Monographs in Electrical and Electronic Engineering

Series editors: P. Hammond, T.J.E. Miller, and T. Kenjo

Monographs in Electrical and Electronic Engineering

Series editors: P. Hammond, T.J.E. Miller, and T. Kenjo

25. *Electrical machines and drives: a space-vector theory approach* (1992) Peter Vas
26. *Spiral vector theory of a.c. circuits and machines* (1992) Sakae Yamamura
27. *Parameter estimation, condition monitoring, and diagnosis of electrical machines* (1993) Peter Vas
28. *An introduction to ultrasonic motors* (1993) S. Sashida and T. Kenjo
29. *Ultrasonic motors: theory and applications* (1993) S. Ueha and Y. Tomikawa
30. *Linear induction drives* (1993) J.F. Gieras
31. *Switched reluctance motors and their control* (1993) T.J.E. Miller
32. *Numerical modelling of eddy currents* (1993) Andrzej Krawczyk and John A. Tegopoulos
33. *Rectifiers, cycloconverters, and AC controllers* (1994) Thomas H. Barton
34. *Stepping motors and their microprocessor controls. Second edition* (1994) T. Kenjo and A. Sugawara
35. *Inverse problems and optimal design in electricity and magnetism* (1995) P. Neittaanmäki, M. Rudnicki, and A. Savini
36. *Electric drives and their control* (1995) Richard M. Crowder
37. *Design of brushless permanent magnet motors* (1995) J.R. Hendershott and T.J.E. Miller
38. *Reluctance synchronous machines and drives* (1995) I. Boldea
39. *Geometry of electromagnetic systems* (1996) D. Baldomir and P. Hammond
40. *Permanent-magnet d.c. linear motors* (1996) Amitava Basak
41. *Vector control and dynamics of a.c. drives* (1996) D.W. Novotny and T.A. Lipo
42. *Sensorless vector and direct torque control* (1998) Peter Vas
43. *Intelligent control: power electronic systems* (1998) Yasuhiko Dote and Richard G. Hoft
44. *Time domain wave-splittings and inverse problems* (1998) Sailing He, Staffen Ström, and Vaughan H. Weston
45. *AI-based electrical machines and drives: power electronics, fuzzy logic, and neural networks* (1998) Peter Vas
46. *Finite element methods in electrical power engineering* (2000) A.B.J. Reece and T.W. Preston
47. *Modern electric vehicle technology* (2001) C.C. Chan and K.T. Chau

Modern Electric Vehicle Technology

C.C. Chan

and

K.T. Chau

Department of Electrical and Electronic Engineering,
International Research Centre for Electric Vehicles,
The University of Hong Kong

OXFORD
UNIVERSITY PRESS

Great Clarendon Street, Oxford OX2 6DP

Oxford University Press is a department of the University of Oxford.
It furthers the University's objective of excellence in research, scholarship,
and education by publishing worldwide in

Oxford New York

Athens Auckland Bangkok Bogotá Buenos Aires Cape Town
Chennai Dar es Salaam Delhi Florence Hong Kong Istanbul Karachi
Kolkata Kuala Lumpur Madrid Melbourne Mexico City Mumbai
Nairobi Paris São Paulo Shanghai Singapore Taipei Tokyo Toronto Warsaw

and associated companies in Berlin Ibadan

Oxford is a registered trade mark of Oxford University Press
in the UK and in certain other countries

Published in the United States
by Oxford University Press Inc., New York

First published 2001

A catalogue record for this title is available from the British Library

Library of Congress Cataloging in Publication Data

Chan, C.C.
Modern electric vehicle technology/C.C. Chan and K.T. Chau.
(Monographs in electrical and electronic engineering; 47)
Includes bibliographical references.
1. Electric vehicles. I. Chau, K.T. II. Title. III. Series.
TL220 .C45 2001 629.22′93–dc21 2001033913

ISBN 0 19 850416 0

10 9 8 7 6 5 4 3 2 1

Typeset by Integra Software Services Pvt. Ltd, Pondicherry, India
www.integra-india.com

Printed in Great Britain
on acid-free paper by
Biddles Ltd,
Guildford & King's Lynn

To our parents, families, colleagues and friends worldwide

Contents

Preface

First of all, I have a confession to make. Perhaps I have written this book on electric vehicles under false pretences. I am not the inventor of electric vehicles. Electric vehicles had been in existence before I was born. The only main motivation for me to write this book is that electric vehicles have been my bread and butter in the latter half of my life. Some 20 years ago, with the encouragement of my colleagues, I started my research work in electric vehicles. I published with my research students among the earliest papers on the Multi-Microprocessor Based Adaptive Decoupling Controlled Induction Motor Drives for Electric Vehicles and won the IEEE Best Paper Award. Another sister paper also awarded the Alcan Prize at the Ninth International Electric Vehicle Symposium, since then I was promoted from Lecturer to Senior Lecturer, Reader then to the Honda Chair Professor of Engineering and the Head of Department of Electrical & Electronic Engineering (1994–2000) at the University of Hong Kong. Indeed it is the electric vehicles that have made me what I am today. I have devoted myself into the Engineering Philosophy of Electric Vehicles, System Integration and Optimization, Development of Advanced Propulsion Systems and Intelligent Energy Management Systems. I have been involved in major international electric vehicle projects, served as Technical Consultant to a number of companies and Technical Advisor to Governments. Believe it or not, in many dreams I dream of electric vehicles. In other words, I am simply obsessed by electric vehicles. Therefore, it is my great privilege to initiate and co-author this book, to share my passion with you on the subject of electric vehicles.

Since the 1980s, with the support from colleagues and friends in electric vehicle community, I have been active in International Electric Vehicle Symposium, deeply involved in the founding of the World Electric Vehicle Association (WEVA). The WEVA was inaugurated in Hong Kong on the 3rd December 1990 during the 10th International Electric Vehicle Symposium (EVS-10) where I served as the General Chairman. I have been made as President of the WEVA and President of the Electric Vehicle Association of Asia Pacific (EVAAP). Since then, I have been invited to deliver keynote speeches at a number of international conferences and to give lectures on electric vehicles at many universities around the world. Last year, I was awarded the 2000 International Lecture Medal by the Institution of Electrical Engineers (IEE), UK, and delivered the International Lectures on Electric Vehicles in four continents. Due to the lack of comprehensive reference book on modern electric vehicle technology, many audiences enthusiastically and eagerly urged me to consider to write a book on Modern Electric Vehicles. Although I have commenced the outline of book in early 1990s, it was only possible to implement my thought after my former Ph.D. student, Dr. K.T. Chau joined my Department at the University of Hong Kong in January 1995 and became the co-author. The inspiration and support from my colleagues and friends in international electric vehicle community always enlightened me whenever I faced difficulties. The first manuscript was completed in 1998 and used as lecture handouts for our course on 'Electric Vehicle Technology' for final-year students of B.Eng. Degree Programme in Electrical Energy System Engineering and 'Advanced Electric Vehicle Technology' for M.Sc.(Eng.) Programme in Electrical and Electronic Engineering at the University of Hong Kong. Feedback from the students was sought and major revision was made due to rapid development in electric/hybrid vehicle technology during the past few years.

What has it been like to an Academic and Engineer in Electric Vehicle Technology over the past two decades? In two words, exciting and frustrating. Exciting, because I have seen great technical advances. Frustrating, because I have experienced difficulties and barriers both in technology and society. But overall, I have lived to see the modern electric/hybrid technology and industry develop from gleams to a technology and industry that is universally acknowledged as the necessity for the clean, efficient, intelligent and sustainable transportation means for the 21st Century. I am also pleased that my two sons followed my footsteps to be interested in EVs. My elder son Lam participated the Luciole electric vehicle project in Japan National Institute of Environment Studies in 1995. My younger son Ting expressed keen interest in EVs when I delivered a seminar on EVs in his university—the University of Washington in 1998.

I understand that this book is not perfect. However, I hope that this book will stimulate the engineering philosophy of electric vehicle development and to contribute to the advancement of the state of the art of electric vehicle technology so that the commercialization of electric vehicles can be accelerated in the right direction.

In my Presidential Address of the Hong Kong Institution of Engineers in 1999, I said: 'If Engineering is to have a strong say about the Future, it needs to become more Integrated with Science, Society and Humanistic Concerns. Engineering should venture to develop new capabilities and a broader sense of mission. We should integrate what we know, what we can do and what we should do, that is, to integrate the knowledge, the action and the wisdom'. It is gratifying that we are engaging in part of a programme that has the potential for having such a major impact on future of our society and the welfare of our future generations.

HKU C.C. Chan
18 January 2001

The electric vehicle is tomorrow's car—Clean, Green and Keen. Clean is due to its zero local emissions and very low global emissions. Green is due to its environmental friendliness. Keen is due to its keen intelligence. Despite these obvious benefits, electric vehicles have not been widely used around the world. The key reasons are due to their high price and short driving range. Mass production, which would lower the price of electric vehicles, is being held back by the argument that there is no market needs. Since there is no market, naturally there will not be mass production and the cost thus remains high—it is a vicious cycle. I believe that the essence is simply a lack of devotion or real concern on our environment. Confucius said, 'Everyone has a duty to all mankind, namely to strive for the realization of a world community exemplifying the principle of reciprocity in the relations of all mankind to each other'. Governments should have clear policy and regulations on environmental protection while drivers should be proud of using zero-emission vehicles, hence creating a sustainable market for electric vehicles. With the ever improving technology of energy sources such as advanced batteries and fuel cells, the short range problem should be solvable. As an interim, the introduction of hybrid electric vehicles can instantly solve this problem though with a compromise of exhaust fumes. The success of hybrid electric vehicles will definitely accelerate the realization of electric vehicles. In the foreseeable future, it is my vision that internal combustion engine vehicles, hybrid electric vehicles and electric vehicles will coexist. Nevertheless, it is my ultimate dream that all cars are electric vehicles with absolute-zero emissions.

This book is not only a milestone of my work, but also a record of my obsession with electric vehicles. My romance with electric vehicles actually began in 1988, as the final-year undergraduate student in the Department of Electrical & Electronic Engineering, the University of Hong Kong. I chose the battery management system for electric vehicles for my final-year project. At that time, I did not know much about electric vehicle technology, and I was only attracted by the reputation of Professor C.C. Chan. But as soon as I started the project, I loved electric vehicles. Further nurtured by Professor Chan, I earned my M.Phil. and Ph.D. degrees through the developments of permanent-magnet synchronous motor drives and switching power converters for electric vehicles. Thus, my love with electric vehicles was eternized. In 1995, I joined my Alma Mater and was encouraged by C.C. to co-author a reference book on modern electric vehicle technology. I simply said: 'I love it'. Over the years, we have expanded our electric vehicle technology research group from a handful to more than a dozen people, and have created fruitful research outcomes to enrich the academic value of this reference book. Nothing is perfect, we strive to write the preferred reference book on modern electric vehicle technology for students, researchers, engineers and administrators.

While electric vehicles are the driving force for better environment, my family is the driving force for my work on electric vehicles. Especially, I must express my indebtedness to two ladies—my mother, Leung-Chun, who created my life and my wife, Joan Wai-Yi, who lights up my life. I would like to make use of this chance to celebrate my mother's 85th birthday anniversary and my 10th wedding anniversary.

HKU K.T. Chau
8 February 2001

Introduction

We have now entered into the 21st Century. In a world where energy conservation, environmental protection and sustainable development are growing concerns, the development of electric vehicle (EV) technology has taken on an accelerated pace. The dream of having commercially viable EVs is becoming a reality.

This book is published after celebrating the 30th Anniversary of the International Electric Vehicle Symposium (EVS) and the 10th Anniversary of the World Electric Vehicle Association (WEVA), as well as at dawn of commercialization of electric/hybrid vehicles. In these activities, the first author has deeply involved. We hope this book reflects the recent development of electric/hybrid vehicles, contributes to the advancement of electric/hybrid vehicle technology and promotes the commercialization of electric/hybrid vehicles—clean, efficient and intelligent transportation means for the 21st Century.

Electric vehicle technology is the happy marriage of mechanical/chemical and electrical/electronic laws which operate in perfect harmony. This book covers multidisciplinary aspects of EVs, and is written for a wide coverage of readers including students, researchers, engineers and administrators. It is organized in such a way that it provides maximum flexibility without any loss of continuity from one chapter to another. We believe that this approach would facilitate the readers to select reading those chapters that are most interesting to them. The suggestion for reading is as follows:

- Electrical engineering students taking a 35- to 45-h course dedicated to Electric Vehicle Technology may be interested in all chapters.
- Researchers in the field of EVs, EV technology or automotive engineering may be more interested in Chapters 1, 3, 4, 5, 6, 7, 8 and 9.
- Researchers in the field of chemical engineering may be more interested in Chapters 6 and 7.
- Researchers in the field of computer engineering may be more interested in Chapter 8.
- Researchers in the field of power engineering may be more interested in Chapter 9.
- Researchers in the field of environmental sciences may be more interested in Chapter 10.
- Practising engineers for EVs and automobiles may be more interested in Chapters 1, 3, 4, 5, 6, 7 and 9.
- Administrators relating to EVs, hybrid EVs (HEVs), energy conservation, environmental protection and sustainable businesses may be more interested in Chapters 1, 2, 3, 4, 9 and 10.
- General readers may be interested in any chapters.

Chapters 1 and 2 are written to review the important of engineering philosophy and fundamental concepts of EV technology, then to summarize the development of EVs. Chapter 1 gives an overview on the past, present and future of EVs, and reveals the essential engineering philosophy of EVs, which is the guiding ideology of the whole book. Then, Chapter 2 briefly describes the development of both EVs and HEVs, and hence identifies the state-of-the-art EVs and HEVs.

Chapters 3 and 4 present the system configurations of modern EVs and HEVs. New concepts and classifications of both EVs and HEVs are revealed. Chapter 3 includes the variations in EV configurations due to fixed and variable gearing, single- and multiple-motor drives and in-wheel drives. Some unique EV parameters are also discussed. Chapter 4 includes the variations in HEV configurations and the corresponding power flow control.

Chapters 5, 6 and 7 are core chapters of this book. These three chapters are essentially in technical nature. Chapter 5 is devoted to present electric propulsion systems for modern EVs. It involves in-depth discussions of dc motor drives, induction motor drives, permanent-magnet motor drives and switched reluctance motor drives for electric propulsion. Chapter 6 presents different types of energy sources for EVs, including batteries, fuel cells, ultracapacitors and ultrahigh-speed flywheels. Their operating principles, unique features and potentialities are discussed and evaluated. Also, a new concept on the hybridization of multiple energy sources is revealed to solve the problems due to the use of sole energy source. Chapter 7 discusses various auxiliaries for modern EVs, including battery chargers, battery indicators, energy management systems, temperature control units, power steering units, auxiliary power supplies, navigation systems and regenerative braking systems.

Chapter 8 delineates the concept of system level simulation for EVs, hence the development of a dedicated EV simulator. Based on the EV simulator, the deductions of optimal transmission ratio, optimal system voltage and optimal hybridization ratio are exemplified. Furthermore, a case study on the implementation of electric light buses is given.

Chapters 9 and 10 deal with the commercialization and implementation of EVs. Chapter 9 discusses the most essential factor, EV infrastructure, for the commercialization and popularisation of EVs. It includes the discussions on domestic charging infrastructure, public charging infrastructure, standardization and regulations, training and promotion as well as various impacts on the power system. Chapter 10 presents the energy, environment and economy (EEE) benefits resulting from the implementation of EVs.

Acknowledgements

Material presented in this book is a collection of many years of research and development by the authors in the International Research Centre for Electric Vehicles, Department of Electrical and Electronic Engineering, The University of Hong Kong. It is our great pleasure to present this book in celebration of the 90th anniversary of The University of Hong Kong, 30th anniversary of the International Electric Vehicle Symposium, and 10th anniversary of the World Electric Vehicle Association.

We are grateful to all members of our Electric Vehicle Technology Research Group, especially Mr. Y.S. Wong, Dr. Q. Jiang, Mr. S.W. Chan, Dr. M. Cheng, Mr. J.Y. Gan, Dr. J.H. Chen, Mr. S.Z. Jiang, Dr. W.C. Lo, Dr. W. Xia, Dr. H.Q. Wang, Dr. R.J. Zhang, Dr. C.W. Ng, Dr. Y.J. Zhan, Mr. J.M. Yao, Mr. W.X. Shen, Mr. T.W. Chan, and Mr. K.C. Chu for their help in the preparation of this work. We must express our sincere gratitude to all the reviewers, especially Prof. J.F. Eastham, Prof. Z.Q. Zhu, Prof. A. Szumanowski, Prof. J.Z. Jiang, Dr. V. Wouk and Dr. G. Brusaglino, who have helped us at various stages in the revision of this book. We are deeply indebted to our colleagues and friends at the World Electric Vehicle Association (WEVA), the Electric Vehicle Association of Americas (EVAA), the European Road Electric Vehicle Association (AVERE) and the Electric Vehicle Association of Asia Pacific (EVAAP) including the Japan Electric Vehicle Association (JEVA) for their continuous support and encouragement over past decades. We are highly appreciated to the Editors of Oxford University Press for their patience and effective support who made this book possible. We are also obliged to our secretary Ms Clara Chung for her efficient assistance.

Last but not least, we thank our families for their unconditional support and absolute understanding during the writing of this book.

1 Engineering philosophy of EV development

Let us begin with the investigation of the growth of population and vehicles as shown in Fig. 1.1. In the next 50 years, the global population will increase from 6 billions to 10 billions and the number of vehicles will increase from 700 millions to 2.5 billions. If all these vehicles are propelled by internal combustion engines, where will the oil come from? and where should the emissions be disseminated? Under that circumstance, our sky will be permanently grey. Therefore, we should strive for sustainable road transportation for the 21st Century.

In a world where environmental protection and energy conservation are growing concerns, the development of electric vehicle (EV) technology has taken on an accelerated pace to fulfil these needs. Concerning the environment, EVs can provide emission-free urban transportation. Even taking into account the emissions from the power plants needed to fuel the vehicles, the use of EVs can still significantly reduce global air pollution. From the energy aspect, EVs can offer a secure, comprehensive and balanced energy option that is efficient and environmentally friendly, such as the utilization of various kinds of the renewable energies. Furthermore, EVs will be more intelligent to improve traffic safety and road utilization, and will have the potential to have a great impact on energy, environment and transportation as well as hi-tech promotion, new industry creation and economic development (Chan, 2000).

1.1 Past, present and future of EVs

Energy and environment issues have pushed the development of EVs. At the beginning of this new millennium, it is timely to review what we have achieved in the development of EVs over the past 30 years, where we are today, and to predict where we will be in the next 30 years.

1.1.1 PAST 30 YEARS DEVELOPMENT

About 30 years ago, some countries started the rekindling of interests in EVs, since EVs had almost vanished from the scene in the 1930s after its invention in 1834. The major driving force for the development of EVs at that time was the energy issue due to energy crisis in the early 1970s. In 1976, the USA launched the Electric and Hybrid Vehicle Research, Development and Demonstration Act, Public Law 94-413. At that time, the main question to be answered was 'Can EVs do the job in our modern society?', although EVs did work well in the late 1800s and early 1900s. The development of EVs for over 30 years has answered the above question—yes. For example, an experimental EV in 1968 racing from California

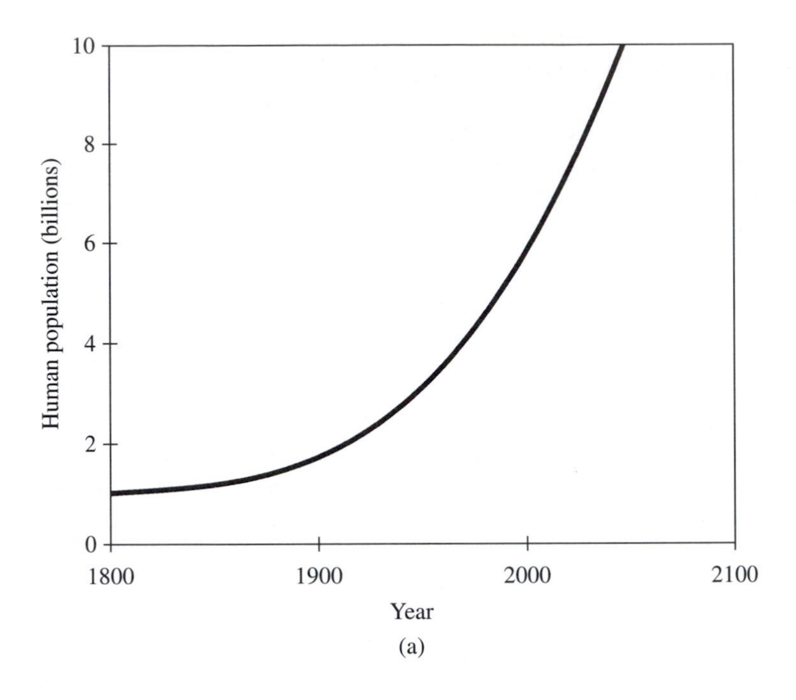

(a)

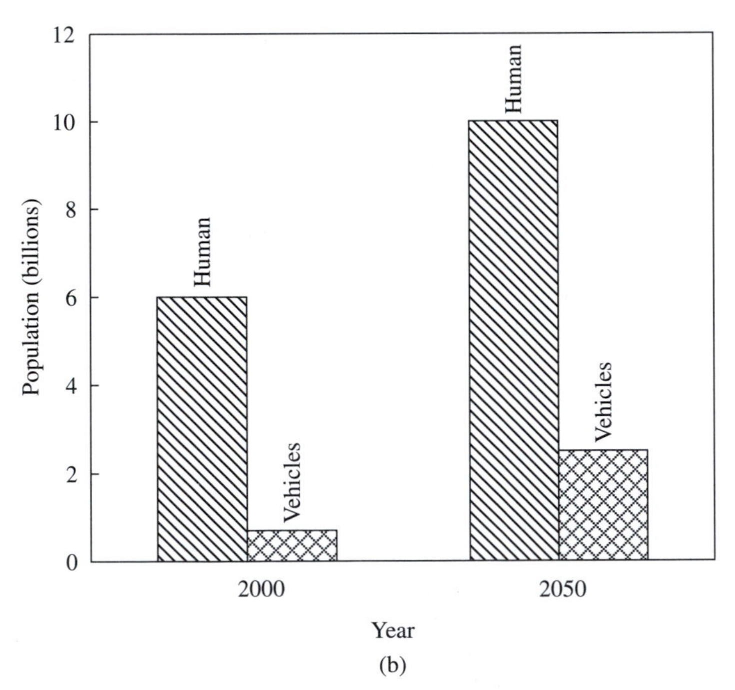

(b)

Fig. 1.1. Growth of population and vehicles.

Institute of Technology (Caltech) to Massachusetts Institute of Technology (MIT) suffered from failures in virtually every critical component; whereas a commercially built EV in 1998 running from Los Angeles to Detroit exhibited a success with no component failures. Within the 1970s, EVs were still in research and development stage, and most of them were conversion of internal combustion engine vehicles (ICEVs). Today, major automobile manufacturers are offering EVs for sale or lease. Most of them are the purpose-built EV, not conversion EV.

1.1.2 PRESENT MAJOR ISSUES

At present, the major driving force for EVs is the environment issue, such as mandate by California rule, rather than the previous energy issue. Thus, the main question to be answered becomes 'Can EVs be made affordable?'. The major factors that make EV affordable are the range and cost. To tackle the range, the development of advanced batteries such as the nickel–metal hydride, zinc/air and lithium–ion are in progress. However, since both specific energy and energy density of batteries are much lower than that of petrol, the development of fuel cells for EVs has taken on an accelerated pace in recent years, focusing on solid polymer fuel cells. Meanwhile, the development of commercial hybrid electric vehicles (HEVs) is also going on rapidly. HEVs essentially improve the range and performance of EVs at higher complexity and cost because of the additional energy source, engine and other accessories. To tackle the cost, efforts are being made to improve various EV subsystems, such as electric motors, power converters, electronic controllers, energy management units, battery chargers, batteries and other EV auxiliaries, as well as EV system level integration and optimization.

Besides subsystem level improvement and system level optimization, effort should be made at the global level. We need to combine the strength of East and West to solve those EV problems, especially the production cost. For instance, the eastern countries such as China, India, Thailand and Malaysia can readily offer low labour cost while the western countries such as the USA, Germany and France can provide fundraising and high technologies. Japan can also play an important role in the global development of EVs. Hence, by combining low labour cost and hi-tech facilities, the production cost of EVs can be remarkably reduced. Moreover, the development of EVs is a global issue—how to restructure the global business environment to create a sustainable global market for EVs. For instance, the energy-storage device of ICEVs, the fuel tank, represents only a minor fraction of the total vehicle cost; whereas the energy-storage device of EVs, the battery pack, is the most expensive EV subsystem. Thus, it is natural to sell the fuel tank as part of an ICEV, while it may be more appropriate to lease the battery pack of an EV so that the user just needs to pay for the rental cost and electricity cost. In fact, power utilities can own the batteries, lease them for EV users, and run battery storage business to leverage their electricity generation business, thereby maximizing their profits on storage and generation of electricity, while making EVs more affordable.

1.1.3 DEVELOPMENT TRENDS

In order to see the development trends of various EV aspects, a survey has been made with respect to the number of papers on various topics in the Proceedings of International Electric Vehicle Symposium (EVS) from 1984 to 2000. It should be noted that the EVS were held biennially from 1984 to 1996, and then held annually to keep abreast with the fast-changing development of EVs. Figure 1.2 shows the development trend of various motor drives, including dc motor drives (DC), induction motor drives (IM), permanent-magnet brushless motor drives (PM) and switched reluctance motor drives (SR). It can be observed that the research papers on IM and PM are highly dominant, whereas those on DC are drooping while those on SR are still in a crawling stage. On the other hand, Fig. 1.3 shows the

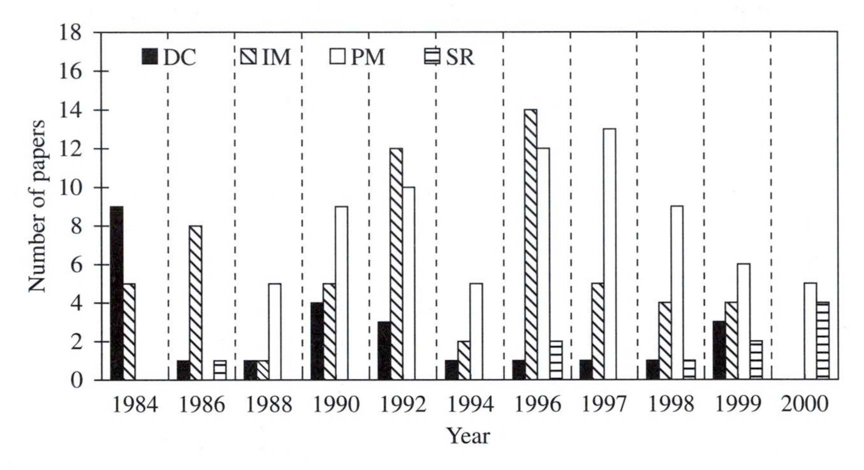

Fig. 1.2. Survey on motor drives in EVS proceedings.

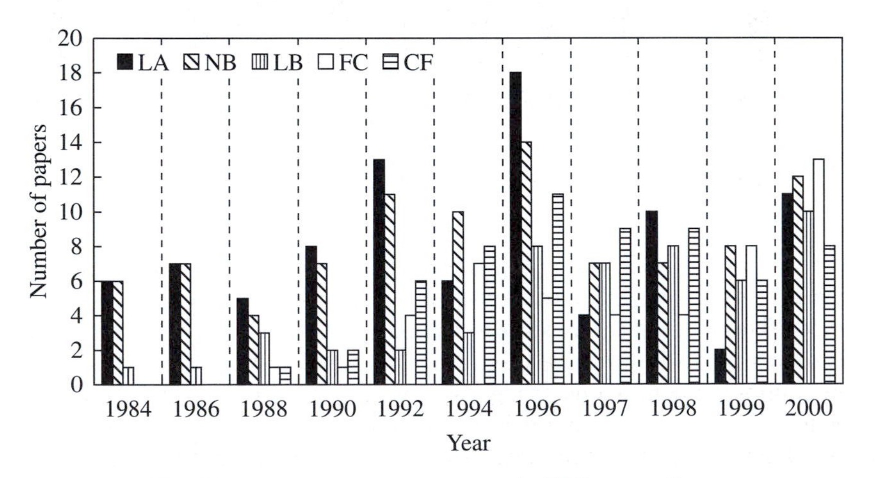

Fig. 1.3. Survey on energy sources in EVS proceedings.

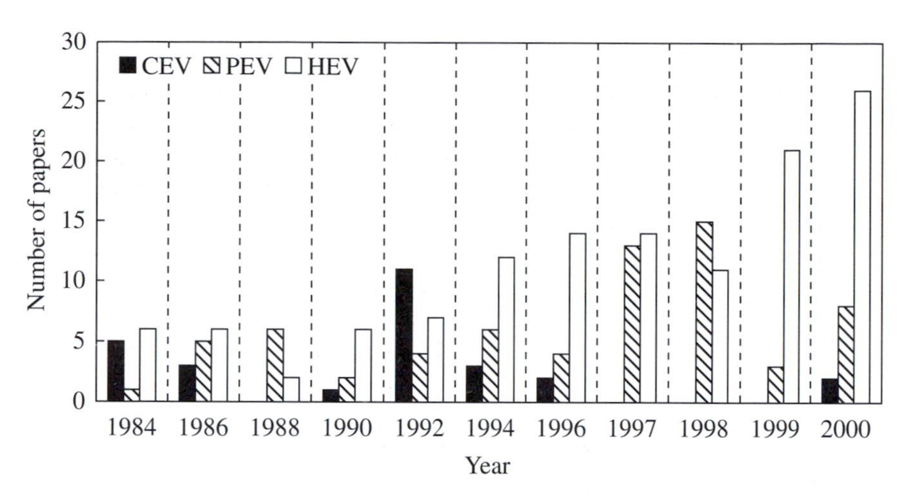

Fig. 1.4. Survey on EV types in EVS proceedings.

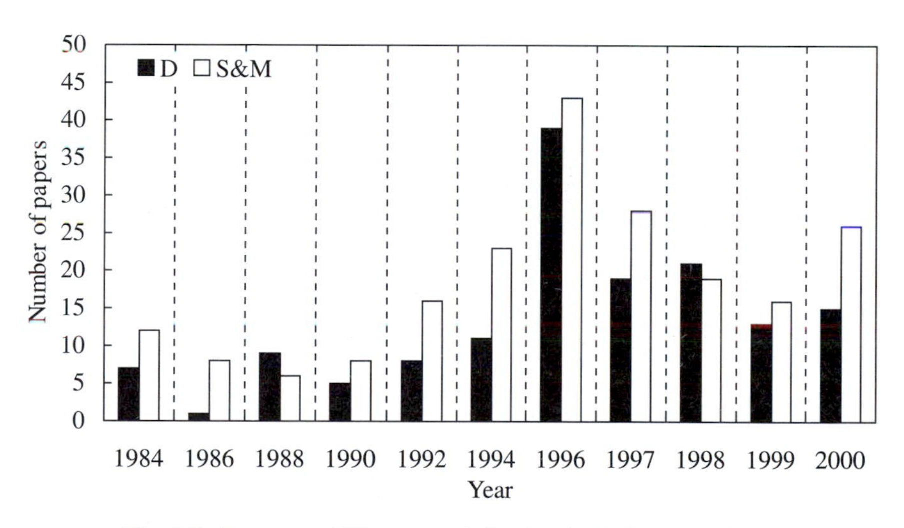

Fig. 1.5. Survey on EV commercialization in EVS proceedings.

development trend of various energy sources, including lead–acid batteries (LA), nickel-based batteries (NB), lithium-based batteries (LB), fuel cells (FC) and capacitors/flywheels (CF). As shown, the number of papers in LB, FC and CF are becoming more and more attractive, though LA and NB are still undergone continual improvement. Concerning the types of EVs, Fig. 1.4 illustrates that the conversion EV (CEV) is becoming less attractive than the purpose-built EV (PEV) while the HEV is of growing interests for the coming EV markets. Finally, Fig. 1.5 shows that EVs are on the verge of commercialization, since more and more

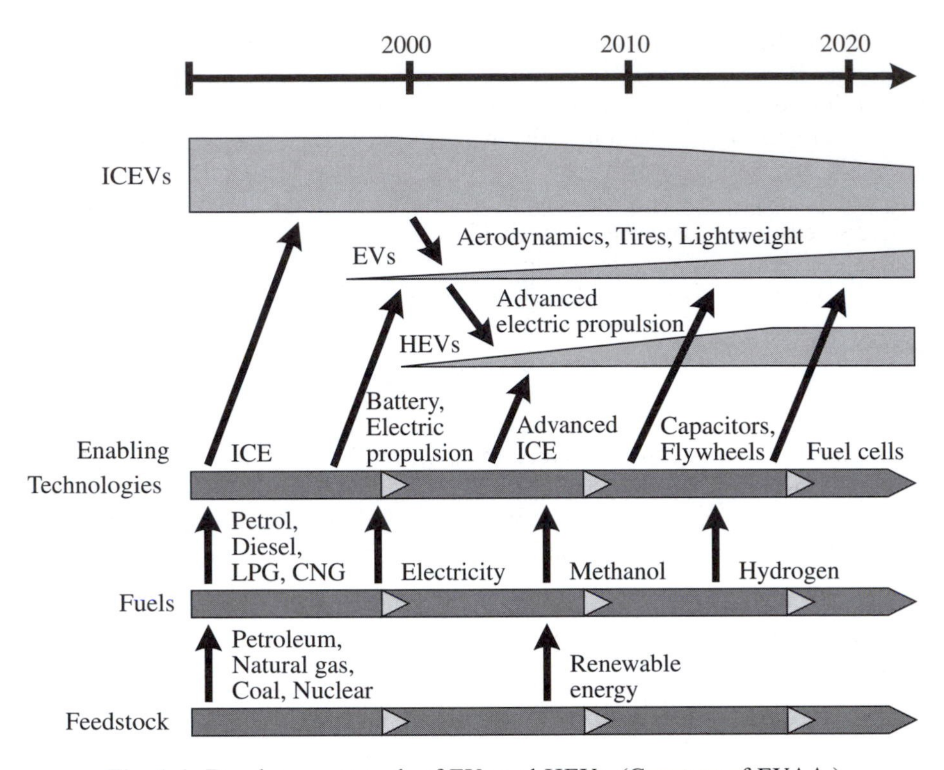

Fig. 1.6. Development trends of EVs and HEVs. (Courtesy of EVAA.)

papers are published on the topics of demonstration (D) as well as standardization and marketing (S&M) of EVs.

In the next 30 years, it is anticipated that both EVs and HEVs will be commercialized, and they will have their market shares. EVs will be well accepted by some niche markets, namely the users for community transportation, the places where electricity is cheap and ease of access, and the cities with zero-emission mandate. On the other hand, HEVs will have a niche market for those users desiring long driving ranges. The ultimate penetration of EVs and HEVs will mainly depend on their respective costs. Particularly, the commercialization of fuel cell EVs will be accelerated in the next two decades, since they have the greatest potential to deliver the same range and performance as our ICEVs.

In summary, electric propulsion and energy sources will still be the key technologies to be addressed, EVs and HEVs will still be coexistent, while energy, environment and economy will still be the key issues for EV commercialization. Figure 1.6 illustrates the development trends of EVs and HEVs. It should be noted that some core technologies can be shared among ICEVs, EVs and HEVs. Our ultimate goal is the use of clean, efficient and intelligent energy to achieve sustainable transportation system for the 21st Century.

1.2 Engineering philosophy of EVs

Engineering integrates science, technology and management to solve real world problems effectively and economically, while usually asks for 'how' and 'what for', pursues 'best solution' and emphasizes 'teamwork'. As EV engineering is multidisciplinary, specialists in those areas of engineering must work together; electrical engineers, electronic engineers, mechanical engineers, chemical engineers and mainstream automotive engineers must pool their knowledge in the main areas that must be integrated: body design, electric propulsion, energy sources, energy refuelling and energy management.

1.2.1 EV CONCEPT

Although the EV was around before the turn of the 20th Century, the modern EV is a completely new machine that is totally different from the classical EV. It is not only a transportation vehicle, but also a new type of electric equipment. The modern EV concept is summarized as follows:

- The EV is a road vehicle based on modern electric propulsion which consists of the electric motor, power converter and energy source, and it has its own distinct characteristics;
- The EV is not just a car but a new system for our society, realizing clean and efficient road transportation;
- The EV system is an intelligent system which can readily be integrated with modern transportation networks;
- EV design involves the integration of art and engineering;
- EV operating conditions and duty cycles must be newly defined;
- EV users' expectations must be studied, hence appropriate education must be conducted.

The system architecture of EVs is very different from that of ICEVs, similar to the fact that the system architecture of quartz-based electronic watches is very different from that of spring-based mechanical watches. In short, their appearances are very similar whereas their principles are very different. The EV system architecture consists of mechanical subsystems, electrical and electronic subsystems, and information subsystems. Concerning those mechanical subsystems, namely the body and chassis, propulsion structure and transmission as well as energy source frame, relevant factors include road characteristics, crash worthiness, interior space, assembly time, serviceability and cost. Concerning those electrical and electronic subsystems, including the power network, motors, controllers and energy source system, relevant factors are safety, regulation, standards, efficiency, reliability, weight and cost. Concerning those information subsystems, handling the driver's desire, vehicle operating status, energy source status, motor status, controller status and charger status, relevant factors are communication network, data processing algorithms as well as communication links for diagnostic and charging control.

1.2.2 EV ENGINEERING PHILOSOPHY

The EV engineering philosophy essentially is the integration of automobile engineering and electrical engineering. Thus, system integration and optimization are prime considerations to achieve good EV performance at affordable cost. Since the characteristics of electric propulsion are fundamentally different from those of engine propulsion, a novel design approach is essential for EV engineering. Moreover, advanced energy sources and intelligent energy management are key factors to enable EVs competing with ICEVs. Of course, the overall cost effectiveness is the fundamental factor for the marketability of EVs.

The design approach of modern EVs should include state-of-the-art technologies from automobile engineering, electrical and electronic engineering and chemical engineering, should adopt unique designs particularly suitable for EVs, and should develop special manufacturing techniques particularly suitable for EVs. Every effort should be made to optimise the energy utilisation of EVs. The following points are those typical considerations for EV design:

- To identify the niche market and environment;
- To determine the design philosophy;
- To determine the technical specifications including the driving cycle;
- To determine the infrastructure required including the recycling of batteries;
- To determine the overall system configuration—EV, HEV or fuel cell EV configurations;
- To determine the chassis and body;
- To determine the energy source—generation or storage, single or hybrid;
- To determine the propulsion system—motor, converter and transmission types, single or multiple motors, gearless or geared, mounting methods, and ICE systems in case of an HEV;
- To determine the specifications of electric propulsion (power, torque, speed) and energy source (capacity, voltage, current) according to various driving cycles; for example, Fig. 1.7 shows that the torque-speed requirement of Federal Urban Driving Schedule (FUDS) is very different from that of Federal Highway Driving Schedule (FHDS);
- To adopt intelligent energy management system;
- To analyse the interaction of EV subsystems by using the quality function matrix, hence understanding the degree of interaction that affects the cost, performance and safety;
- To optimize the efficiency of the motor drive according to the selected driving pattern and operating conditions;
- To optimize the overall system using computer simulation.

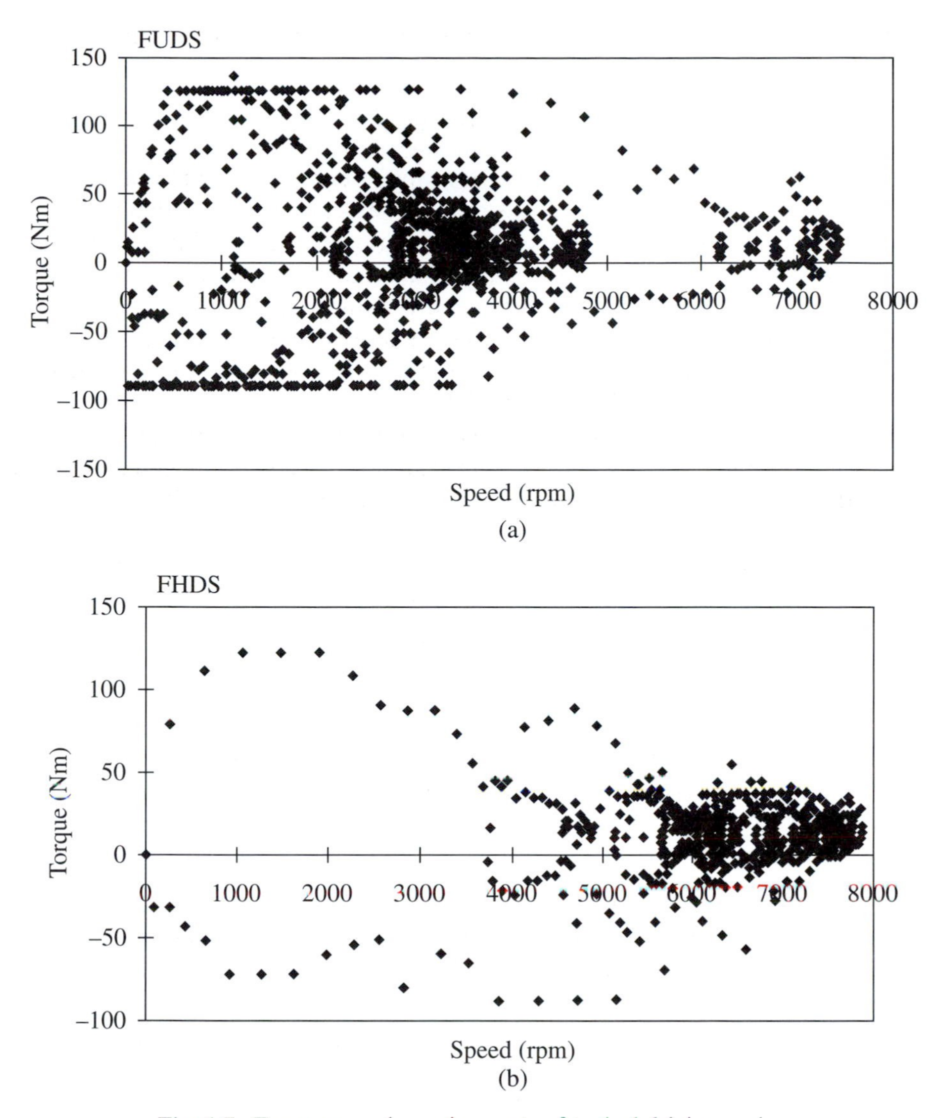

Fig. 1.7. Torque-speed requirements of typical driving cycles.

1.2.3 KEY EV TECHNOLOGIES

The key technologies of EVs include automotive technology, electrical technology, electronic technology, information technology and chemical technology. Although the energy source is the most crucial area, body design, electric propulsion, energy management and system optimization are equally important. In fact, the integration of all these areas is the key to success (Chan, 1993).

Body design

There are two basic approaches for producing EVs—either conversion or purpose-built. For the conversion EV, the engine and associated equipment of an existing ICEV are replaced by the electric motor, power converter and battery. This offers some economy for a small volume production because the existing ICEV chassis can be utilized. However, in most conversions, the resulting EV suffers from a greater curb weight, a higher centre of gravity and an unbalanced weight distribution. Therefore, this approach is gradually fading out. At present, the modern EV is mostly purpose-built, sometimes called ground-up design. This purpose-built EV takes the definite advantage over the conversion one because it allows the engineers having the flexibility to coordinate and integrate various EV subsystems so that they can work together efficiently.

There are some design concepts for purpose-built EVs so that the overall performances such as range, gradeability, acceleration and top speed can be improved. These concepts include the consistent weight-saving design, low drag coefficient body design, and low rolling resistance concept. Firstly, the vehicle weight directly affects the performance of EVs, especially the range and gradeability. To reduce the curb weight, the use of lightweight materials such as aluminium and composite material for the body and chassis can be adopted. Secondly, low drag coefficient body design can effectively reduce the vehicle aerodynamic resistance, which has a significant effect on extending the range of EVs in highway driving or cruising. In general, the aerodynamic resistance can be reduced by tapering front and rear ends, adopting undercover and flat under-floor design, optimizing airflow around the front and rear windows, using rear spats, providing airflow streaks along the front and rear tires, and employing slanted front nose design. Thirdly, low rolling resistance tires are particularly effective in reducing running resistance at low and medium driving speeds, and play an important role in extending the range of EVs in city driving. This can be achieved through the use of a newly developed blended tire polymer, together with an increase in tire pressure.

Electric propulsion

The electric propulsion system of EVs is responsible for converting electrical energy to mechanical energy in such a way that the vehicle is propelled to overcome aerodynamic drag, rolling resistance drag and kinetic resistance. Since the torque-speed characteristic of an engine covers only a narrow range, the required torque-speed performance of the vehicle has to be achieved through gear changing. On the other hand, in a modern motor drive, high-torque low-speed and constant-power high-speed regions can be achieved through electronic control. Moreover, the EV propulsion design can be more flexible, namely single or multiple motors, with or without reduction gearing, with or without differential gearing, and axle or wheel motors.

The electric propulsion system consists of the motor drive, transmission device and wheels. The transmission device sometimes is optional. In fact, the motor drive, comprising of the electric motor, power converter and electronic controller, is the core of the EV propulsion system. The major requirements of the EV motor drive are summarized as follows:

- High instant power and high power density;
- High torque at low speeds for starting and climbing as well as high speed at low torques for cruising;
- Very wide speed range including constant-torque and constant-power regions;
- Fast torque response;
- High efficiency over wide speed and torque ranges;
- High efficiency for regenerative braking;
- High reliability and robustness for various vehicle operating conditions;
- Reasonable cost.

To satisfy these special requirements, the power rating and torque-speed requirements of the motor drive should be determined on the basis of driving cycles and system-level simulation. New motor design technologies and control strategies are being pursued to extend the speed range, to optimize the system efficiency and to enlarge the high-efficiency region. Newly developed electronic products are also adopted to improve the system performance and to reduce the total cost.

Based on the technological growth of electric motors, power electronics, microelectronics and control strategies, more and more kinds of motor drives become available for EVs. Dc motor drives have been traditionally used for EV propulsion because of their ability to achieve high torque at low speeds and easy to control. However, the dc motor needs careful maintenance due to its commutator and brushes. Recent technological developments have enabled a number of advanced motor drives to offer definite advantages over those dc motor drives, namely high efficiency, high power density, efficient regenerative braking, robust, reliable, and maintenance free. Among them, the vector controlled induction motor drive is most popular and mature, though it may suffer from low efficiency at light-load ranges. On the other hand, permanent magnet (PM) brushless motors possess the highest efficiency and power density over the others, but may suffer from a difficulty in flux weakening control for the constant-power high-speed region. The PM hybrid motor is a special type of PM brushless motors. In this motor, an auxiliary dc field winding is so incorporated that the air-gap flux is a resultant of the PM flux and field-winding flux. By adjusting the field-winding excitation current, the air-gap flux can be varied flexibly, hence offering optimal efficiency over a wide speed range. Switched reluctance (SR) motors offer promising features for EV applications because of their simplicity and reliability in both motor construction and power converter configuration, wide speed range, favourable thermal distribution, 'limp-home' capability, and efficient regenerative braking. However, they may suffer from torque ripples and acoustic noise problems.

In summary, for EV motor drives, dc motor drives have been gradually superseded by induction motor drives, PM brushless motor drives with various configurations and SR motor drives. These advanced motor drives must be specially designed to meet the EV special requirements. For transmission devices, conventional gearing can no longer satisfy the needs of EVs. Recently, planetary gearing has been accepted as the transmission device of the latest EVs.

Energy sources

At present, the main obstacles of the commercialization of EVs are the relatively high initial cost and short driving range. The EV energy source has been identified to be the major cause of these problems. Thus, the present and foreseeable future most important EV development issue is on how to develop various EV energy sources. Those development criteria are summarized as follows:

- High specific energy and energy density;
- High specific power and power density;
- Fast charging and deep discharging capabilities;
- Long cycle and service lives;
- Low self discharging rate and high charging efficiency;
- Safety and cost effectiveness;
- Maintenance free;
- Environmental sound and recyclable.

Currently, there is no sole EV energy source that can fully satisfy all of these criteria. When batteries are selected, there are various compromises among those criteria. For examples, the lead–acid battery offers the merits of relatively low cost and high specific power, and the demerits of relatively short cycle life and low specific energy; whereas the nickel–metal hydride battery exhibits the relatively high specific energy but with relatively high cost. In general, all batteries face a compromise among the specific energy, specific power and cost. In the foreseeable future, the lithium-based batteries such as lithium-ion and lithium-polymer should have good prospects for modern EVs. On the other hand, emerging energy sources including ultracapacitors and ultrahigh-speed flywheels provide promising EV applications because of their exceptionally high specific power. Recently, fuel cells have been identified as one of the most important EV energy sources that can fundamentally solve the key EV problem—short driving range. Provided that the high initial cost of fuel cells can be significantly reduced, it is anticipated that fuel cell EVs can directly compete with the existing ICEVs in the next generation of road transportation.

Rather than based on one energy source, the use of multiple energy sources, so-called hybridization of energy sources, can eliminate the compromise between the specific energy and specific power. For the hybridization of two energy sources, one is selected for high specific energy while the other for high specific power. For examples, there are the battery & battery hybrid, battery & ultracapacitor hybrid, battery & ultrahigh-speed flywheel hybrid, and fuel cell & battery hybrid. In fact, the HEV is a special case of this hybridisation, namely the petrol is of high specific energy for the long driving range while the battery is of high specific power for assisting fast acceleration and providing emission-free operation.

Energy management

Compared with ICEVs, EVs offer a relatively short driving range. Thus, in order to maximize the utilization of on-board stored energy, an intelligent energy management system (EMS) needs to be adopted. Making use of sensory inputs from various EV subsystems, including sensors for temperatures of outside and inside air, current and voltage of the energy source during charging and discharging, current and voltage of the electric motor, vehicle speed and acceleration as well as external climate and environment, the EMS can realise the following functions:

- to optimize the system energy flow;
- to predict the remaining available energy and hence the residual driving range;
- to suggest more efficient driving behaviour;
- to direct regenerative energy from braking to receptive energy sources such as batteries;
- to modulate temperature control in response to external climate;
- to adjust lighting brightness in response to external environment;
- to propose a suitable battery charging algorithm;
- to analyse the operation history of the energy source, especially the battery;
- to diagnose any incorrect operation or defective components of the energy source.

When the EMS is coupled with a navigation system, it can plan energy efficient routes, locate charging facilities for extended trips, and modify range predictions on the basis of traffic conditions. In summary, the EMS has the distinct features of integrated multi-functions, flexibility and adaptability (just like the brain of EVs) such that the limited on-board energy can be used wisely.

System optimization

As mentioned before, the EV system has a complex architecture that contains multidisciplinary technologies. Since the EV performance can be affected by many multidisciplinary interrelated factors, computer simulation is the most important technology to carry out the optimization for performance improvement and cost reduction. Also, EV simulation can help those manufacturers to minimize prototyping cost and time, and to provide rapid concept evaluation. Since the whole EV system consists of various subsystems clustered together by mechanical link, electrical link, control link and thermal link, the simulation should be based on the concept of mixed-signal simulation. Hence, the system optimization can be carried out in the system level in which there are many trade-offs among various subsystem criteria. Generally, numerous iterative processes are involved for the preferred system criteria.

In summary, the system-level simulation and optimization of EVs should consider the following key issues:

- As the interactions among various subsystems greatly affect the performance of EVs, the significance of those interactions should be analysed and taken into account.
- As the model accuracy is usually coherent with the model complexity but may be contradictory to the model usability, trade-offs among the accuracy, complexity and usability as well as simulation time should be considered.
- As the system voltage generally causes contradictory issues for EV design, including the battery weight, motor drive voltage and current ratings, acceleration performance, driving range and safety, it should be optimized on the system level.
- In order to increase the driving range, multiple energy sources may be adopted for modern EVs. The corresponding combination and hybridization ratio should be optimized on the basis of the vehicle performance and cost.
- Since EVs generally adopt fixed gearing, the gear ratio can greatly affect the vehicle performance and driveability. An optimal ratio should be determined through iterative optimization under different driving profiles.

References

Chan, C.C. (1993). An overview of electric vehicle technology. *Proceedings of IEEE*, **81**, 1201–13.

Chan, C.C. (2000). *The 21st Century Green Transportation Means—Electric Vehicles*. National Key Book Series in Chinese. Tsing Hua University Press, Beijing.

Proceedings of the 1st International Electric Vehicle Symposium, 1969.

Proceedings of the 2nd International Electric Vehicle Symposium, 1971.

Proceedings of the 3rd International Electric Vehicle Symposium, 1974.

Proceedings of the 4th International Electric Vehicle Symposium, 1976.

Proceedings of the 5th International Electric Vehicle Symposium, 1978.

Proceedings of the 6th International Electric Vehicle Symposium, 1981.

Proceedings of the 7th International Electric Vehicle Symposium, 1984.

Proceedings of the 8th International Electric Vehicle Symposium, 1986.

Proceedings of the 9th International Electric Vehicle Symposium, 1988.

Proceedings of the 10th International Electric Vehicle Symposium, 1990.

Proceedings of the 11th International Electric Vehicle Symposium, 1992.

Proceedings of the 12th International Electric Vehicle Symposium, 1994.
Proceedings of the 13th International Electric Vehicle Symposium, 1996.
Proceedings of the 14th International Electric Vehicle Symposium, 1997.
Proceedings of the 15th International Electric Vehicle Symposium, 1998.
Proceedings of the 16th International Electric Vehicle Symposium, 1999.
Proceedings of the 17th International Electric Vehicle Symposium, 2000.

2 EV and HEV developments

In 1801, Richard Trevithick built a steam-powered carriage, opening the era of horseless transportation. After tolerating over 30 years of noise and dirtiness due to steam engines, the first battery-powered electric vehicle (EV) was built in 1834. Over 50 years later, the first petrol-powered internal combustion engine vehicle (ICEV) was built in 1885. So, the EV is not new and is already over 166 years old. It was better than the ICEV in the early 1900s. Having slept for almost 70 years, it began to recover in the 1970s. Now, the EV intends to strike back (Daniels, 2000). Detailed history of the development of EVs can be found in Wakefield (1994, 1998) and Shacket (1979).

2.1 Historical development

In 1834, the first battery-powered EV, actually a tricycle, was built by Thomas Davenport. It was powered by a non-rechargeable battery and used on a short track. Four years later, Robert Davidson also built a non-rechargeable battery-powered electric locomotive. After the invention of the lead–acid battery, Sir David Salomons successfully built a rechargeable battery-powered EV in 1874. Twelve years later, the first electric trolley system was set by Frank Sprague in 1886. Then, the EV quickly became popular and played the main role in automotive transportation. In 1900, among an annual sale of 4200 automobiles in America, 38% were EVs, 22% were ICEVs, and 40% steam-powered vehicles. At that time, EVs were the preferred method of transportation among the wealthy elite. Their cost was equivalent to a Rolls Royce of today.

During the last decade of the 19th century, there were a number of companies producing EVs in America, Britain and France. One of the earliest EV manufacturer was the Electric Carriage and Wagon Company owned by Morris and Salom. As shown in Fig. 2.1 (The Horseless Age, December 1895: **1**(2), 15; Scientific American, November 16, 1895: **LXXIII**(20), 315), its 'Electrobat' was converted from a delivery wagon, and ran on the streets of Philadelphia in 1894. Another early EV maker was the Pope Manufacturing Company. By the end of 1898, Pope had produced about 500 EVs of the 'Columbia' model. From 1896 to 1902, the Riker Electric Motor Company also produced a variety of EVs. Its Victoria was an example of the better designs in 1897. Apart from those EV manufacturers in America, the London Electrical Cab Company in England was inaugurated in 1897 with 15 taxis. This electric taxi is shown in Fig. 2.2 (Scientific American Supplement, November 13, 1897: **XLIV**(1141), p. 1897). Moreover, Bouquet, Garcin & Schivre (BGS) in France built different types of commercial EVs, including cars, trucks, buses and limousines, from 1899 to 1906.

Fig. 2.1. Morris & Salom's Electrobat. (Courtesy of The Horseless Age; Courtesy of Scientific American; Photo courtesy of History of the Electric Automobile by Ernest H. Wakefield.)

As it designed and made batteries specially for its own EVs, the BGS EV of 1900 held the world's electric distance record of almost 290 km per charge. It is interesting to note that the first vehicle running over the 100 km/h barrier was an EV, namely the 'Jamais Contente' (Never Satisfied), which was driven by Camille Jenatzy, a Belgian. It was a bullet-shaped electric racing car and captured the record of 110 km/h on the first of May in 1899.

After the era of horseless carriages, EVs entered the era of commercial development. They had the features of wire-spoken wheels, pneumatic tires, comfortable springs and rich upholstery. By 1912, nearly 34 000 EVs were registered in America. From 1899 to 1916, the Baker Electric Company was one of the most important EV makers in America. From 1901 to 1920, the British Electromobile Company of London also produced rear-wheel-driven EVs with motors on the rear wheels, inclined wheel steering and pneumatic tires. From 1907 to 1938, the Detroit Electric not only offered the merits of being silent, clean and reliable, but also had a top speed of 40 km/h and a range of 129 km per charge.

Fig. 2.2. London Electrical Cab Company's taxi. (Courtesy of Scientific American Supplement; Photo courtesy of History of the Electric Automobile by Ernest H. Wakefield.)

People like to say that one's enemy is one's partner. This saying applied to EVs because the electric motor, being a key element for EV propulsion, also helped ICEVs in their fight against EVs. In 1911, Kettering invented the automobile starter motor which made the ICEVs more attractive to the drivers who had relied on the easy-to-drive EVs. This eliminated a major part of the EV market. Another man with an idea which finished off the EVs for good was Ford. His mass-produced Ford Model T, originally priced for US$850 in 1909, was selling for US$260 in 1925. This accelerated the disappearance of EVs. Moreover, the ICEVs could offer a range double or triple that of the EVs but at only a fraction of the cost. This virtually removed the possibility of EV manufacturers upholding any market share. By the 1930s, the EVs had almost vanished from the scene.

2.2 Recent development

The rekindling of interests in EVs started at the outbreak of the energy crisis and oil shortage in the 1970s. In the early 1970s, many countries throughout the world, such as America, Britain, France, Germany, Italy and Japan, were developing

EVs. In 1976, the USA issued the Electric and Hybrid Vehicle Research, Development and Demonstration Act, Public Law 94-413. Since late 1970s, companies in many other nations and regions, including Australia, Belgium, Brazil, Bulgaria, Canada, China, Denmark, France, Germany, Hong Kong, Holland, India, Italy, Japan, Mexico, Sweden, Switzerland, UK, USA and USSR, were building EVs. However, oil prices fell at the end of 1970s. The energy crisis and oil shortage thus faded out before an EV acceptable for commercial production could be developed. The impetus to commercialise EVs therefore lost momentum and those activities related to EV development slowed down dramatically.

Due to the growing concern over air quality and the possible consequences of the greenhouse effect in the 1980s, EV development activities were re-triggered. In the early 1990s, some countries and cities started to enforce stricter emissions regulations. In 1990, the California Air Resources Board (CARB) established rules that 2% of all vehicles sold in California in 1998 should be zero emission vehicles (ZEVs), and by 2003, the ZEV sales should be 10%. Being stimulated by the California rules, many other states in America and some other countries throughout the world set similar regulations. Since EVs were identified to be the only available technology meeting the criteria of ZEVs, the development of EVs was accelerated. Although the CARB's 1998 target was not realized, the California rules did mobilize the EV development.

Automotive makers lead the development and begin the commercialization of EVs. Throughout the world, especially in America, Japan and Europe, many automotive makers have produced their EVs or been involved in EV activities. In America, General Motors (GM), Ford, Chrysler, U.S. Electricar and Solectria played active roles in the development of EVs in response to the California rules. In Japan, nearly all automotive makers, including Toyota, Nissan, Honda, Mazda, Daihatsu, Mitsubishi, Suzuki, Isuzu and Subaru, had their programmes to evaluate or commercialize EVs. Similarly, many countries in Europe, especially France, Germany, Italy and Britain, also launched their projects to target the coming EV market. Some of these active European automotive makers were PSA Peugeot Citroën Group, Renault, BMW, Mercedes-Benz, Audi, Volvo, Opel, Volkswagen, Fiat and Bedford. Apart from automotive makers, power utilities and battery manufacturers also play an active role in the demonstration of EVs. Both of them aim to enhance the commercialization of rechargeable battery-powered EVs so as to benefit their businesses. Usually, they either collaborate with automotive makers to develop their EVs or simply purchase available EVs for evaluation and demonstration. In response to the inherent benefits of energy efficiency, energy diversification and environmental friendliness resulting from the use of EVs, both energy and environment agencies are also actively involved in advancing EV technologies and promoting EV commercialization. Last, but not the least, research laboratories and universities continually fuel EV technologies in such a way that EVs can be a real competitor to ICEVs.

The work of GM with EVs began in 1916 when GMC Truck produced a number of electric trucks using lead–acid batteries (Rajashekara, 1994). Over the last three decades, GM has built a number of experimental EVs, such as the Electrovair in 1966, the Electrovan in 1968, the Stir-Lec I in 1968, the Series 512 in 1969 and the Electrovette in 1979. In January 1990, GM announced the first EV intended for production called the Impact. This Impact demonstrated that electric propulsion technology is available to produce practical EVs that may be competitive with today's ICEVs in performance. In 1993, GM started limited production of the Impact 3, a revised version of the 1990 Impact, for evaluation by potential customers of EVs. In 1996, GM began to lease its EV1 which was a production version of the Impact coupe, and to market an electric conversion of the Chevrolet S-10 pickup truck for use in commercial fleets. Recently, GM has worked hard on hybrid EV (HEV) and fuel cell EV (FCEV) concepts. Its HEV, called the Precept, features a dual-axle regenerative hybrid propulsion system which combines an electric front-wheel drive and a hybrid rear-wheel drive. In this first instance, its FCEV, called the Zafira, runs on liquid hydrogen stored in a cryogenic fuel tank.

Ford's involvement in modern EVs began in the early 1960s when Ford of Europe built its first complete research EV, the Comuta, which was a very compact EV for carrying two adults plus two children. Subsequently, it conducted a number of EV projects, such as the Cortina Station Wagon EV in the late 1960s, the Econoline Van Hybrid in the 1970s as well as the Fiesta EV, the Escort EV, the ETX-I (LN7) and the ETX-II (Aerostar) in the 1980s. Based on the Ford's European Escort van, the Ecostar was its demonstration EV in the 1990s. As of May 1995, 103 Ecostars had been placed into service and driven a total of 800 000 km. In 1995, Ford also announced the sale of an electric conversion pickup truck, the Ranger. In 1998, Ecostar became a company name as a result of the alliance among Ford, Ballard and Daimler-Benz (now DaimlerChysler), and supplied electric drive systems for both EVs and HEVs. Recently, Ford has sought to generate an EV brand, called the Th!nk, which includes the City and the Neighbor. Also, Ford has demonstrated its FCEV, called the P2000, and announced its HEV, called the Escape. The P2000 is a four-door sedan as fuelled by compressed hydrogen gas, whereas the Escape is a sport-utility vehicle (SUV).

Fundamental development work on EVs was under way at Nissan from 1970. Its first concept EV, the City, was exhibited at the Tokyo Motor Show in 1970. Then, Nissan produced the EV-4 in 1976, the March EV in 1983, the EV Resort in 1985, the March EV-II in 1987, and the Garbage Collecting Truck in 1988. In 1991, it produced the President EV, the Cedric EV and the FEV. The FEV stood for the future electric vehicle that was designed as a concept car. In 1995, its second generation, the FEV II, was produced, and was firstly powered by lithium-ion batteries. In 1994, Nissan launched domestic sales of the Cedric EV (a four-seater sedan) for use by government agencies, and the Avenir EV (a commercial van) for use by utility companies. In 1997, it produced the Prairie Joy and the Altra EV for the marketing of multi-purpose wagons. Recently, Nissan has demonstrated its HEV, called the Tino, which is on sale in Japan on the internet. Also, it has participated in the California Fuel Cell Partnership Program, and introduced its direct hydrogen FCEV, called the FCV.

Toyota started its involvement in EV development in 1971 when it joined a programme sponsored by Japan's Ministry of International Trade and Industry. From 1983 to 1989, Toyota produced a series of EVs, from the EV-10 to the EV-40. Its development work led to the TownAce EV in 1991, the Crown Majesta EV in 1992, the EV-50 in 1993, and the Coaster HEV in 1994. In 1995, it produced the RAV4-EV which won the FIA-sanctioned Scandinavian Electric Car Rally '95, testifying to its performance and capabilities. Also in 1995, it successfully developed a practical HEV, the Prius, which was commercially sold in 1997. So far, over 40 000 Prius HEVs have been sold. In addition, Toyota began to develop its FCEV, the RAV4, aiming to greatly extend the driving range of EVs. In 1996, it developed its hydrogen-absorbing alloy storage device to enable the RAV4 to travel up to 202 km on a single charge. In 1997, it also developed its methanol-to-hydrogen reformer to enable it achieving a range of 500 km per tank of liquid methanol. Recently, Toyota has demonstrated a new Post-Prius hybrid system, called the THS-C, which is an electric four-wheel drive system combining a hybrid front-wheel drive and an electric rear-wheel drive.

In the early 1990s, the PSA Peugeot Citroën Group began a commitment to adapt its vehicles to the environment by the launch of EVs. In 1990, it produced the Peugeot J5 and the Citroën C25 for the marketing of utility vehicles. In 1995, the Peugeot 106 and the Citroën AX were produced for the marketing of small sedans. Following the Citroën Citela and the Peugeot Ion, the Tulip presented a new concept for the urban EV, which was purposely designed to offer subscribers the use of rented EVs. Recently, PSA has collaborated with Renault in a FCEV programme, and exhibited its ICE-heavy HEV, called the Xsara Dynactive.

Since the 1960s, the Fiat Group began to develop a set of technologies for the manufacturing of EVs. Subsequently, throughout the 1970s, this activity was concentrated on the integration of these technologies into the vehicle structure. Thus, Fiat developed its first experimental EV, the X1/23, in 1974. In 1978, it cooperated with Pininfarina to produce a city EV, the Ecos. In 1989, its electric race car, the Y10, successfully participated in the 2nd Gran Premio 4E in Turin under the E3 class (vehicles of weight lower than 1000 kg). In 1990, Fiat presented its Panda Elettra which was realized for practical use in urban areas. Then, it also produced the Cinquecento Elettra for the marketing of passenger cars and the Ducato Elettra for the marketing of delivery vans. Consequently, it developed a two-seater EV, the Zic, in 1994, and a four-seater EV, the Seicento Elettra, in 1998. Recently, Fiat has exhibited interests in both HEVs and FCEVs, and presented an ICE-heavy HEV, the Multipla.

Like most automotive makers, BMW started its EV development by converting production cars. The first generation of BMW's conversion EV, namely the E30E, was built in 1989. Its second generation EV, the E36E, was produced in 1991. It offered an optimal conversion design, hence enhancing the suitability for small scale series production, especially for fleet tests. In 1991, BMW launched its first conceptual study for a purpose-design EV, called the E1, which was a two-door four-seater EV. In 1992, BMW unveiled its four-door four-seater version, the E2.

Recently, BMW has focused on the development of hydrogen-fuelled ICEVs, rather than HEVs or FCEVs.

2.3 State-of-the-art EVs and HEVs

Besides the commercialization of EVs and HEVs, the development has been focused on the advancement of their technologies. Examples of this are the use of advanced induction motors or permanent-magnet brushless motors to improve the electric propulsion system, the employment of advanced batteries, fuel cells, and/or engines to improve the on-board energy source, and the adoption of inductive charging or variable temperature seats to improve the EV auxiliaries.

Showcasing the most advanced propulsion system, the 1997 two-seater GM EV1 is shown in Fig. 2.3. It had a front-wheel drive which adopted a 102-kW 3-phase induction motor and a single-speed transaxle with dual-reduction of 10.946:1. It contained 26-module 312-V valve-regulated lead–acid (VRLA) batteries which were inductively charged by a 6.6-kW off-board charger or an 1.2-kW on-board charger. This EV1 could offer an axle torque of 1640 Nm from 0 to 7000 rpm and a propulsion power of 102 kW from 7000 to 14 000 rpm, leading to achieve a top speed of 128 km/h (electronically limited) and an acceleration from 0 to 96 km/h in less than 9 s. For city driving, it could provide a range of 112 km per charge; whereas on highway operation, it offered 144 km per charge. In 1999, the EV1 adopted nickel–metal hybrid batteries, hence reaching 220 km per charge.

Figure 2.4 shows the 1997 four-seater Altra EV which was the flagship of Nissan. It used a 62-kW permanent-magnet synchronous motor which weighed only 39 kg, the highest power-to-weight ratio (1.6 kW/kg) for any EV motor available. Making use of maximum efficiency control, the total efficiency of the propulsion system was more than 89%. Power came from the cobalt-based lithium-ion batteries which had a specific energy of 90 Wh/kg, a specific power of

Fig. 2.3. GM EV1. (Photo courtesy of General Motors.)

Fig. 2.4. Nissan Altra EV. (Photo courtesy of Nissan.)

300 W/kg and a long cycle life of about 1200 recharges. This battery pack could be charged up by an on-board inductive charging system within 5 h. It could achieve a top speed of 120 km/h and a range of 192 km for city driving. In 1999, the Altra adopted the manganese-based lithium-ion batteries to further increase both specific energy and specific power to 91 Wh/kg and 350 W/kg, respectively.

The Ford P2000 symbolized the dedication of Ford in the development of fuel cell EVs. Figure 2.5 shows this four-door sedan which was launched in the year 2000. It was powered by the Ford's Th!nk fuel cell system, namely the proton-exchange-membrane (PEM) fuel cells, which was fuelled by compressed hydrogen gas stored at 25 MPa and oxygen gas simply from the air. It adopted a three-phase induction motor, offering a peak power of 67 kW, a peak torque of 190 Nm and a peak efficiency of 91%. With the curb weight of 1514 kg, the P2000 could achieve a top speed of 128 km/h and a driving range of 160 km per charge.

Daimler-Benz, now DaimlerChrysler, presented its first methanol-fuelled fuel cell EV in 1997—the NECAR 3. It used proton-exchange-membrane (PEM) fuel

Fig. 2.5. Ford P2000. (Photo courtesy of Ford Motor Company.)

Fig. 2.6. DaimlerChrysler NECAR 5. (Photo courtesy of DaimlerChrysler.)

cells to generate a power of 50 kW for propulsion. The hydrogen fuel was directly extracted from methanol via a mini reformer, thus bypassing the problem of having compressed gas canisters on-board the vehicle. The fuel cells were stored beneath the floor, while the reformer, methanol tank and control systems were located in the boot. Based on this first generation methanol-fuelled fuel cell propulsion system, the NECAR 3 could travel over 400 km on 38 l of liquid methanol. As shown in Fig. 2.6, the NECAR 5 launched in 2000 was the technological successor of the NECAR 3, while reducing the size of the drive system by half and the weight of the vehicle by 300 kg. It also boosted up the power to 75 kW to reach speeds over 150 km/h.

The world's first mass-production HEV was the Toyota Prius as shown in Fig. 2.7. Its motive power was sourced from both a four-cylinder ICE (52 kW at 4500 rpm) and a permanent-magnet synchronous motor (33 kW at 1040–5600 rpm). Since it was an ICE-heavy HEV, a power split device, namely the planetary gear, sent part of the ICE power to the wheels and part to a generator. The generated electrical energy could supply the electric motor to increase the motive power, or could be stored in the 38-module nickel–metal hybrid batteries. The Prius could offer a top speed of 160 km/h, an acceleration from 0 to 96 km/h in 12.7 s, and a fuel economy of 20 km/l for combined city and highway operation. Both of its fuel economy and exhaust emissions were much better than that of any conventional ICEVs.

The Honda Insight, shown in Fig. 2.8, went on sale in December 2000. It employed an ICE-heavy hybrid system, combining a three-cylinder ICE (50 kW at 5700 rpm) and a permanent-magnet synchronous motor (10 kW at 3000 rpm). The electric motor was powered by a 144-V nickel–metal hydride battery pack which

Fig. 2.7. Toyota Prius. (Photo courtesy of Toyota.)

Fig. 2.8. Honda Insight. (Photo courtesy of Honda.)

was recharged by regenerative braking during normal cruising and downhill driving. The Insight was claimed to be the most fuel-efficient HEV with the fuel economy of 26–30 km/l. Also, it satisfied the stringent ultra-low-emission vehicle (ULEV) standard in California.

To simultaneously address the problems of air pollution, wasteful energy consumption and traffic safety, the National Institute for Environmental Studies

Fig. 2.9. NIES Luciole. (Photo courtesy of Japan National Institute for Environmental Studies.)

(NIES) in Japan presented a high-performance lightweight EV, namely the Luciole (formerly called Eco-Vehicle), in 1996 for convenient city commuting. The first author was invited to visit NIES as Distinguished Fellow in 1996 and partly participated the development of Luciole. As shown in Fig. 2.9, it adopted a tandem two-seater layout so that the seats could be kept comfortable, and the safety in side crushes could be improved by thickening the doors. It was a rear-wheel drive which was powered by two in-wheel permanent-magnet synchronous motors with the total output of 72 kW and 154 Nm. The battery pack contained 224-V VRLA batteries (which were mounted inside the square holes of the purpose-built chassis) could be charged up by normal charging within 5 h, by fast charging within 15 min or even partially charged by solar charging (Shimizu *et al.*, 1997). The Luciole could achieve a top speed of 130 km/h, a range on the Japan 10.15 Mode driving cycle of 130 km, and an acceleration from 0 to 40 km/h in 3.9 s.

Figure 2.10 shows an EV, the U2001, which was developed by the University of Hong Kong (HKU) in 1993. It was a four-seater EV which adopted a 45-kW permanent-magnet hybrid motor and a 264-V nickel–cadmium battery pack. This specially designed EV motor could offer high efficiencies over a wide operating range. It also incorporated a number of advanced EV technologies, such as the adoption of thermoelectric variable temperature seats to minimize the energy used for air-conditioning, the use of an audio navigation system to facilitate safe and user-friendly driving, and the use of an intelligent energy management system to optimize the energy flow within the vehicle. The U2001 could offer a top speed of 110 km/h, an acceleration from 0 to 48 km/h in 6.3 s, and a range of 176 km at 88-km/h operation.

Fig. 2.10. HKU U2001.

Apart from the USA, Europe and Japan, India also plays an active role to commercialize EVs. Figure 2.11 shows a two-door hatchback EV, the Reva EV, which was launched in the year 2001 and would be India's first mass-produced

Fig. 2.11. Reva EV. (Photo courtesy of Reva Electric Car Company.)

EV. It adopted a separately excited dc motor (70 Nm, 13 kW peak) and a 48-V tubular lead–acid battery pack. Its on-board charger (220 V, 2.2 kW) could provide 80% charge within 3 h and 100% within 6 h. With the curb weight of 650 kg, the Reva EV could achieve a top speed of 65 km/h and a range of 80 km per charge. The most attractive feature was its incredibly low initial and running costs—the selling price is about US$5000 and the running cost is less than one US cent per kilometer.

References

Daimler-Benz NECAR 3. Daimler-Benz, 1997.

Daniels, J. (2000). Millennium motors. *Electric & Hybrid Vehicle Technology International*, pp. 6–10.

Ford P2000. Ford Motor Company, 2000.

GM EV1. General Motors, 1997.

Honda Insight. Honda, 2000.

Http://www.media.daimlerchrysler.com/.

NIES Luciole. National Institute for Environmental Studies, 1996.

Nissan Altra EV. Nissan, 1999.

Powering the Future. Ballard Automotive, 1999.

Prius Product Information. Toyota, 2000.

Rajashekara, K. (1994). History of electric vehicles in General Motors. *IEEE Transactions on Industry Applications*, **30**, 897–904.

Reva: The Ideal City Car. Reva Electric Car Company, 2000.

Shacket, S.R. (1979). *The Complete Book of Electric Vehicles*. Domus Books, Chicago.

Shimizu, H., Harada, J., Bland, C., Kawakami, K. and Chan, L. (1997). Advanced concepts in electric vehicle design. *IEEE Transactions on Industrial Electronics*, **44**, 14–18.

Toyota Electric & Hybrid Vehicles. Toyota, 2000.

Wakefield, E.H. (1994). *History of the Electric Automobile: Battery-Only Powered Cars*. Society of Automotive Engineers, Warrendale, PA.

Wakefield, E.H. (1998). *History of the Electric Automobile: Hybrid Electric Vehicles*. Society of Automotive Engineers, Warrendale, PA.

3 EV systems

The conventional ICEV employs a combustion engine for propulsion. Its energy source is liquid petrol or diesel. In contrast, the EV employs an electric motor and the corresponding energy sources are batteries, fuel cells, capacitors and/or flywheels. Notice that the presently achievable specific energy of capacitors and flywheels precludes them from being the sole energy sources for EVs. The key difference between the ICEV and EV is the device for propulsion (ICE vs. electric). There are many alternatives for configuring EVs. Although EVs borrow most of the system parameters from ICEVs, they have some parameters that are absent in ICEVs (Chan, 1993; Unnewehr and Nasar, 1982; Shacket, 1979).

3.1 EV configurations

Previously, the EV was mainly converted from the ICEV, simply replacing the combustion engine by the electric motor while retaining all the other components. This converted EV has been fading out because of the drawback of heavy weight, loss of flexibility and degradation of performance. At present, the modern EV is purposely built. This purpose-built EV is based on original body and frame designs to satisfy the structural requirements unique to EVs and to make use of the greater flexibility of electric propulsion.

Compared with the ICEV, the configuration of the EV is particularly flexible. This flexibility is due to several factors unique to the EV. Firstly, the energy flow in the EV is mainly via flexible electrical wires rather than bolted flanges or rigid shafts. Thus, the concept of distributed subsystems in the EV is really achievable. Secondly, different EV propulsion arrangements (such as independent four-wheel and in-wheel drives) involve a significant difference in the system configuration. To a lesser degree, different EV propulsion devices (such as dc and induction motor drives) also have different weights, sizes and shapes. Thirdly, different EV energy sources (such as batteries and fuel cells) have different weights, sizes and shapes. The corresponding refuelling systems also involve different hardware and mechanism. For example, the batteries can be electrically recharged via conductive or inductive means, or can also be mechanically exchanged (so-called mechanically recharged) and then recharged centrally.

Figure 3.1 shows the general configuration of the EV, consisting of three major subsystems—electric propulsion, energy source and auxiliary. The electric propulsion subsystem comprises the electronic controller, power converter, electric motor, mechanical transmission and driving wheels. The energy source subsystem involves the energy source, energy management unit and energy refuelling unit.

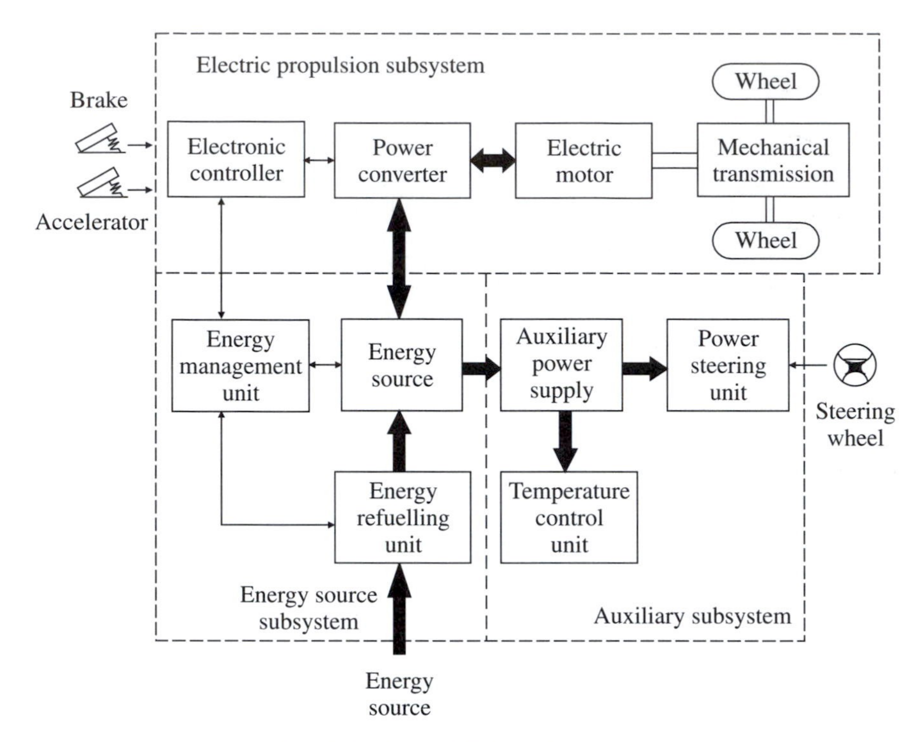

Fig. 3.1. General EV configuration.

The auxiliary subsystem consists of the power steering unit, temperature control unit and auxiliary power supply. In the figure, a mechanical link is represented by a double line, an electrical link by a thick line and a control link by a thin line. The arrow on each line denotes the direction of electrical power flow or control information communication. Based on the control inputs from the brake and accelerator pedals, the electronic controller provides proper control signals to switch on or off the power devices of the power converter which functions to regulate power flow between the electric motor and energy source. The backward power flow is due to regenerative braking of the EV and this regenerative energy can be stored provided the energy source is receptive. Notice that most available EV batteries (except some metal/air batteries) as well as capacitors and flywheels readily accept regenerative energy. The energy management unit cooperates with the electronic controller to control regenerative braking and its energy recovery. It also works with the energy refuelling unit to control refuelling and to monitor usability of the energy source. The auxiliary power supply provides the necessary power with different voltage levels for all EV auxiliaries, especially the temperature control and power steering units. Besides the brake and accelerator, the steering wheel is another key control input of the EV. Based on its angular position, the power steering unit can determine how sharply the vehicle should turn.

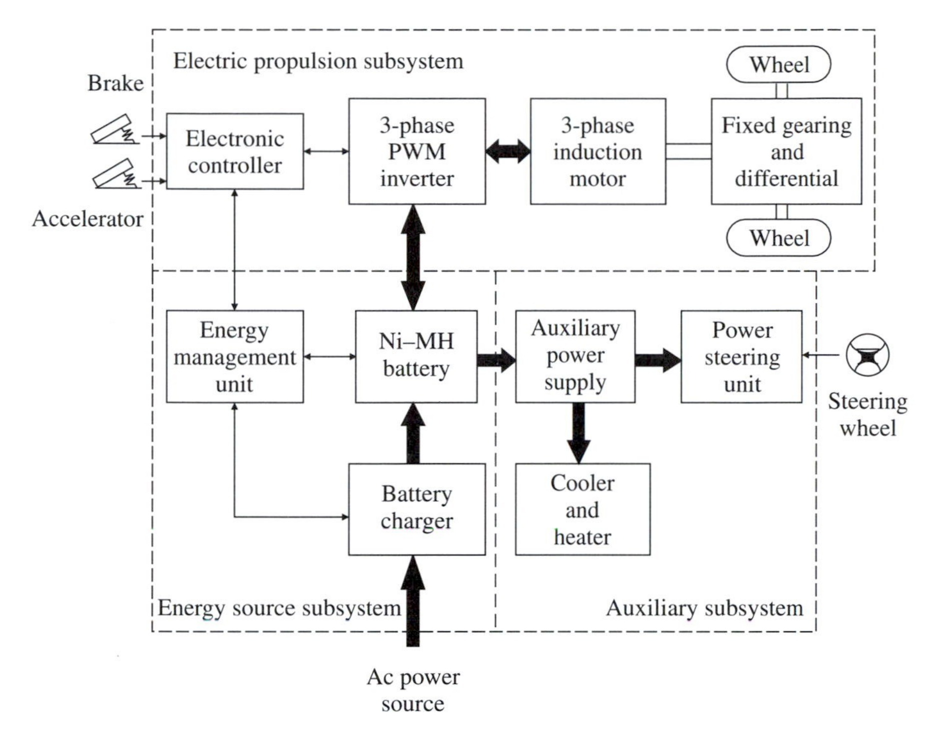

Fig. 3.2. Typical EV configuration.

For a modern EV, a three-phase induction motor is typically selected. The corresponding power converter is a three-phase PWM inverter. In general, the mechanical transmission is based on fixed gearing and a differential. Also, a nickel–metal hydride (Ni–MH) battery is also typically selected as the energy source. The corresponding refuelling unit becomes a battery charger. The temperature control unit generally consists of a cooler and/or a heater, depending on the climate of a particular country. This typical set-up is shown in Fig. 3.2.

At present, there are many possible EV configurations due to the variations in electric propulsion and energy sources. Focusing on those variations in electric propulsion, there are six typical alternatives as shown in Fig. 3.3:

(1) Figure 3.3(a) shows the first alternative which is a direct extension of the existing ICEV adopting longitudinal front-engine front-wheel drive. It consists of an electric motor, a clutch, a gearbox and a differential. The clutch is a mechanical device which is used to connect or disconnect power flow from the electric motor to the wheels. The gearbox is another mechanical device which consists of a set of gears with different gear ratios. By incorporating both clutch and gearbox, the driver can shift the gear ratios and hence the torque going to the wheels. The wheels have high torque low speed in the lower gears and high speed low torque in the higher gears. The differential is a mechanical

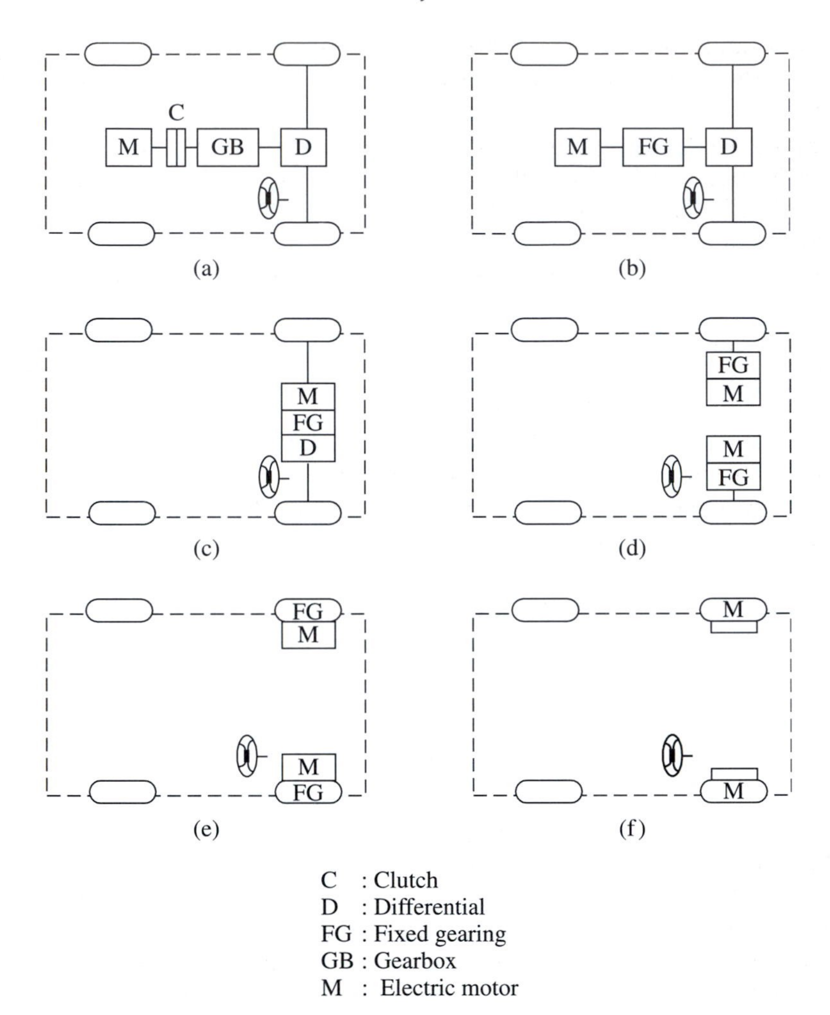

Fig. 3.3. EV configuration due to variations in electric propulsion.

device which enables the wheels to be driven at different speeds when cornering—the outer wheel covering a greater distance than the inner wheel.

(2) By replacing the gearbox with fixed gearing and hence removing the clutch, both the weight and size of the mechanical transmission can be greatly reduced. Figure 3.3(b) shows this arrangement which consists of an electric motor, fixed gearing and a differential. Notice that this EV configuration is not suitable for the ICEV as the engine by itself, without the clutch and gearbox, cannot offer the desired torque-speed characteristics.

(3) Similar to the concept of transverse front-engine front-wheel drive of the existing ICEV, the electric motor, fixed gearing and differential are integrated

into a single assembly, while both axles point at both driving wheels. Figure 3.3(c) show this configuration which is in fact most commonly adopted by modern EVs.

(4) Besides the mechanical means, the differential action of an EV when cornering can be electronically provided by two electric motors operating at different speeds. Figure 3.3(d) shows this dual-motor configuration in which two electric motors separately drive the driving wheels via fixed gearing.

(5) In order to further shorten the mechanical transmission path from the electric motor to the driving wheel, the electric motor can be placed inside a wheel. This arrangement is the so-called in-wheel drive. Figure 3.3(e) shows this configuration in which fixed planetary gearing is employed to reduce the motor speed to the desired wheel speed. It should be noted that planetary gearing offers the advantages of a high speed-reduction ratio as well as an in-line arrangement of input and output shafts.

(6) By fully abandoning any mechanical gearing, the in-wheel drive can be realized by installing a low-speed outer-rotor electric motor inside a wheel. Figure 3.3(f) shows this gearless arrangement in which the outer rotor is directly mounted on the wheel rim. Thus, speed control of the electric motor is equivalent to the control of the wheel speed and hence the vehicle speed.

Apart from the variations in electric propulsion, there are other EV configurations due to the variations in energy sources (batteries, fuel cells, capacitors and flywheels). Six typical alternatives are shown in Fig. 3.4.

(1) Figure 3.4(a) shows a basic battery-powered configuration that is almost exclusively adopted by existing EVs. The battery may be distributed around the vehicle, packed together at the vehicle back or located beneath the vehicle chassis. This battery should be able to offer reasonable specific energy and specific power as well as being able to accept regenerative energy during braking. Notice that both high specific energy and high specific power are desirable for EV applications as the former governs the driving range while the latter dictates the acceleration rate and hill-climbing capability. A battery having a design compromised between specific energy and specific power is generally adopted in this configuration.

(2) Instead of using a compromised battery design, two different batteries (one is optimized for high specific energy while another for high specific power) can be used simultaneously in an EV. Figure 3.4(b) shows the basic arrangement of this battery & battery hybrid energy source. This arrangement not only decouples the requirements on energy and power but also affords an opportunity to use those mechanically rechargeable batteries which cannot accept regenerative energy during braking or downhill.

(3) Differing from the battery which is an energy storage device, the fuel cell is an energy generation device. The operating principle of fuel cells is a reverse process of electrolysis—combining hydrogen and oxygen gases to form

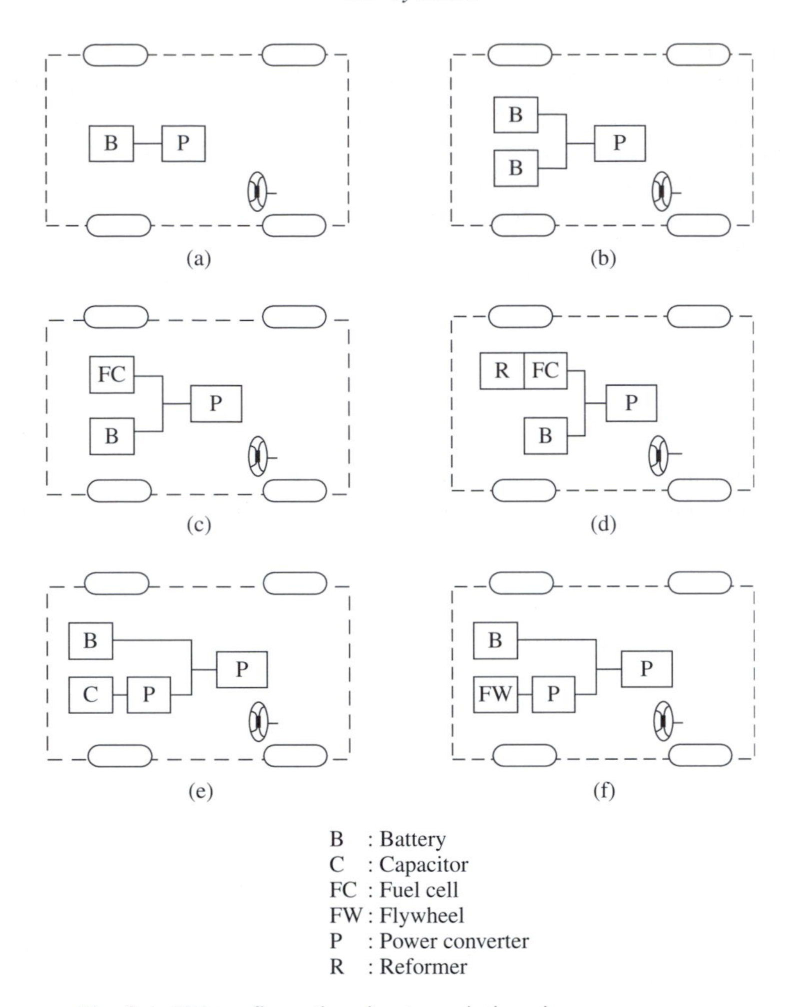

Fig. 3.4. EV configuration due to variations in energy sources.

electricity and water. Hydrogen gas can be stored in an on-board tank whereas oxygen gas is simply extracted from air. Since the fuel cell can offer high specific energy but cannot accept regenerative energy, it is preferable to combine it with a battery with high specific power and high energy receptivity. Figure 3.4(c) shows this arrangement which is denoted as a fuel cell & battery hybrid energy source.

(4) Rather than storing it as a compressed gas, a liquid or a metal hydride, hydrogen can be on-board generated from ambient-temperature liquid fuels such as methanol or even petrol. As shown in Fig. 3.4(d), a mini reformer is installed in the EV to produce on line the necessary hydrogen gas for the fuel cell.

(5) In contrast to the fuel cell & battery hybrid in which the battery is purposely selected to offer high specific power and high energy receptivity, the battery in the battery & capacitor hybrid is aimed to have high specific energy. This is because a capacitor can inherently offer a much higher specific power and energy receptivity than a battery. Since the available capacitors for EV application, usually termed as ultracapacitors, are of relatively low voltage level, an additional dc–dc power converter is needed to interface between the battery and capacitor terminals. Figure 3.4(e) shows this configuration.

(6) Similar to the capacitor, the flywheel is another emerging energy storage device which can offer high specific power and high energy receptivity. It should be noted that the flywheel for EV applications is different from the conventional design which is characterized by low speed and massive size. In contrast, it is lightweight and operates at ultrahigh speeds under a vacuum environment. This ultrahigh-speed flywheel is incorporated into the rotor of an electric machine which operates at motoring and generating modes when converting electrical energy to and from kinetic energy, respectively. The corresponding configuration is shown in Fig. 3.4(f) in which the battery is selected to offer high specific energy. Since this flywheel is preferably incorporated into an ac machine which is brushless and can offer a higher efficiency than that of a dc machine, an additional ac–dc converter is needed to interface between the battery and flywheel terminals.

3.1.1 FIXED AND VARIABLE GEARING

Fixed gearing means that there is a fixed gear ratio between the propulsion device (ICE or electric motor) to the driving wheels. In contrast, variable gearing involves shifting between different gear ratios, this can be accomplished by using a combination of clutch and gearbox. The purpose of variable gearing is to provide multiple-speed transmission (achieving wide ranges of speed and torque using different gear ratios). Generally, four- or five-speed transmission is used for passenger cars, and up to 16-speed transmission for trucks. When the clutch is engaged, the propulsion device and the gearbox are coupled together and power transmission is enabled. When it is disengaged manually or automatically, the power transmission is interrupted so that the gear ratio in the gearbox can be shifted.

For ICEVs, there is no alternative to the use of variable gearing as the ICE cannot offer the desired torque-speed characteristics (such as high torque for hill climbing and high speed for cruising) without using multiple-speed transmission. Figure 3.5 shows typical force-speed characteristics of an ICE with five-speed transmission. For EVs, the employment of variable gearing to achieve multiple-speed transmission used to be controversial. For EVs converted from ICEVs, the use of variable gearing was claimed to be natural because both gearbox and clutch are already present and their maintenance costs are minor. However, the concept of converted EVs is almost obsolete as it cannot fully utilize the flexibility and

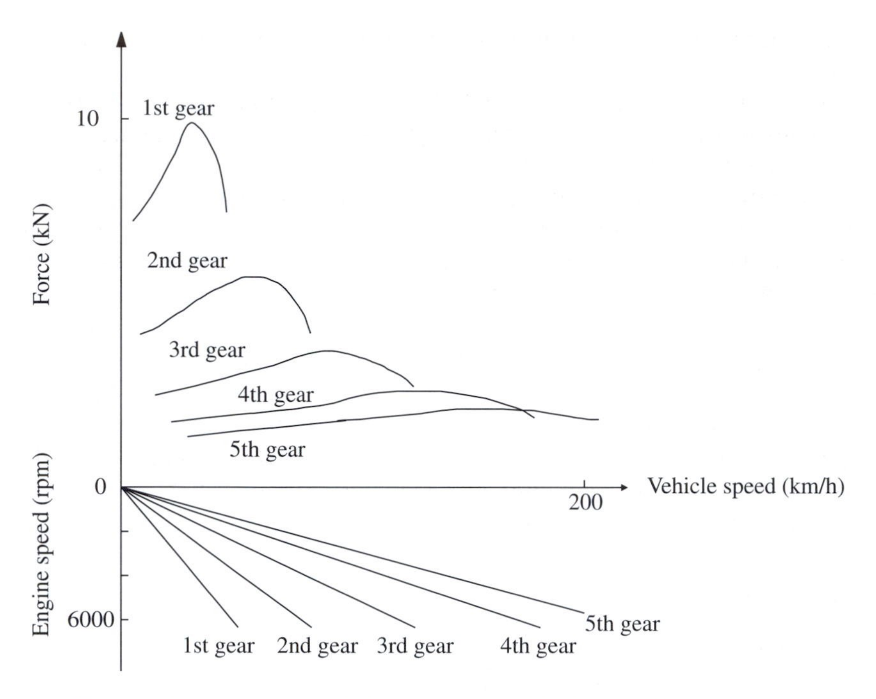

Fig. 3.5. ICEV force-speed characteristics with five-speed transmission.

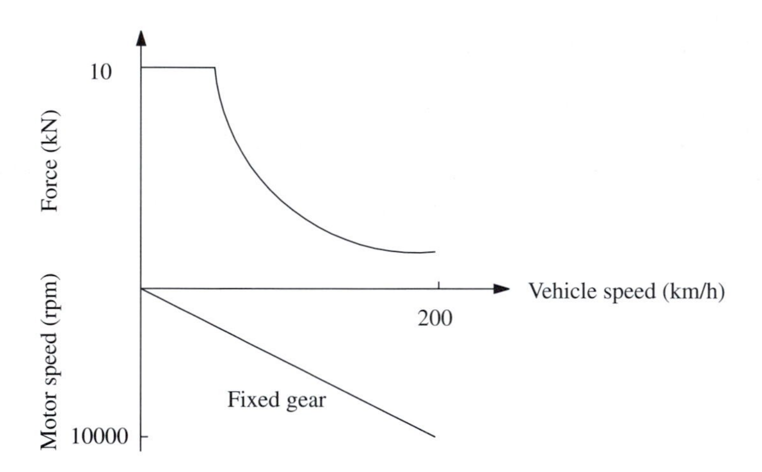

Fig. 3.6. EV force-speed characteristiscs with fixed gearing.

potentiality offered by EVs. It was also claimed that the use of variable gearing can enhance the electric motor achieving regenerative braking and high efficiency operation over a wide speed range. With the advances of power electronics and control algorithms, both regenerative braking and high efficiency operation of electric motors can be easily achieved by electronic means rather than mechanical means.

Fixed-gearing transmission is usually based on planetary gearing. A planetary gearset consists of a sun gear, several planet gears, a planet gear carrier and a ring gear. It takes the advantages of strong, compact, high efficiency, high speed-reduction ratio and in-line arrangement of input and output shafts over the conventional parallel-shaft variable gearset. Thus, the removal of this variable gearing can significantly reduce the overall complexity, size, weight and cost of the transmission. Moreover, modern electric motors with the use of fixed gearing can readily offer the desired torque-speed characteristics for vehicular operation. Figure 3.6 shows typical force-speed characteristics of an EV with fixed gearing, consisting of constant-torque operation for acceleration and hill climbing as well as constant-power operation for high-speed cruising. Moreover, the absence of gear changing (irrespective of whether it is manual or automatic) can greatly enhance smooth driving and transmission efficiency. Therefore, modern EVs almost exclusively adopt fixed gearing rather than variable gearing.

3.1.2 SINGLE- AND MULTIPLE-MOTOR DRIVES

A differential is a standard component for conventional vehicles and this technology can be carried forward to the EV field. When a vehicle is rounding a curved road, the outer wheel needs to travel on a larger radius than the inner wheel. Thus, the differential adjusts the relative speeds of the wheels; otherwise, the wheels will slip which causes tire wear, steering difficulties and poor road holding. For all ICEVs, whether front- or rear-wheel drive, a differential is mandatory. Figure 3.7 shows a typical differential in which pinion spider gears can rotate on their shaft, allowing axle side gears to turn at different speeds.

For EVs, it is possible to dispense with a mechanical differential. By separately coupling two or even four electric motors to the driving wheels, the speed of each wheel can be independently controlled in such a way that the differential action

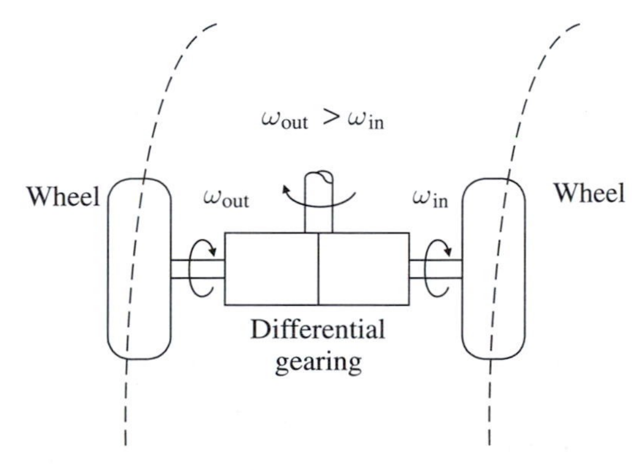

Fig. 3.7. Mechanical differential.

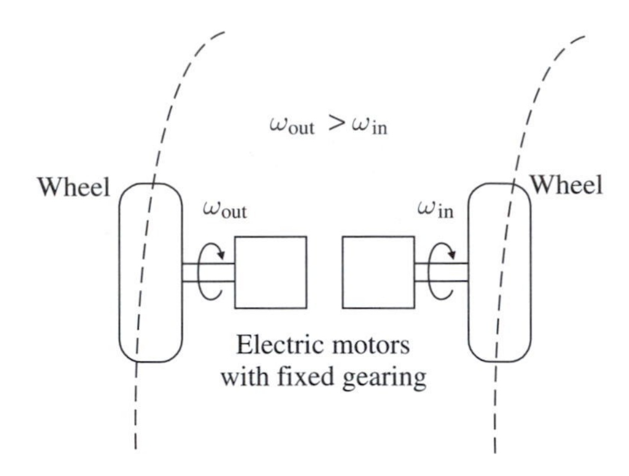

Fig. 3.8. Electronic differential.

can be electronically achieved when cornering. Figure 3.8 shows a typical dual-motor drive with an electronic differential. This arrangement is smaller and lighter than the mechanical counterpart. Unlike the choice between variable gearing and fixed gearing, the selection of either a single-motor drive with a differential or a multiple-motor drive without a differential is still controversial. Positively, the removal of a mechanical differential can reduce the overall size and weight while the electronic differential can accurately control the wheel speeds so as to achieve better performance during cornering. Negatively, the use of an additional electric motor and power converter causes an increase in the initial cost while the reliability of the electronic controller to accurately control two electric motors at various driving conditions is to be observed. In recent years, the reliability of this electronic controller has been greatly improved by incorporating the capability of fault tolerance. For example, the electronic controller consists of three microprocessors. Two of them are used to separately control the motor speeds for the left and right wheels while the remaining one is used for system control and coordination. All of them watch one another by using a watchdog to improve the reliability.

3.1.3 IN-WHEEL DRIVES

By placing an electric motor inside the wheel, the in-wheel motor has the definite advantage that the mechanical transmission path between the electric motor and the wheel can be minimized or even eliminated, depending whether the electric motor is a high-speed inner-rotor type or a low-speed outer-rotor type. When it is a high-speed inner-rotor motor, a fixed speed-reduction gear becomes necessary to attain a realistic wheel speed. In general, a high speed-reduction planetary gearset is adopted which is mounted between the motor shaft and the wheel rim. Typically, this motor is purposely designed to operate up to about 10 000 rpm so as to give a higher power density. This maximum speed is limited by the friction and windage losses as well as the transmission tolerance. Thus, the corresponding

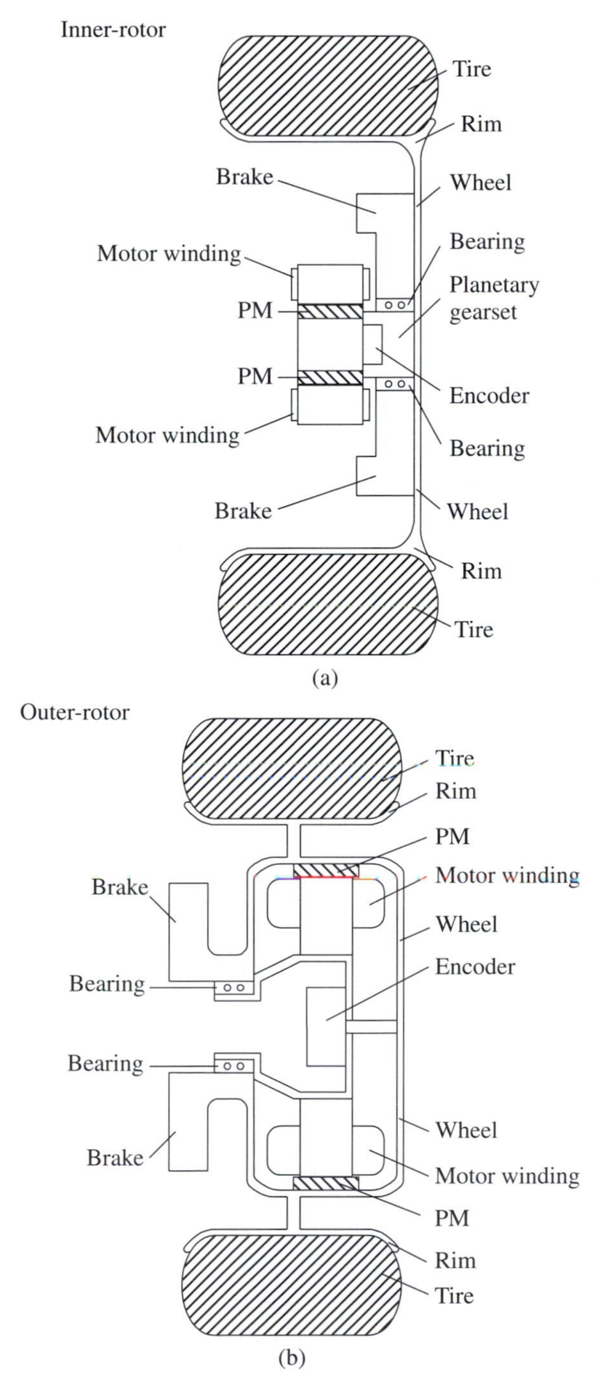

Fig. 3.9. In-wheel drives. (Courtesy of Hiroshi Shimizu.)

planetary gear ratio is of about 10:1 to provide the wheel speed range from zero to about 1000 rpm. On the other hand, the transmission can be totally removed when a low-speed outer-rotor motor is used. The key is that this outer rotor itself is the wheel rim and the motor speed is equivalent to the wheel speed, and no gears are required. Figure 3.9 shows these two in-wheel drives, both employing a permanent-magnet brushless motor. Although different types of electric motors can be adopted, the permanent-magnet brushless machine is most attractive because of its outstanding power density.

The high-speed inner-rotor motor has the advantages of smaller size, lighter weight and lower cost, but needs an additional planetary gearset. On the other hand, the low-speed outer-rotor motor has the definite advantage of simplicity and is gearless, but the motor suffers from the drawbacks of increased size, weight and cost because of the low-speed design. Both types of in-wheel motors have been applied to modern EVs (Shimizu, 1995).

3.2 EV parameters

In general, most of EV parameters are borrowed from the well-established definitions used for ICEVs (Adler *et al.*, 1986; Fenton, 1996; Lucas, 1996). Nevertheless, there are some parameters particularly defined for EVs. For example, the battery weight and the regenerative braking efficiency are absent in the ICEV's handbooks (Duffy *et al.*, 1988; Newton *et al.*, 1996).

3.2.1 WEIGHT AND SIZE PARAMETERS

Vehicle weights are key parameters in EVs because they seriously affect the driving range and vehicle performance. Their typical definitions are listed below:

- Curb weight—the weight excluding payload;
- Gross weight—the weight including payload;
- Payload—the weight of passengers and cargo;
- Inertia weight—the curb weight plus a standard payload;
- Maximum weight—the maximum gross weight for safety operation;
- Drivetrain weight—the weight of the whole drivetrain in the EV;
- Battery weight—the weight of the whole battery pack in the EV.

Another set of important parameters for EVs are the vehicle sizes which are identical to those of ICEVs. They are listed below:

- Vehicle dimensions—the length, width, height and ground clearance;
- Frontal area—the equivalent frontal area affecting the vehicle aerodynamic drag;
- Seating capacity—the number of passengers, sometimes adult or child is also specified;
- Cargo capacity—the volume of cargo.

3.2.2 FORCE PARAMETERS

The force that a vehicle must overcome to travel is known as road load. As shown in Fig. 3.10, this road load F_l consists of three main components—aerodynamic drag force F_d, rolling resistance force F_r and climbing force F_c as given by:

$$F_l = F_d + F_r + F_c.$$

The aerodynamic drag force is due to the drag upon the vehicle body when moving through air. Its composition is due to three aerodynamic effects—the skin friction drag due to the air flow in the boundary layer, the induced drag due to the downwash of the trailing vortices behind the vehicle, and the normal pressure drag (proportional to the vehicle frontal area and speed) around the vehicle. The skin friction drag and the induced drag are usually small compared to the normal pressure drag, and are generally neglected. Thus, the aerodynamic force can be expressed as:

$$F_d = 0.5\rho C_d A(v + v_0)^2,$$

where C_d is the aerodynamic drag coefficient (dimensionless), ρ is the air density in kg/m^3, A is the frontal area in m^2, v is the vehicle velocity in m/s, and v_0 is the head wind velocity in m/s. Since both v and v_0 can also be expressed in km/h, the corresponding aerodynamic force is written as:

$$F_d = 0.0386\rho C_d A(v + v_0)^2.$$

In general, ρ is taken as 1.23 kg/m^3 although it is dependent on the altitude. On the other hand, C_d varies significantly, ranging from 0.2 to 1.5. For examples, purposely streamlined cars have C_d from 0.2 to 0.3, passenger cars from 0.3 to 0.5, vans from 0.5 to 0.6, buses from 0.6 to 0.7, and trucks from 0.8 to 1.5.

The rolling resistance force is due to the work of deformation on the wheel and road surface. The deformation on the wheel heavily dominates the rolling

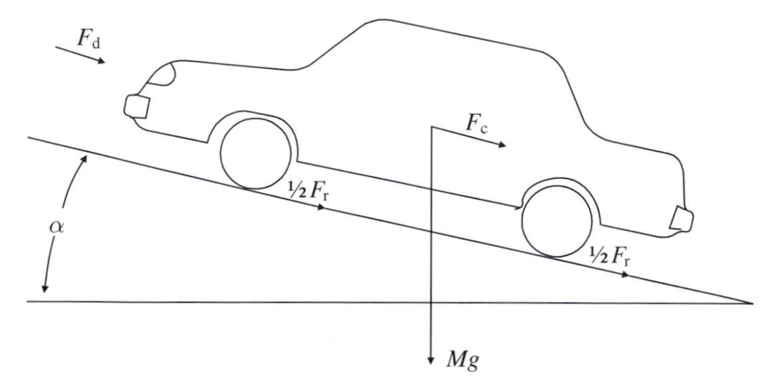

Fig. 3.10. Road load.

resistance while the deformation on the road surface is generally insignificant. Factors that affecting the rolling resistance are tire types, tire pressure, tire temperature, vehicle speed, tread thickness, the number of plies, the mix of the rubber and the level of torque transmitted. Among them, the tire types and tire pressure are relatively more significant. This rolling resistance force is normally expressed as:

$$F_r = MgC_r,$$

where M is the vehicle mass in kg, g is the gravitational acceleration (9.81 m/s^2), and C_r is the rolling resistance coefficient (dimensionless). In general, the C_r of radial-ply tires (around 0.013) is lower than that of cross-ply tires (around 0.018). It also varies inversely with the tire air pressure.

The climbing force is simply the climbing resistance or downward force for a vehicle to climb up an incline. This force is given by:

$$F_c = Mg \sin \alpha,$$

where α is the angle of incline in radian or degree. Usually, the incline is expressed as a percentage gradeability p:

$$p = (h/l)100\%,$$

where h is the vertical height over horizontal distance l. So, p and sin α are related by:

$$\alpha = \tan^{-1}(p/100).$$

For example, in a 20% gradeability, the angle of incline is 11.3°. The maximum gradeability denotes the maximum incline that a vehicle can climb at essentially zero speed.

The motive force F available at the wheels is required to overcome the above road load F_l and to drive the vehicle with an acceleration a. If F_l is greater than F, it becomes a deceleration and a is a negative value. It is expressed as:

$$a = \frac{(F - F_l)}{k_r M},$$

where k_r denotes a correction factor that there is an apparent increase in vehicle mass due to the inertia of rotational masses (such as wheels, gears and shafts), ranging from about 1.01 to 1.40.

3.2.3 ENERGY PARAMETERS

In transportation, the unit of energy is kWh rather than J because the latter is too small for such an application. To assess the energy consumption of a vehicle, the energy per unit distance in kWh/km is generally used. This unit can be applied to both ICEVs and EVs. However, an ICEV's driver generally has no idea about the kWh and prefers a physical unit of fuel volume such as the litre or gallon. So,

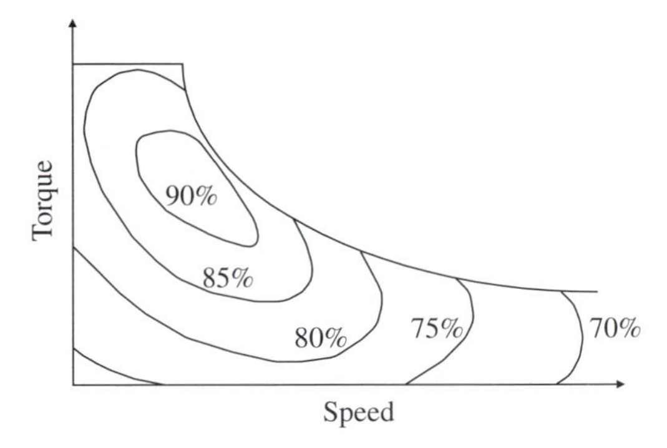

Fig. 3.11. Typical power efficiency map of an EV induction motor.

the energy consumption unit of ICEVs is usually in l/km. On the other hand, the energy parameter of ICEVs can also be expressed as the distance per unit volume of fuel, so-called the fuel economy. The corresponding SI unit is in km/l. In the USA, it is also expressed as the miles per gallon (mpg) of fuel. For battery-powered EVs, the original energy consumption unit in kWh/km becomes suitable because the fuel for recharging EV batteries is electricity which has already been measured by kWh. The corresponding fuel economy is expressed in km/kWh. When the EV is fed by fuel cells, the corresponding fuel may be compressed gaseous hydrogen, liquid hydrogen or even liquid methanol, hence the energy consumption unit in l/km and the fuel economy unit in km/l become applicable.

The energy efficiency η_{e} is the ratio of energy output to energy input, while the power efficiency η_{p} is the ratio of power output to power input. So, they can simply be expressed as:

$$\eta_{\mathrm{e}} = \frac{E_{\mathrm{out}}}{E_{\mathrm{in}}}$$

$$\eta_{\mathrm{p}} = \frac{P_{\mathrm{out}}}{P_{\mathrm{in}}}.$$

For industrial operation, these two efficiencies may not be necessarily distinguishable. On the contrary, for vehicular operation, there is a significant difference because the power efficiency varies continually during the operation of most vehicles. Thus, it is necessary to delineate the power efficiency associated with the speed and torque conditions. Instead of using a particular operating point (such as rated power at rated torque and rated speed) to describe the power efficiency of a vehicle subsystem or component, an efficiency map is generally adopted. Figure 3.11 shows a typical efficiency map of a three-phase induction motor for propelling an EV. Hence, the energy efficiency can be derived by summing powers over a given time period.

Regenerative braking is a definite advantage of EVs over ICEVs. During braking, the motor operates in the regenerative mode which converts the reduction of kinetic energy during braking into electrical energy, hence recharging the batteries. On average, the amount of convertible energy is only about 30–50% as there is significant dissipation in road load. Assuming that the in/out efficiency of the drivetrain and energy source is about 70%, the amount of energy actually stored in the batteries is about 21–35%. This is known as the regenerative braking efficiency. Similar to the previous case, in order to depict the value at different loads, a regenerative braking efficiency map should be adopted.

3.2.4 PERFORMANCE PARAMETERS

For ICEVs, the maximum speed and the acceleration rate are essentially the most important parameters to assess vehicle performance. The corresponding range per refuelling is of relatively less concern. In contrast, the range per charge for EVs is of the utmost importance. The reason is simply that the range of a 1500-cc ICEV can be around 500 km per tank, whereas the range of an EV is only about 100–200 km per charge which may not satisfy a normal driver's expectation. Hence, the commonly used performance parameters for EVs are listed below:

- Range per charge—the driving range in km of an EV that has been fully charged up. It can also be extended to describe the range per refuelling for those EVs adopting other energy sources such as fuel cells.
- Acceleration rate—it is usually expressed as the minimum time required to accelerate the vehicle from zero to a specified speed such as 40, 60 or 80 km/h.

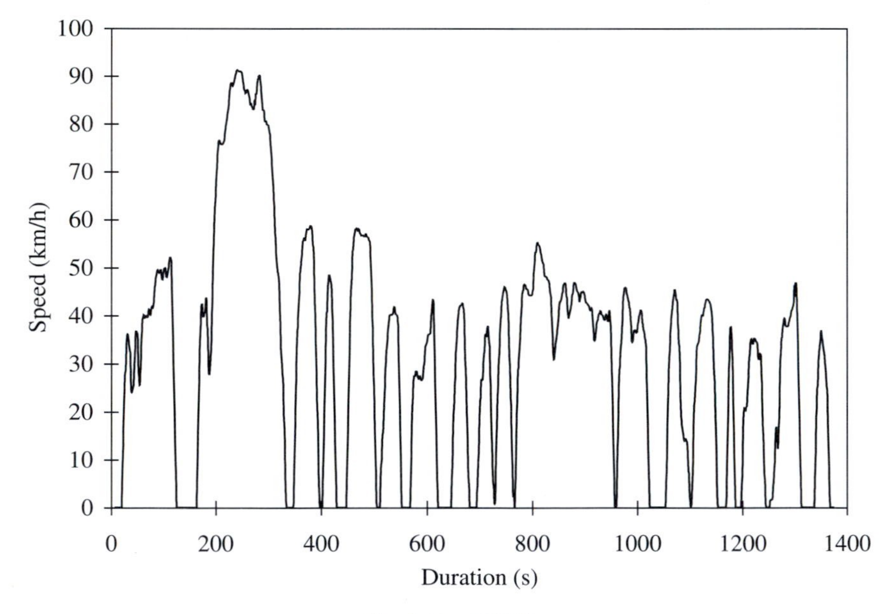

Fig. 3.12. FUDS.

- Maximum speed—it is simply the quoted maximum safe speed in km/h that a vehicle can attain.

Notice that the range per charge can be remarkably varied as it is claimed in many different ways. It can be based on a specified constant-speed operation on a level road, or for a specified type of driving cycle. For example, an EV can be claimed to offer a range of 200 km per charge at a constant speed of 40 km/h or a range of 120 km per charge under FUDS. Sometimes, some loose references may be used such as urban or highway situations.

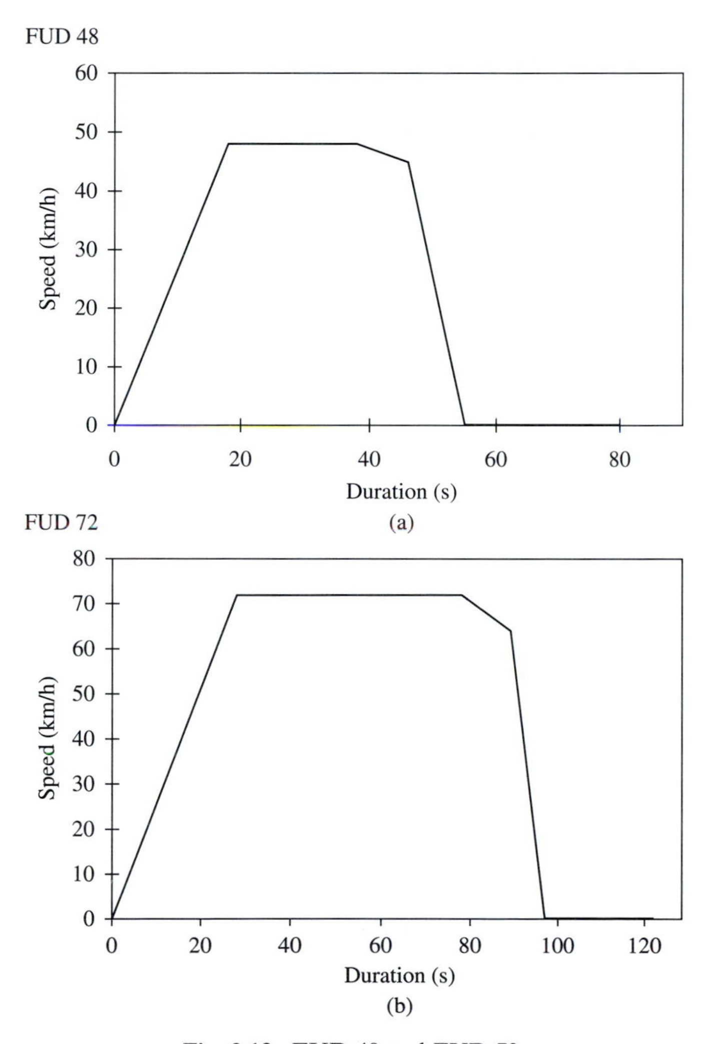

Fig. 3.13. FUD 48 and FUD 72.

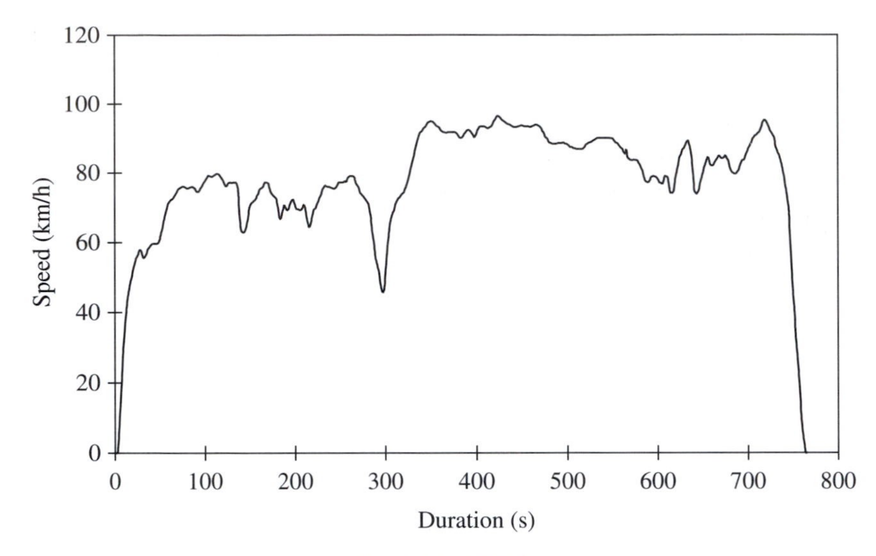

Fig. 3.14. FHDS.

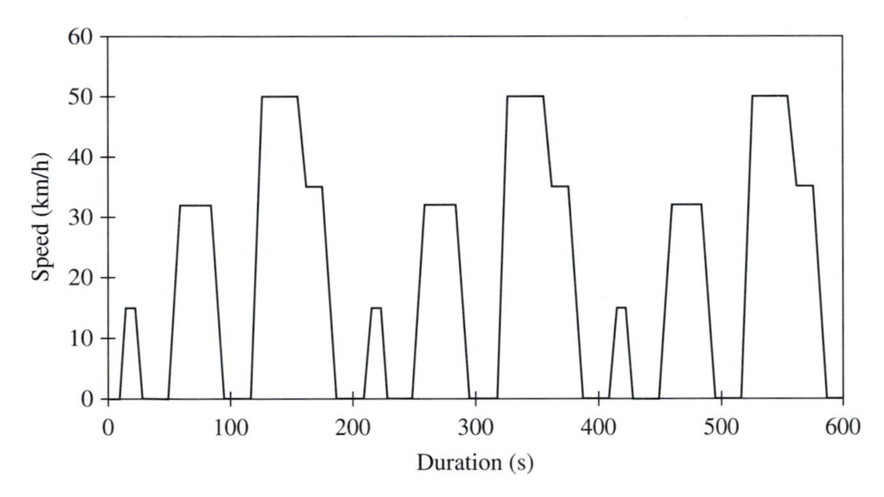

Fig. 3.15. ECE 15 Urban Cycle.

In recent years, several standard driving cycles have been developed. They are basically used to characterize various driving modes in different regions or countries.

(1) Federal Urban Driving Schedule (FUDS) and Federal Highway Driving Schedule (FHDS)—the FUDS is the most common driving cycles in the USA. It was developed originally to evaluate the noxious emissions of ICEVs and was based on a cycle derived from the statistical flow of traffic patterns

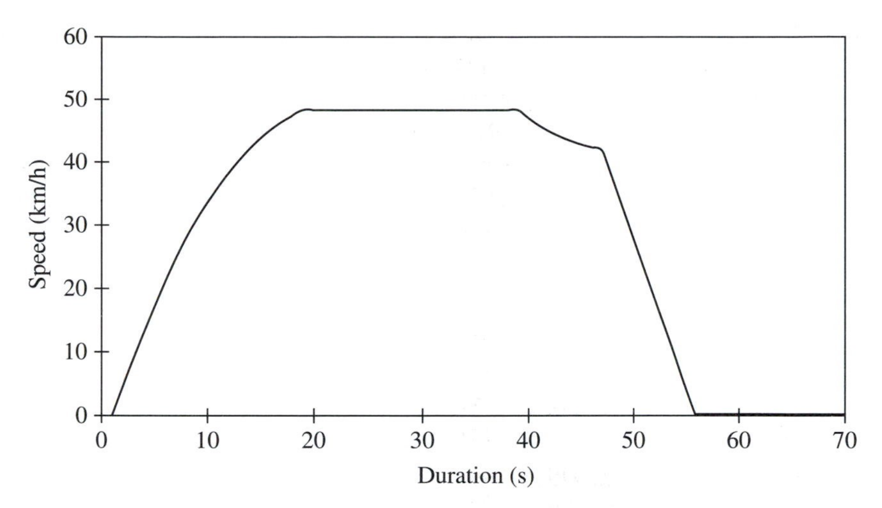

Fig. 3.16. J227a-C Cycle.

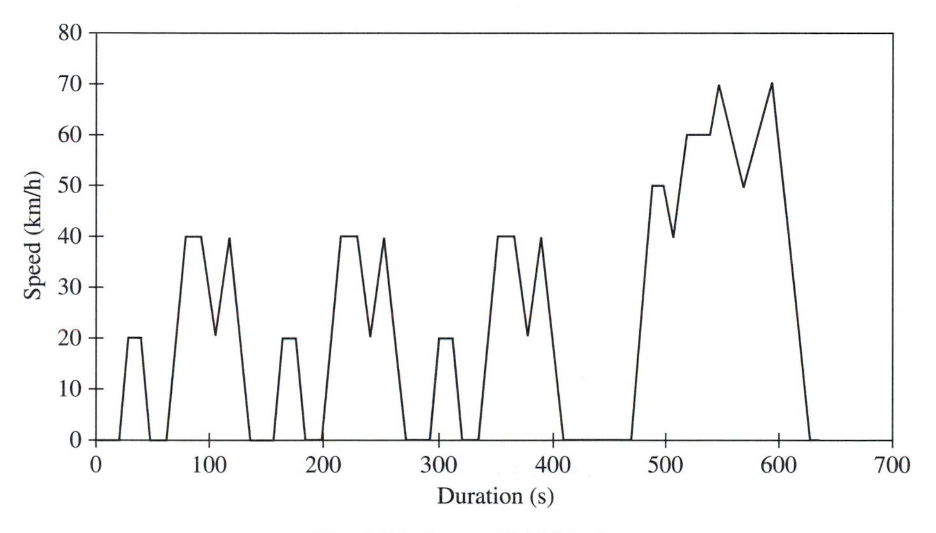

Fig. 3.17. Japan 10.15 Mode.

in Los Angeles. Subsequently, it is widely used to evaluate the fuel economy of urban or city driving. Two of its simplified versions, namely the FUD 48 and FUD 72 with their maximum speeds of 48 km/h and 72 km/h, respectively, are also widely used for computer simulation. In contrast, the FHDS was developed to typify rural or cross-country driving in the USA. Figures 3.12–3.14 show the corresponding speed–time profiles.

(2) European Driving Cycle—it is known as the ECE Cycle and is a composite of two driving modes—urban and highway. It is widely adopted in Europe to

evaluate fuel economy and emissions. There is a reduced version called the ECE 15 Urban Cycle, shown in Fig. 3.15, focusing on the driving mode in urban areas.

(3) SAE J227 Cycle—in the early 1970s, the Society of Automotive Engineers (SAE) developed some standard driving cycles for EVs. The SAE J227 cycle was designed to give approximately the same road-load energy demand as the FUDS but with a lower peak road-load power. Since many EVs in the 1970s were unable to achieve the required road load power levels, the J227 Cycle was re-issued as a set of four simplified cycles, so-called J227a-A, -B, -C, and -D Cycles. The capital letters refer to four cycles with increasing power requirements and increasing maximum speed requirements. Figure 3.16 shows the J227a-C Cycle which used to be adopted in the design of EVs.

(4) Japan 10.15 Mode—Figure 3.17 shows the profile of the Japan 10.15 Mode driving cycle. As a regulation, the driving range of EVs in Japan should be evaluated by using only this driving cycle.

References

Adler, U. and Bazlen, W. (1986). *Automotive Handbook* (2nd edition), Robert-Bosch, Stuttgart.

Chan, C.C. (1993). An overview of electric vehicle technology. *Proceedings of IEEE*, **81**, 1201–13.

Duffy, J.E., Stockel, M.T. and Stockel, M.W. (1988). *Automotive Mechanics Fundamentals: How and Why of the Design, Construction, and Operation of Modern Automotive Systems and Units*. Gregory's Automotive Publications, Sydney.

Fenton, J. (1996). *Handbook of Vehicle Design Analysis*. Society of Automotive Engineers, Warrendale, PA.

Lucas, G.G. (1996). *Road Vehicle Performance: Methods of Measurement and Calculation*. Gordon and Breach, New York.

Newton, K., Steeds, W. and Garrett, T.K. (1996). *The Motor Vehicle*. Butterworth-Heinemann, Oxford.

Shacket, S.R. (1979). *The Complete Book of Electric Vehicles*. Domus Books, Chicago.

Shimizu, H. (1995). *Electric Vehicle* (2nd edition), In Japanese, Japan Industry News Agency, Tokyo.

Unnewehr, L.E. and Nasar, S.A. (1982). *Electric Vehicle Technology*. John Wiley & Sons, New York.

4 HEV systems

When incorporating both of the ICE and electric motor, the vehicle is usually known as the hybrid EV (HEV). This HEV is actually not a new idea. A patent by Pieper in 1905 delineated that a battery-powered electric motor was used to boost the acceleration of an ICEV (Wouk, 1995; Bates, 1995).

What exactly is a HEV? The definition available is so general that it anticipates future technologies of energy sources. As proposed by Technical Committee 69 (Electric Road Vehicles) of the International Electrotechnical Commission, a HEV is a vehicle in which propulsion energy is available from two or more kinds or types of energy stores, sources or converters, and at least one of them can deliver electrical energy. Based on this general definition, there are many types of HEVs, such as the petrol ICE & battery, diesel ICE & battery, battery & fuel cell, battery & capacitor, battery & flywheel, and battery & battery hybrids. However, the above definition is not well accepted. Ordinary people have already borne in mind that a HEV is simply a vehicle having both an ICE and electric motor. To avoid confusing readers or customers, specialists also prefer not using the HEV to represent a vehicle adopting energy source combinations other than the ICE & battery hybrid. For examples, they prefer to call a battery & fuel cell HEV simply as a fuel cell EV and a battery & capacitor HEV as an ultracapacitor-assisted EV. As we prefer general perception to loose definition, the term HEV in this book refers only to the vehicle adopting the ICE and electric motor.

4.1 HEV configurations

Due to the limitations of available energy sources, the EV cannot offer a compatible driving range to the ICEV. People may not buy an EV, no matter how clean, if their range between charges is only 100–200 km. The HEV, using an ICE and electric motor, has been introduced as an alternative solution before the full implementation of EVs when there is a breakthrough in EV energy sources. The definite advantage of the HEV is to greatly extend the original EV driving range by 2–4 times, and to offer rapid refuelling of liquid petrol or diesel. An important plus is that it requires only little changes in the energy supply infrastructure. The key drawbacks of the HEV are the loss of zero-emission concept and the increased complexity. Nevertheless, the HEV is vastly less polluting and has less fuel consumption than the ICEV while having the same range. These merits are due to the fact that the ICE of the HEV can always operate in its most efficient mode, yielding low emissions and low fuel consumption. Also, the HEV may be purposely operated as an EV in the zero-emission zone.

Traditionally, HEVs were classified into two basic kinds—series and parallel. Recently, with the introduction of some HEVs offering the features of both the series and parallel hybrids, the classification has been extended to three kinds—series, parallel and series–parallel. In the year 2000, it is interesting to note that some newly introduced HEVs cannot be classified into these three kinds. Hereby, HEVs are newly classified into four kinds:

- series hybrid
- parallel hybrid
- series–parallel hybrid and
- complex hybrid.

Figure 4.1 shows the corresponding functional block diagrams, in which the electrical link is bidirectional, the hydraulic link is unidirectional and the mechanical

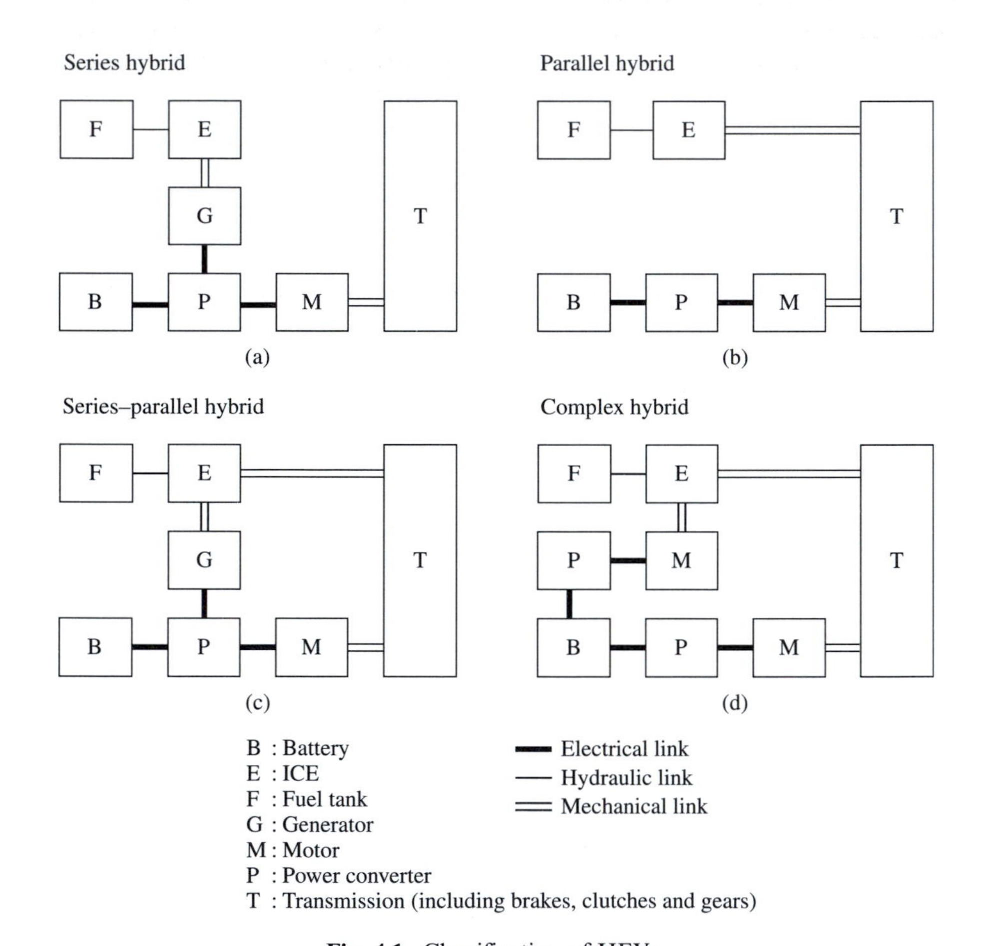

Fig. 4.1. Classification of HEVs.

link (including the clutches and gears) is also bidirectional. It can be found that the key feature of the series hybrid is to couple the ICE with the generator to produce electricity for pure electric propulsion, whereas the key feature of the parallel hybrid is to couple both the ICE and electric motor with the transmission via the same drive shaft to propel the wheels. The series–parallel hybrid is a direct combination of both the series and parallel hybrids. On top of the series–parallel hybrid operation, the complex hybrid can offer additional and versatile operating modes.

In order to achieve the series–parallel hybrid or even the complex hybrid, the ICE needs to mechanically link with the generator/motor and together with the transmission. Such mechanical device can be a planetary gear. For example, Fig. 4.2 shows a planetary gear for the series–parallel hybrid where the sun gear is connected to the generator, the ring gear (also called the crown gear) to the transmission, and the carrier (also called the cage or yoke) to the ICE. Thus, the ICE transfers power outwards via the planet gears (also called the pinion gears) to the ring gear which is coupled to the transmission that drives the wheels, and also inwards to the sun gear which then generates electrical energy via the generator. The angular velocities of all gear shafts are governed by a constraint equation:

$$\omega_1 + k_p\omega_2 - (1 - k_p)\omega_3 = 0,$$

where $k_p = Z_2/Z_1$ is the base gear ratio, Z_1 is the number of teeth of the sun gear, Z_2 is the number of teeth of the ring gear, and ω_1, ω_2, ω_3 are the angular velocities

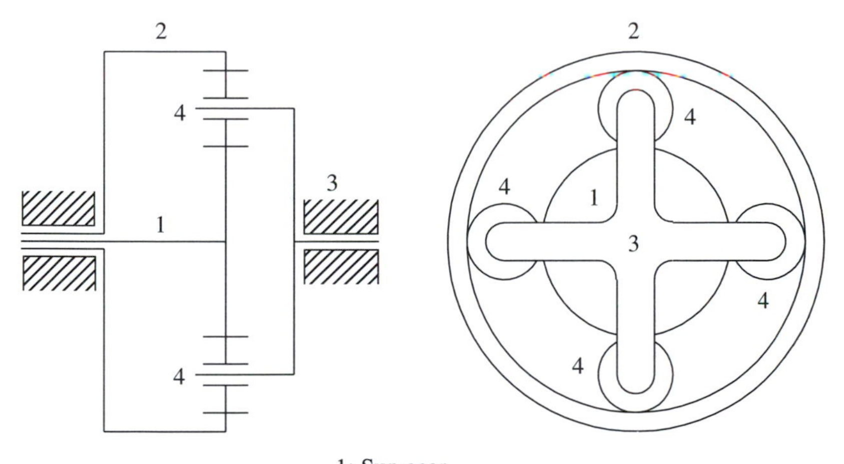

1: Sun gear
2: Ring gear
3: Carrier
4: Planet gears

Fig. 4.2. Planetary gear.

of the sun gear, ring gear and carrier, respectively. The motion equations can be expressed as:

$$J_1\dot{\omega}_1 = \eta_1 M_1 - \frac{1}{k_p}\eta_2 M_2,$$

$$J_3\dot{\omega}_3 = M_3 + \frac{k_p + 1}{k_p}\eta_3 M_2,$$

where J_1 is the total inertial torque of the sun gear and its connecting elements reduced to the sun gear shaft, J_3 is the total inertial torque reduced to the carrier gear shaft, M_1, M_2, M_3 are the external torques acting on the sun gear shaft, ring gear shaft and carrier gear shaft, respectively, and η_1, η_2, η_3 are the coefficients of internal power losses in the sun gear shaft, ring gear shaft and carrier gear shaft, respectively.

4.1.1 SERIES HYBRID SYSTEM

The series hybrid is the simplest kind of HEV. Its ICE mechanical output is first converted into electricity using a generator. The converted electricity either charges the battery or can bypass the battery to propel the wheels via the same electric motor and mechanical transmission. Conceptually, it is an ICE-assisted EV which aims to extend the driving range comparable with that of the ICEV. Due to the absence of clutches throughout the mechanical link, it has the definite advantage of flexibility for locating the ICE-generator set. Although it has an added advantage of simplicity of its drivetrain, it needs three propulsion devices—the ICE, the generator and the electric motor. Another disadvantage is that all these propulsion devices need to be sized for the maximum sustained power if the series HEV is designed to climb a long grade. On the other hand, when it is only needed to serve such short trips as commuting to work and shopping, the corresponding ICE-generator set can adopt a lower rating.

4.1.2 PARALLEL HYBRID SYSTEM

Differing from the series hybrid, the parallel HEV allows both the ICE and electric motor to deliver power in parallel to drive the wheels. Since both the ICE and electric motor are generally coupled to the drive shaft of the wheels via two clutches, the propulsion power may be supplied by the ICE alone, by the electric motor or by both. Conceptually, it is inherently an electric assisted ICEV for achieving lower emissions and fuel consumption. The electric motor can be used as a generator to charge the battery by regenerative braking or absorbing power from the ICE when its output is greater than that required to drive the wheels. Better than the series HEV, the parallel hybrid needs only two propulsion devices—the ICE and the electric motor. Another advantage over the series case is that a smaller ICE and a smaller electric motor can be used to get the same performance until the battery is depleted. Even for long-trip operation, only the

ICE needs to be rated for the maximum sustained power while the electric motor may still be about a half.

4.1.3 SERIES–PARALLEL HYBRID SYSTEM

In the series–parallel hybrid, the configuration incorporates the features of both the series and parallel HEVs, but involving an additional mechanical link compared with the series hybrid, and also an additional generator compared with the parallel hybrid. Although possessing the advantageous features of both the series and parallel HEVs, the series–parallel HEV is relatively more complicated and costly. Nevertheless, with the advances in control and manufacturing technologies, some modern HEVs prefer to adopt this system.

4.1.4 COMPLEX HYBRID SYSTEM

As reflected by its name, this system involves a complex configuration which cannot be classified into the above three kinds. As shown in Fig. 4.1, the complex hybrid seems to be similar to the series–parallel hybrid, since the generator and electric motor are both electric machinery. However, the key difference is due to the bidirectional power flow of the electric motor in the complex hybrid and the unidirectional power flow of the generator in the series–parallel hybrid. This bidirectional power flow can allow for versatile operating modes, especially the three propulsion power (due to the ICE and two electric motors) operating mode which cannot be offered by the series–parallel hybrid. Similar to the series–parallel HEV, the complex hybrid suffers from higher complexity and costliness. Nevertheless, some newly introduced HEVs adopt this system for dual-axle propulsion.

4.2 Power flow control

Due to the variations in HEV configurations, different power control strategies are necessary to regulate the power flow to or from different components. These control strategies aim to satisfy a number of goals for HEVs (Beretta, 2000; Van Mierlo, 2000). There are four key goals:

- maximum fuel economy
- minimum emissions
- minimum system costs and
- good driving performance.

The design of power control strategies for HEVs involves different considerations. Some key considerations are summarized below:

- Optimal ICE operating point—The optimal operating point on the torque-speed plane of the ICE can be based on the maximization of fuel economy,

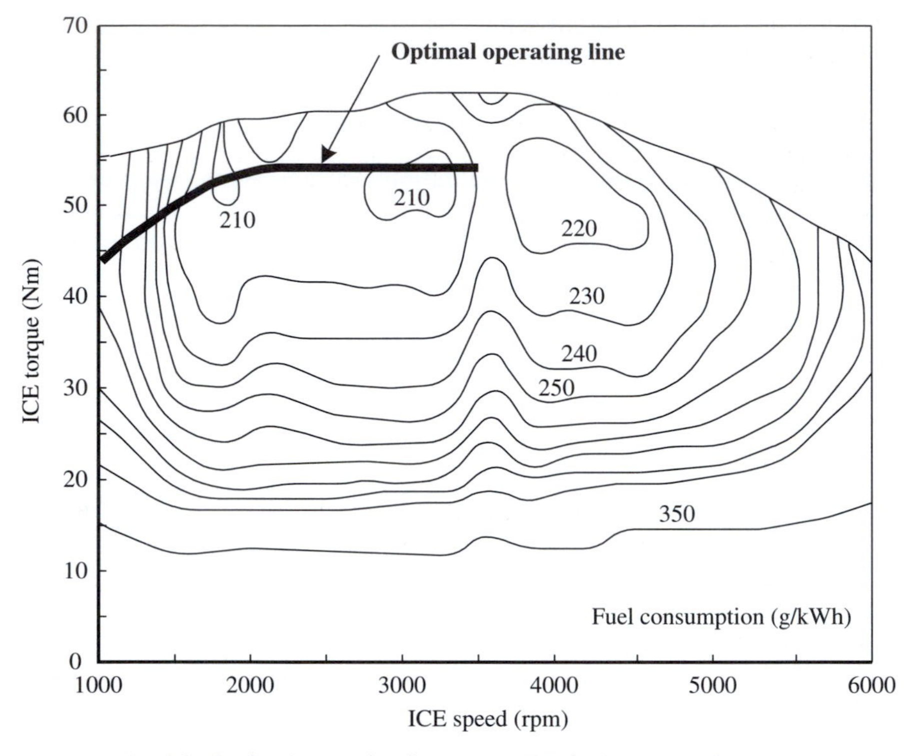

Fig. 4.3. Optimal operating line on an ICE fuel consumption map.

the minimization of emissions, or even a compromise between fuel economy and emissions.

- Optimal ICE operating line—in case the ICE needs to deliver different power demands, the corresponding optimal operating points constitute an optimal operating line. Figure 4.3 shows a typical optimal operating line of an ICE, in which the optimization is based on the minimum fuel consumption which is equivalent to the maximum fuel economy.
- Optimal ICE operating region—the ICE has a preferred operating region on the torque-speed plane, in which the fuel efficiency remains optimum.
- Minimum ICE dynamics—the ICE operating speed needs to be regulated in such a way that any fast fluctuations are avoided, hence minimizing the ICE dynamics.
- Minimum ICE speed—when the ICE operates at low speeds, the fuel efficiency is very low. The ICE should be cut off when its speed is below a threshold value.
- Minimum ICE turn-on time—the ICE should not be turned on and off frequently; otherwise, it results in additional fuel consumption and emissions. A minimum turn-on time should be set to avoid such drawbacks.
- Proper battery capacity—the battery capacity needs to be kept at a proper level so that it can provide sufficient power for acceleration and can accept

regenerative power during braking or downhill. When the battery capacity is too high, the ICE should be turned off or operated idly. When the capacity is too low, the ICE should increase its output to charge the battery as fast as possible.

- Safety battery voltage—the battery voltage may be significantly altered during discharging, generator charging or regenerative charging. This battery voltage should not be over-voltage or under-voltage; otherwise, the battery may be permanently damaged.
- Relative distribution—the distribution of power demand between the ICE and battery can be proportionally split up during the driving cycle.
- Geographical policy—in certain cities or areas, the HEV needs to be operated in the pure electric mode. The changeover should be controlled manually or automatically.

4.2.1 SERIES HYBRID CONTROL

In the series hybrid system, the power flow control can be illustrated by four operating modes shown in Fig. 4.4. During startup, normal driving or acceleration of the series HEV, both the ICE (via the generator) and battery deliver electrical energy to the power converter which then drives the electric motor and hence the wheels via the transmission. At light load, the ICE output is greater than that required to drive the wheels so that the generated electrical energy is also used to charge the battery until the battery capacity reaches a proper level. During braking or deceleration, the electric motor acts as a generator which transforms the kinetic energy of the wheels into electricity, hence charging the battery via the power converter. Also, the battery can be charged by the ICE via the generator and power converter, even when the vehicle comes to a complete stop.

4.2.2 PARALLEL HYBRID CONTROL

Figure 4.5 illustrates the four operating modes of the parallel HEV. During startup or full-throttle acceleration, both the ICE and electric motor proportionally share the required power to propel the vehicle. Typically, the relative distribution between the ICE and electric motor is 80–20%. During normal driving, the ICE solely supplies the necessary power to propel the vehicle while the electric motor remains in the off mode. During braking or deceleration, the electric motor acts as a generator to charge the battery via the power converter. Also, since both the ICE and electric motor are coupled to the same drive shaft, the battery can be charged by the ICE via the electric motor when the vehicle is at light load. Recently, the Honda Insight HEV has adopted similar power flow control.

4.2.3 SERIES–PARALLEL HYBRID CONTROL

In the series–parallel hybrid system, it involves the features of series and/or parallel hybrids. Thus, there are many possible operating modes to carry out its

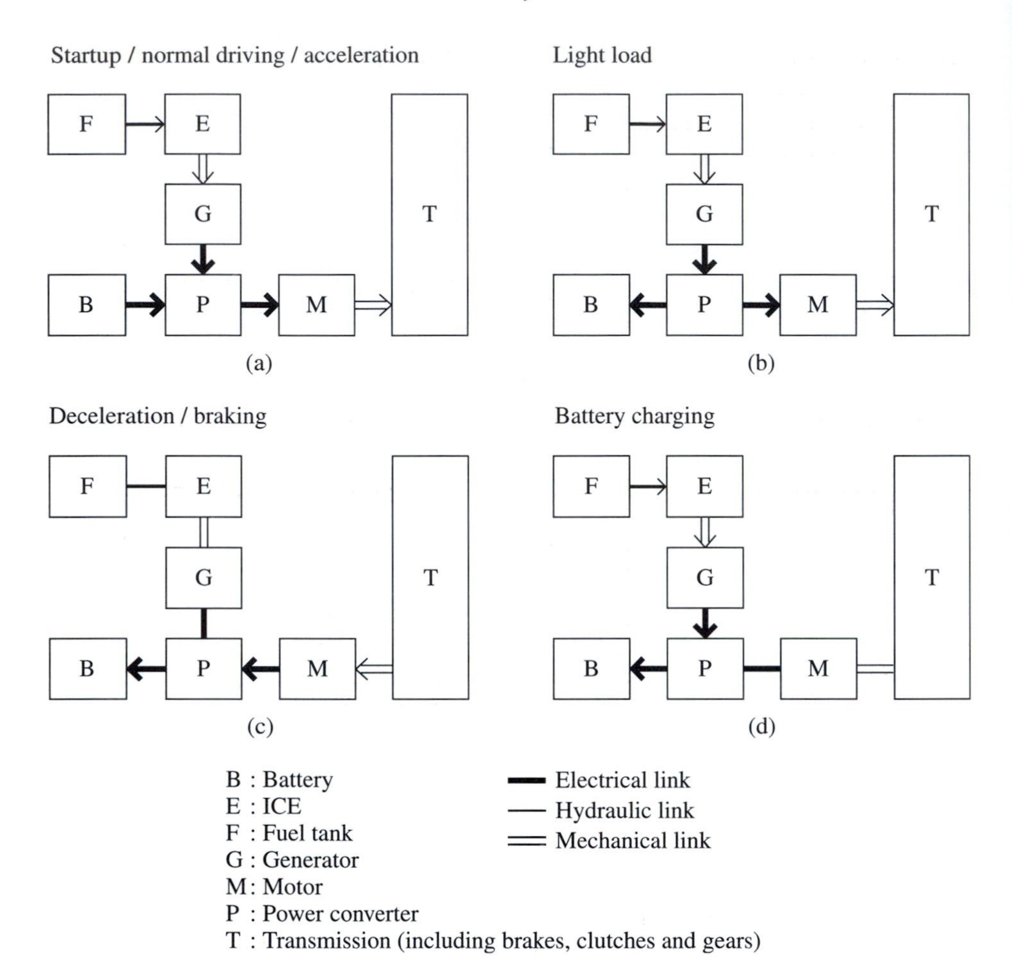

Fig. 4.4. Series hybrid operating modes.

power flow control. Basically, we can identify them into two groups, namely ICE-heavy and electric-heavy. The ICE-heavy one denotes that the ICE is more active than the electric motor for series–parallel hybrid propulsion, whereas the electric-heavy one indicates that the electric motor is more active.

Figure 4.6 shows an ICE-heavy series–parallel hybrid system, in which there are six operating modes. At startup, the battery solely provides the necessary power to propel the vehicle while the ICE is in the off mode. During full-throttle acceleration, both the ICE and electric motor proportionally share the required power to propel the vehicle. During normal driving, the ICE solely provides the necessary power to propel the vehicle while the electric motor remains in the off mode. During braking or deceleration, the electric motor acts as a generator to charge the battery via the power converter. For battery charging during driving, the ICE not only drives the vehicle but also the generator to charge the battery via the

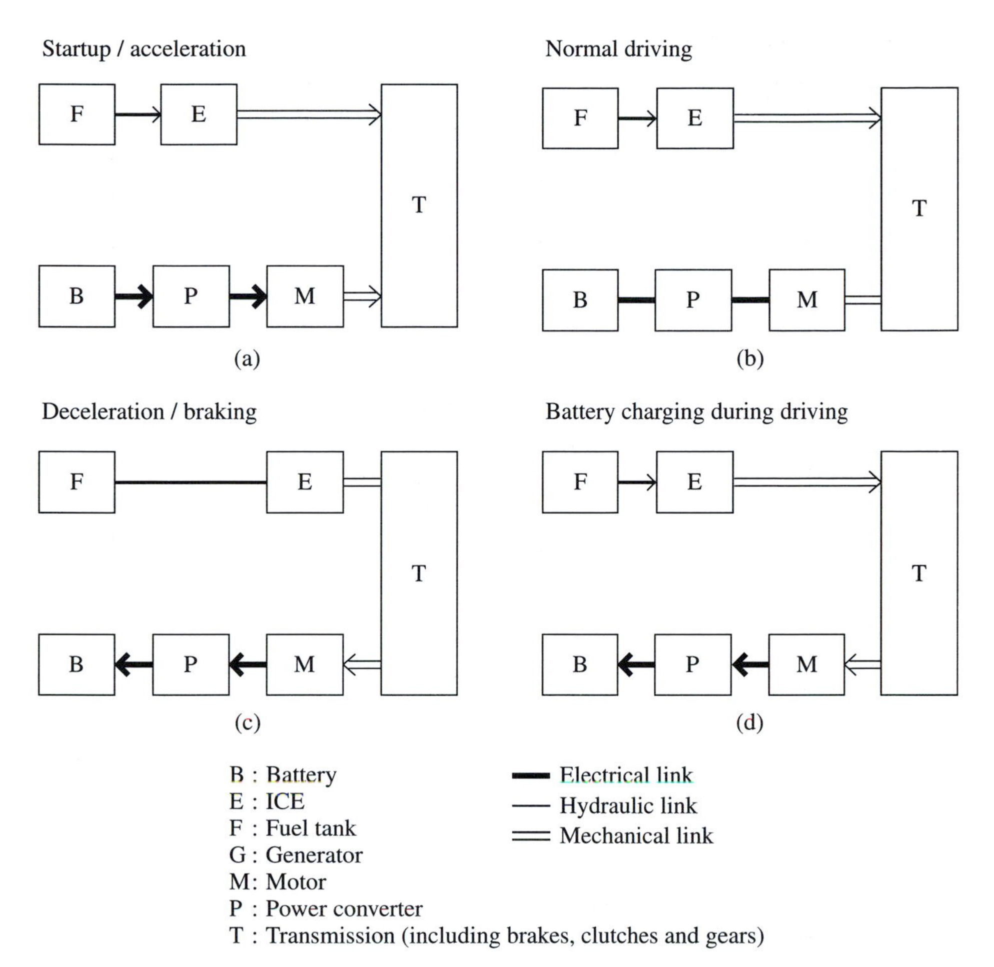

Fig. 4.5. Parallel hybrid operating modes.

power converter. When the vehicle is at a standstill, the ICE can maintain driving the generator to charge the battery. Recently, a similar power flow control system has been applied to the Nissan Tino HEV (Inada, 2000).

Figure 4.7 shows a relatively electric-heavy series–parallel hybrid system, in which there are six operating modes. During startup and driving at light load, the battery solely feeds the electric motor to propel the vehicle while the ICE is in the off mode. For both full-throttle acceleration and normal driving, both the ICE and electric motor work together to propel the vehicle. The key difference is that the electrical energy used for full-throttle acceleration comes from both the generator and battery whereas that for normal driving is solely from the generator driven by the ICE. Notice that a planetary gear is usually employed to split up the ICE output, hence to propel the vehicle and to drive the generator. During

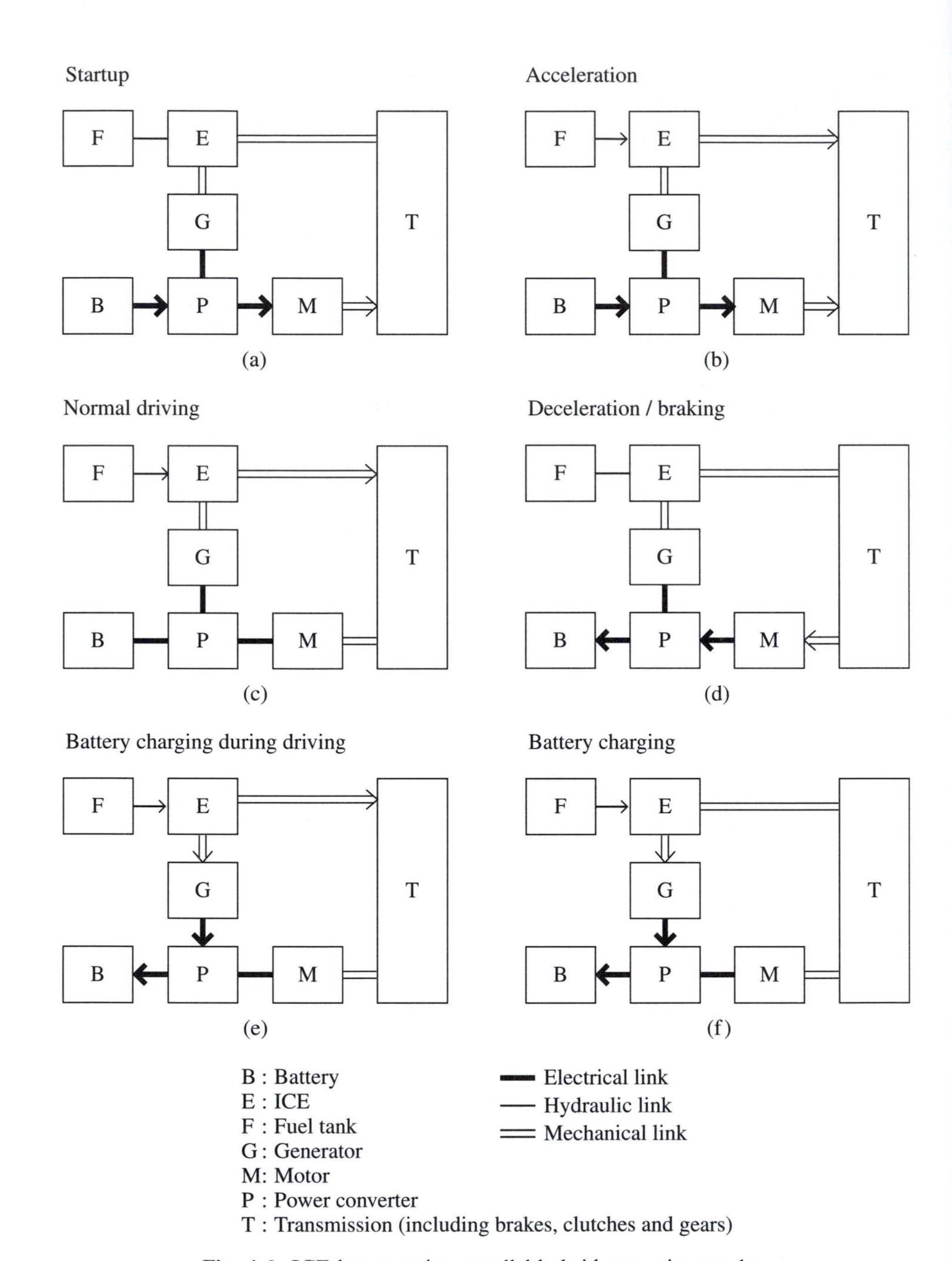

Fig. 4.6. ICE-heavy series–parallel hybrid operating modes.

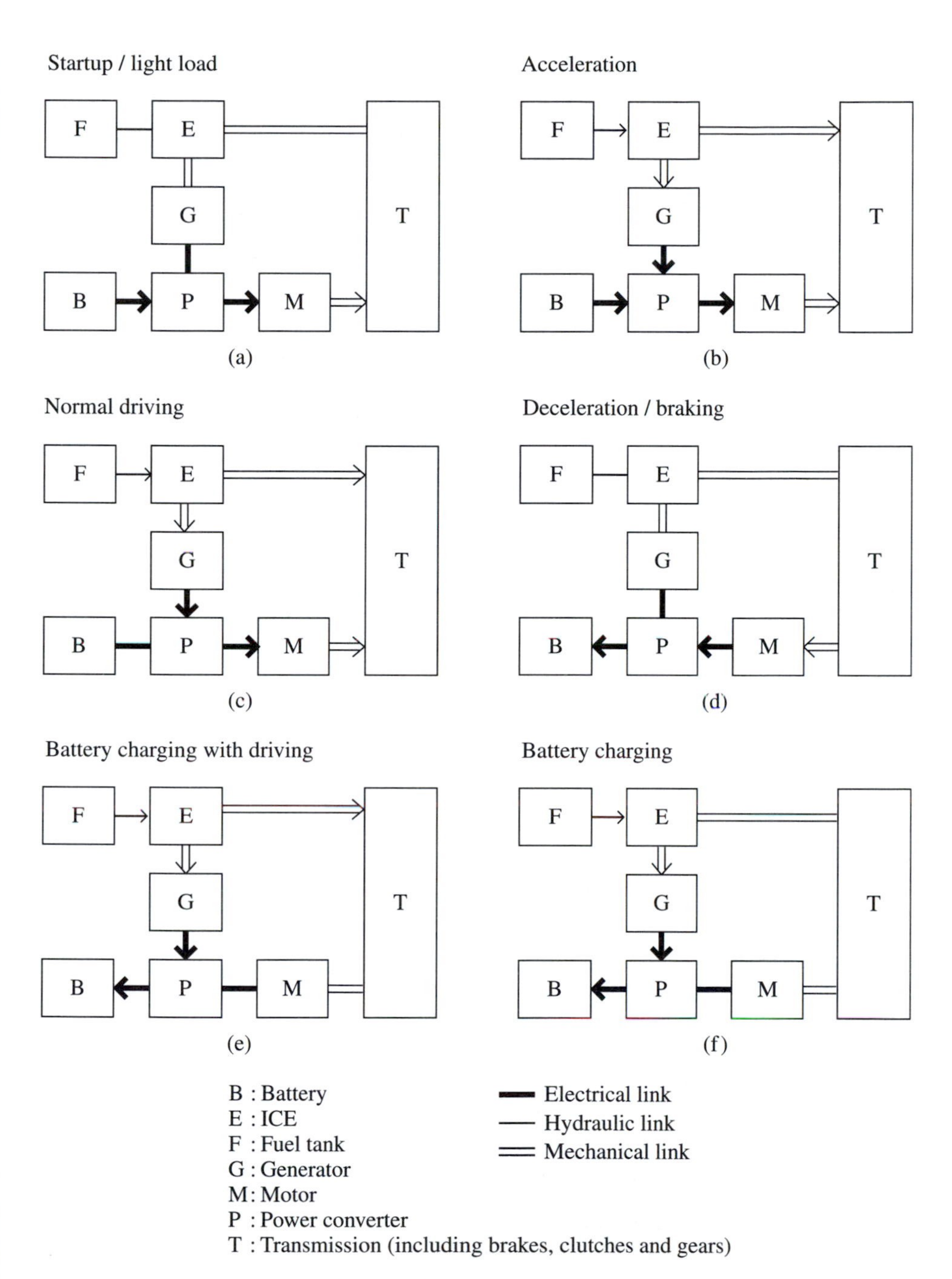

Fig. 4.7. Electric-heavy series–parallel hybrid operating modes.

braking or deceleration, the electric motor acts as a generator to charge the battery via the power converter. Also, for battery charging during driving, the ICE not only drives the vehicle but also the generator to charge the battery. When the vehicle is at a standstill, the ICE can maintain driving the generator to charge the battery. Recently, the Toyota Prius has adopted a similar power flow control system.

4.2.4 COMPLEX HYBRID CONTROL

The development of complex hybrid control has been focused on the dual-axle propulsion system for HEVs. In this system, the front-wheel axle and rear-wheel axle are separately driven. There is no propeller shaft or transfer to connect the front and rear wheels, so it enables a more lightweight propulsion system and increases the vehicle packaging flexibility. Moreover, regenerative braking on all four wheels can significantly improve the vehicle fuel efficiency and hence the fuel economy.

Figure 4.8 shows a dual-axle complex hybrid system, where the front-wheel axle is propelled by a hybrid power train and the rear-wheel axle is driven by an electric motor. There are six operating modes. During startup, the battery delivers electrical energy to feed both the front and rear electric motors to individually propel the front and rear axles of the vehicle whereas the ICE is in the off mode. For full-throttle acceleration, both the ICE and front electric motor work together to propel the front axle while the rear electric motor also drives the rear axle. Notice that this operating mode involves three propulsion devices (one ICE and two electric motors) to simultaneously propel the vehicle. During normal driving and/or battery charging, the ICE output is split up to propel the front axle and to drive the electric motor (which works as a generator) to charge the battery. The corresponding device to mechanically couple the ICE, front electric motor and front axle altogether is usually based on a planetary gear. When driving at light load, the battery delivers electrical energy to the front electric motor only to drive the front axle whereas both the ICE and rear electric motor are off. During braking or deceleration, both the front and rear electric motors act as generators to simultaneously charge the battery. A unique feature of this dual-axle system is the capability of axle balancing. In case the front wheels slip, the front electric motor works as a generator to absorb the change of ICE output power. Through the battery, this power difference is then used to drive the rear wheels to achieve axle balancing. Recently, the Toyota Post-Prius system, termed THS-C, has adopted similar power flow control.

Figure 4.9 shows another dual-axle complex hybrid system, where the front-wheel axle is driven by an electric motor and the rear-wheel axle is propelled by a hybrid power train. Focusing on vehicle propulsion, there are six operating modes. During startup, the battery delivers electrical energy only to the front electric motor which in turn drives the front axle of the vehicle whereas both the ICE and rear electric motor are off. Once the vehicle moves forward, the battery

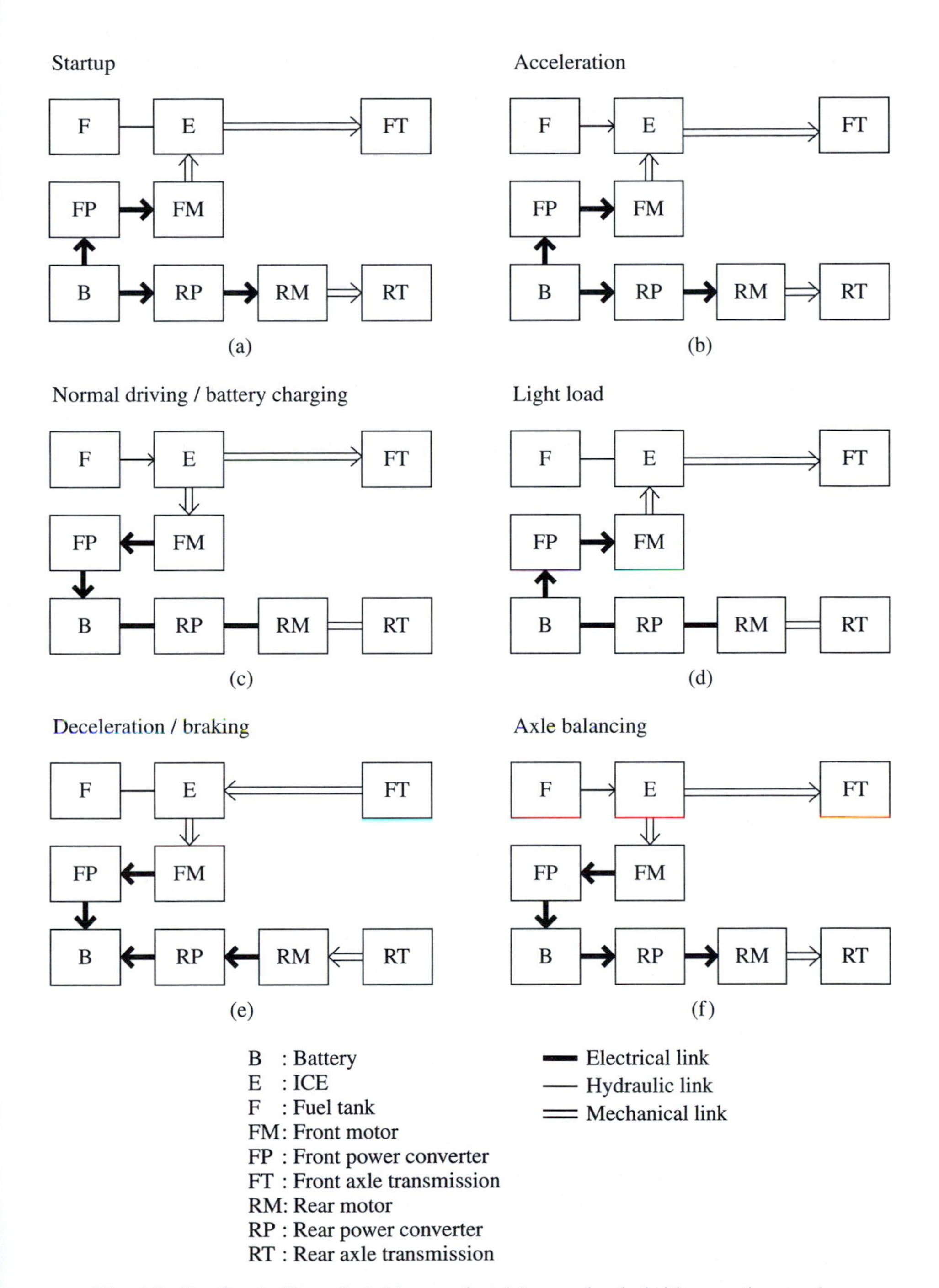

Fig. 4.8. Dual-axle (front-hybrid rear-electric) complex hybrid operating modes.

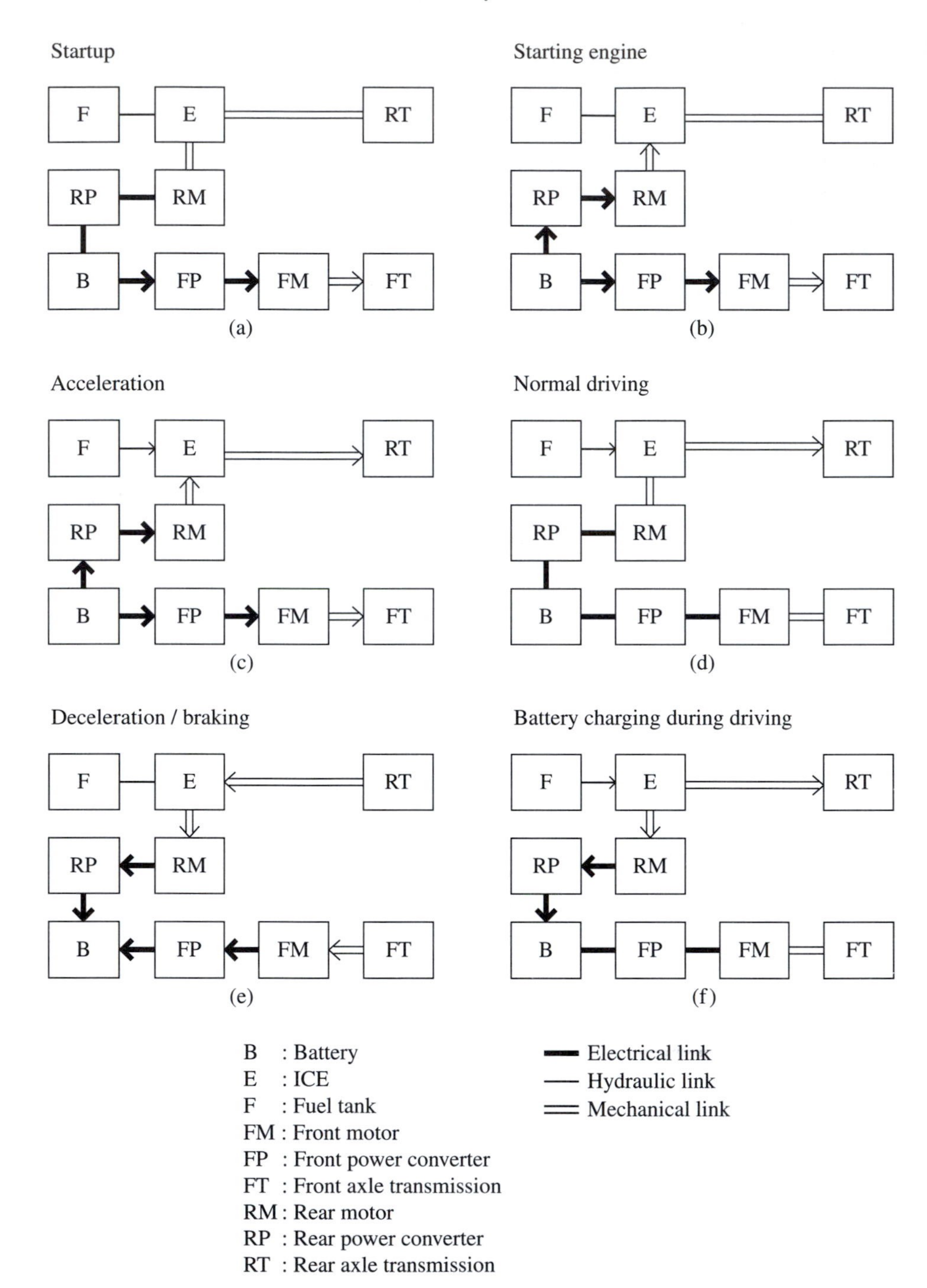

Fig. 4.9. Dual-axle (front-electric rear-hybrid) complex hybrid operating modes.

also delivers electrical energy to the rear electric motor which functions to quickly rise up the ICE speed, thus starting the ICE. For full-throttle acceleration, the front electric motor drives the front axle while both the ICE and rear electric motor work together to propel the rear axle. So, there are three propulsion devices (one ICE and two electric motors) simultaneously propelling the vehicle. During normal driving, the ICE works alone to propel the rear axle of the vehicle. During braking or deceleration, both the front and rear electric motors act as generators to simultaneously charge the battery. For battery charging during driving, the ICE output is split up to propel the rear axle and to drive the rear electric motor (which works as a generator) to charge the battery. Recently, the GM Precept has adopted a similar power flow control system.

4.3 Example of HEV system performances

Figure 4.10 shows a new parallel hybrid HEV system (Szumanowski, 2000). This HEV system not only possesses the features of the parallel hybrid, but

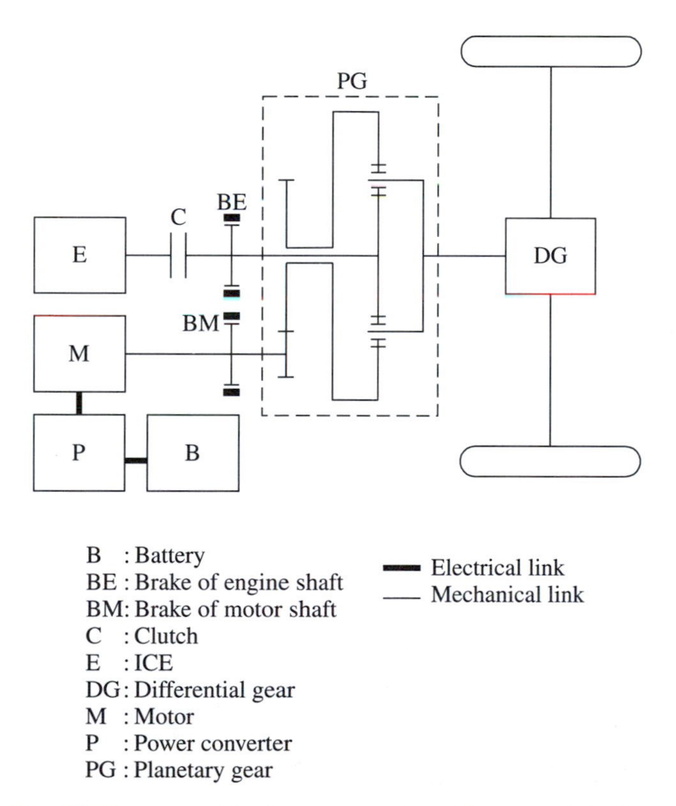

Fig. 4.10. New HEV system with planetary gearing. (Courtesy of A. Szumanowski.)

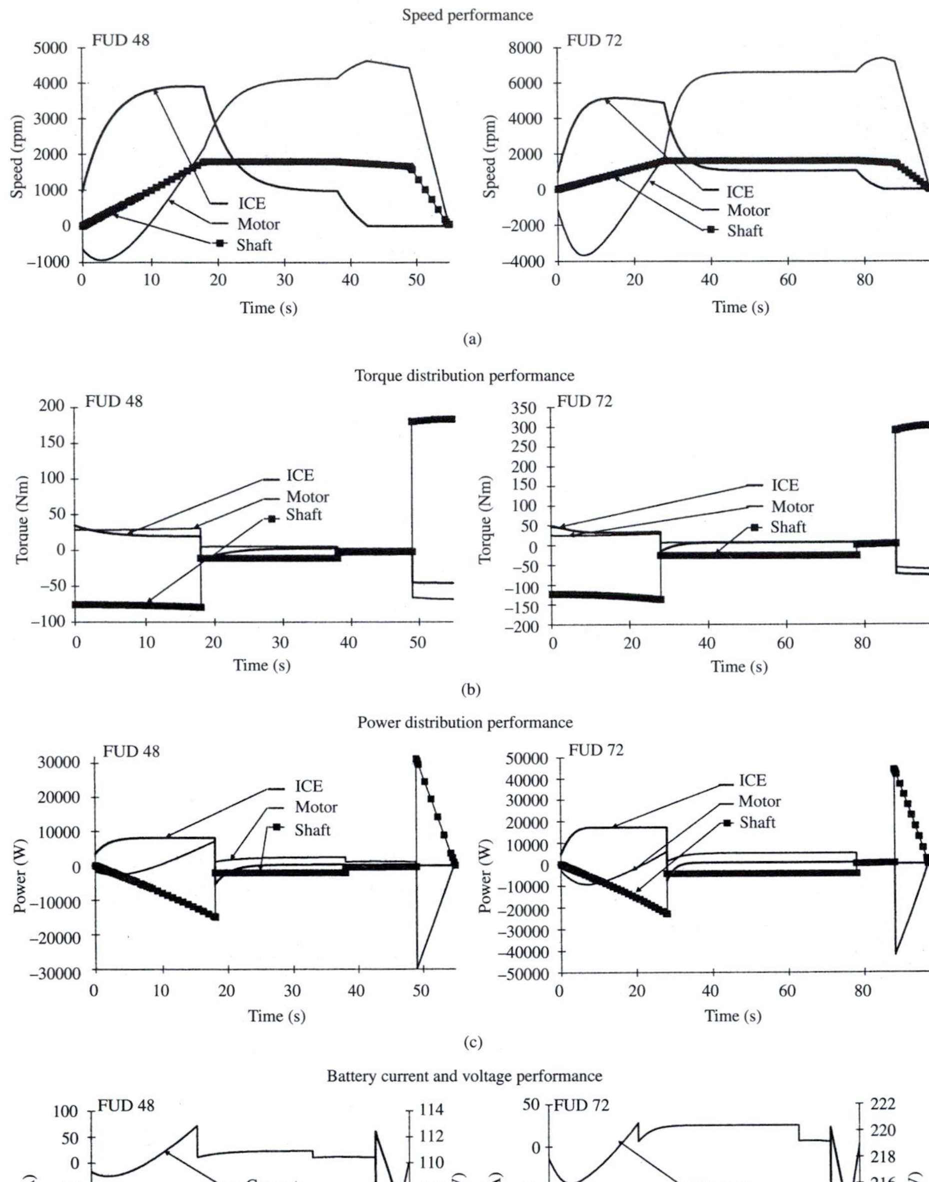

Speed performance
FUD 48
Speed (rpm)
Time (s)
ICE
Motor
Shaft
FUD 72
(a)
Torque distribution performance
Torque (Nm)
(b)
Power distribution performance
Power (W)
(c)
Battery current and voltage performance
Current (A)
Vlotage (V)
Current
Voltage
(d)

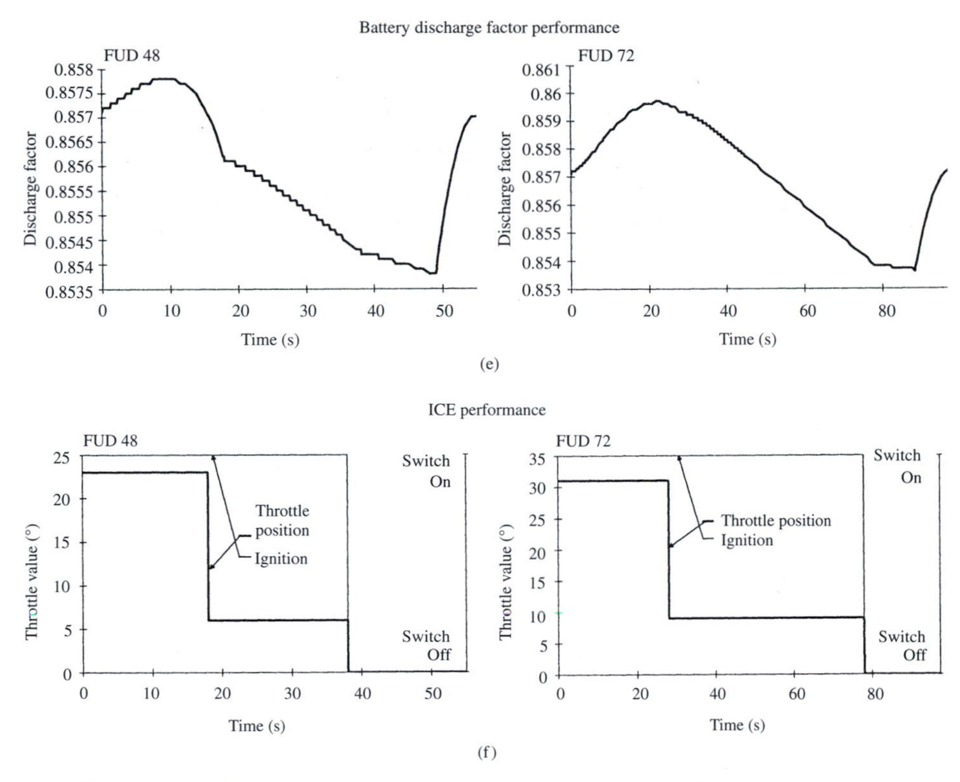

Fig. 4.11. Performances of a new HEV system. (Courtesy of A. Szumanowski.)

also incorporates a unique advantage of the series hybrid (namely the ICE can independently operate at the mode of minimum fuel consumption). The key is to employ a planetary gear which offers two degrees of freedom for mechanical transmission. In fact, A. Szumanowski and C.C. Chan have applied a patent on a similar HEV system with planetary gearing.

In city driving, the HEV system is characterized by the features of a parallel hybrid and the advantage of a series hybrid. When the vehicle is at full-throttle acceleration, the power is simultaneously delivered by the ICE and electric motor. While the vehicle is at normal driving (steady speed operation), the power is collaboratively fed by the ICE and electric motor via the planetary gear with two degrees of freedom in such a way that the fuel consumption of the ICE is minimum. This means that the ICE operates at minimum torque and power, and the majority of power is supplied by the electric motor.

During regenerative braking, the planetary gear operation is reduced to one degree of freedom by disconnecting the clutch and braking the sun gear shaft. Thus, the kinetic energy is converted to electrical energy and hence recharges the battery while the electric motor operates as a generator.

During suburb driving, the HEV operates as an ICEV. In this case, the electric motor is switched off and the ring gear shaft is braked, which means that the planetary gear operation is also reduced to one degree of freedom.

The power distribution using planetary gearing provides the merits of significant torque and power stabilization of ICE operation, hence achieving high efficiency of the whole HEV system. Figure 4.11 shows the simulated performances of this new HEV system and gives comparative results under the driving cycles of FUD 48 (maximum steady speed at 48 km/h) and FUD 72 (maximum steady speed at 72 km/h). It can be seen that the energy level is about the same at the beginning and end of cycle, which implies minimum energy consumption of both the battery and ICE.

References

Bates, B. (1995). Getting a Ford HEV on the road. *IEEE Spectrum*, **32**, 22–5.

Beretta, J. (2000). New tools for energy efficiency evaluation on hybrid system architecture. *Proceedings of the 17th International Electric Vehicle Symposium*, CD-ROM.

GM Precept. General Motors, 2000.

Honda Insight. Honda, 2000.

Inada, E. (2000). Development of a high performance hybrid electric vehicle Tino hybrid. *Proceedings of the 17th International Electric Vehicle Symposium*, CD-ROM.

Szumanowski, A. (2000). *Fundamentals of Hybrid Electric Vehicle Drives*. Warsaw-Radom.

Toyota Electric & Hybrid Vehicles. Toyota, 2000.

Toyota Prius Product Information. Toyota, 2000.

Van Mierlo, J. (2000). Views on hybrid drivetrain power management. *Proceedings of the 17th International Electric Vehicle Symposium*, CD-ROM.

Wouk, V. (1995). Hybrids: then and now. *IEEE Spectrum*, **32**, 16–21.

5 Electric propulsion

An electric propulsion system is the heart of EVs. Its functional block diagram is shown in Fig. 5.1. Its job is to interface batteries with vehicle wheels, transferring energy in either direction as required, with high efficiency, under control of the driver at all times. From the functional point of view, an electric propulsion system can be divided into two parts—electrical and mechanical. The electrical part consists of the subsystems of electric motor, power converter, and electronic controller, whereas the mechanical part includes the subsystems of mechanical transmission (optional), and vehicle wheels. The boundary between the electrical and mechanical parts is the air-gap of the motor, where electromechanical energy conversion is taken place. The electronic controller can be further divided into three functional units—sensor, interface circuitry and processor. The sensor is used to translate the measurable quantities, such as current, voltage, temperature, speed, torque and flux, into electronic signals. Through the interface circuitry, these signals are conditioned to the appropriate level before being fed into the processor. The processor output signals are usually amplified via the interface circuitry to drive power semiconductor devices of the power converter. The converter acts as a power conditioner that regulates the power flow between the energy source and the electric motor for motoring and regeneration. Finally, the motor interfaces with the vehicle wheels via the mechanical transmission. This transmission is optional because the electric motor can directly drive the wheel as in the case of in-wheel drives (Chan, 1993).

5.1 EV considerations

The choice of electric propulsion systems for EVs mainly depends on three factors—driver expectation, vehicle constraint and energy source. The driver expectation is defined by a driving profile which includes the acceleration, maximum speed, climbing capability, braking and range. The vehicle constraint depends on the vehicle type, vehicle weight and payload. The energy source relates with batteries, fuel cells, capacitors, flywheels and various hybrid sources. Thus, the process of identifying the preferred features and packaging options for electric propulsion has to be carried out at the system level. The interactions between subsystems and those likely impacts of system trade-offs must be examined.

The development of electric propulsion systems has been based on the growth of various technologies, especially electric motors, power electronics, microelectronics and control strategies (Chan and Chau, 1997; Chau *et al.*, 1998).

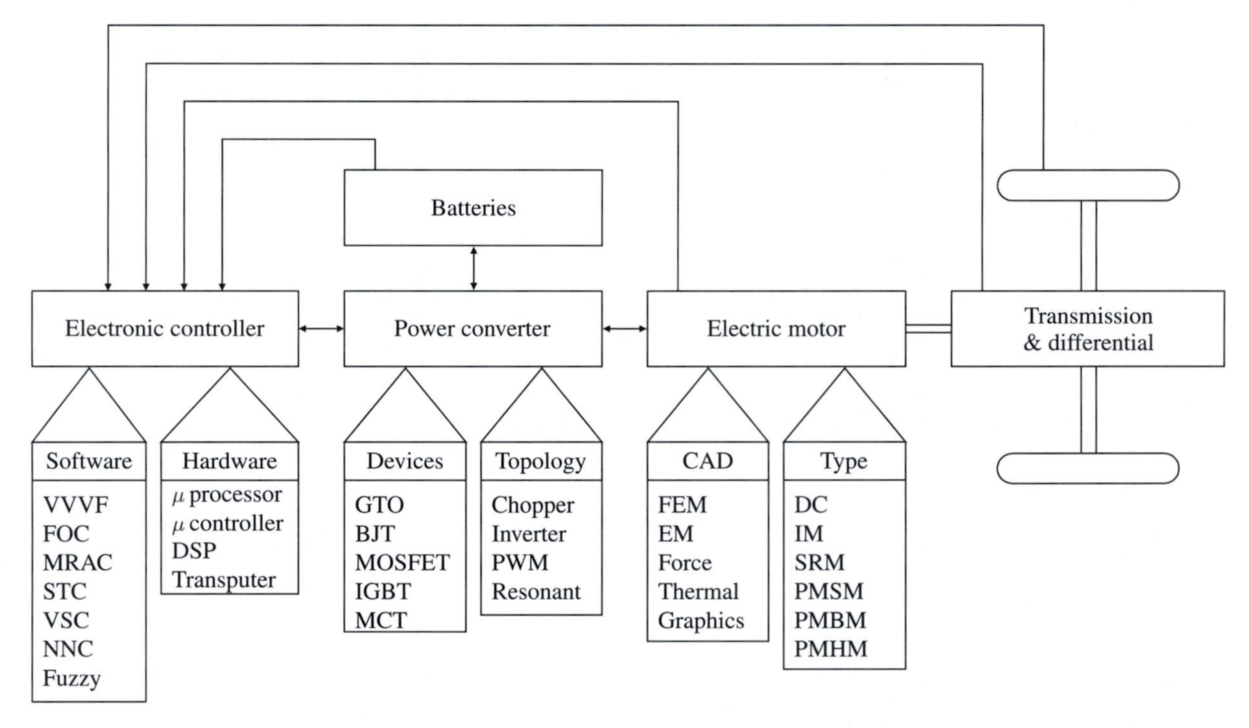

Fig. 5.1. Functional block diagram of an EV propulsion system.

5.1.1 ELECTRIC MOTORS

Concept of EV motors

Some engineers and even researchers may consider EV motors kindred or similar to industrial motors. However, EV motors usually require frequent start/stop, high rate of acceleration/deceleration, high-torque low-speed hill climbing, low-torque high-speed cruising and very wide-speed range of operation, whereas industrial motors are generally optimized at rated conditions. Thus, EV motors are so unique that they are deserved to form an individual class. Their major differences in load requirement, performance specification and operating environment are summarized as follows:

- EV motors need to offer the maximum torque that is four to five times of the rated torque for temporary acceleration and hill-climbing, while industrial motors generally offer the maximum torque that is twice of the rated torque for overload operation.
- EV motors need to achieve four to five times the base speed for highway cruising, while industrial motors generally achieve up to twice the base speed for constant-power operation.
- EV motors should be designed according to the vehicle driving profiles and drivers' habits, while industrial motors are usually based on a typical working mode.
- EV motors demand both high power density and good efficiency map (high efficiency over wide speed and torque ranges) for the reduction of total vehicle weight and the extension of driving range, while industrial motors generally need a compromise among power density, efficiency and cost with the efficiency optimized at a rated operating point.
- EV motors desire high controllability, high steady-state accuracy and good dynamic performance for multiple-motor coordination, while only special-purpose industrial motors desire such performance.
- EV motors need to be installed in mobile vehicles with harsh operating conditions such as high temperature, bad weather and frequent vibration, while industrial motors are generally located in fixed places.

Apart from satisfying the aforementioned special requirements, the design of EV motors also depends on the system technology of EVs. From the technological point of view, the following key issues should be considered:

(1) Single- or multiple-motor configurations—one adopts a single motor to propel the driving wheels, while another uses multiple motors permanently coupled to individual driving wheels. The single-motor configuration has the merit of using only one motor which can minimize the corresponding size, weight and cost. On the other hand, the multiple-motor configuration takes the advantages to reduce the current/power ratings of individual motors and

evenly distribute the total motor size and weight. Also, the multiple-motor one needs additional precaution to allow for fault tolerance during the electronic differential action. For instance, each motor may have its own controller which is controlled by a master controller. The functional block diagrams of single- and dual-motor configurations are shown in Fig. 5.2, while their comparison is listed in Table 5.1. Since these two configurations have their individual merits, both of them have been employed by modern EVs. For examples, the single-motor configuration has been adopted in the GM EV1 while the dual-motor configuration has been adopted in the NIES Luciole. Nevertheless, the use of single-motor configuration is still the majority today.

(2) Fixed- or variable-gearing transmissions—it is also classified as single-speed and multiple-speed transmissions. The former adopts single-speed fixed gearing, while the latter uses multiple-speed variable gearing together with the gearbox

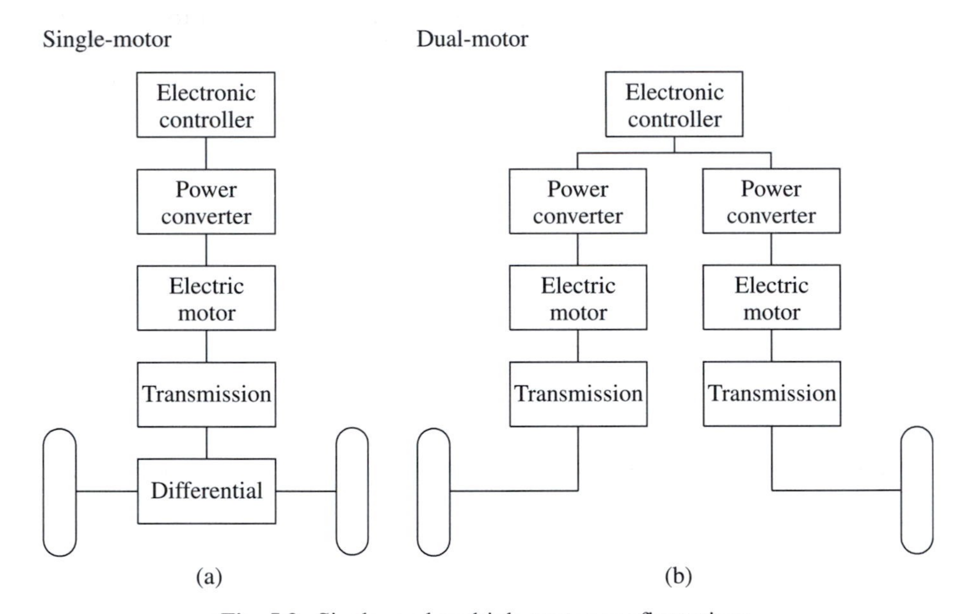

Fig. 5.2. Single- and multiple-motor configurations.

Table 5.1 Comparison of single- and dual-motor configurations

	Single-motor	Dual-motor
Cost	Lower	Higher
Size	Lumped	Distributed
Weight	Lumped	Distributed
Efficiency	Lower	Higher
Differential	Mechanical	Electronic

Table 5.2 Comparison of fixed- and variable-gearing transmissions

	Fixed-gearing	Variable-gearing
Motor rating	Higher	Lower
Inverter rating	Higher	Lower
Cost	Lower	Higher
Size	Smaller	Larger
Weight	Lower	Higher
Efficiency	Higher	Lower
Reliability	Higher	Lower

and clutch. Based on fixed-gearing transmission, the motor should be so designed that it can provide both high instantaneous torque (3 to 5 times the rated value) in the constant-torque region and high operating speed (3 to 5 times the base speed) in the constant-power region. On the other hand, the variable-gearing transmission provides the advantage of using conventional motors to achieve high starting torque at low gear and high cruising speed at high gear. However, there are many drawbacks on the use of variable gearing such as the heavy weight, bulky size, high cost, less reliable and more complex. Table 5.2 gives a comparison of fixed-gearing and variable-gearing transmissions. Actually, almost all the modern EVs adopt fixed-gearing transmission.

(3) Geared or gearless—the use of fixed-speed gearing with a high gear ratio allows EV motors to be designed for high-speed operation, resulting high power density. The maximum speed is limited by the friction and windage losses as well as transaxle tolerance. On the other hand, EV motors can directly drive the transmission axles or adopt the in-wheel drive without using any gearing (gearless operation). However, it results the use of low-speed outer-rotor motors which generally suffer from relatively low power density. The breakeven point is whether this increase in motor size and weight can be outweighed by the reduction of gearing. Otherwise, the additional size and weight will cause suspension problems in EVs. Both of them have been employed by modern EVs. The functional block diagrams of geared and gearless in-wheel motor configurations are shown in Fig. 5.3. For examples, the high-speed geared inner-rotor in-wheel motor has been adopted in the NIES Luciole while the low-speed gearless out-rotor in-wheel motor was adopted in the TEPCO IZA. Nevertheless, with the advent of compact planetary gearing, the use of high-speed planetary-geared in-wheel motors is becoming more attractive than the use of low-speed gearless in-wheel motors.

(4) System voltage—the design of EV motors is greatly influenced by the selection of the EV system voltage level. Reasonable high-voltage motor design can be adopted to reduce the cost and size of inverters. If the desired voltage is too high, a large number of batteries will be connected in series, leading to the

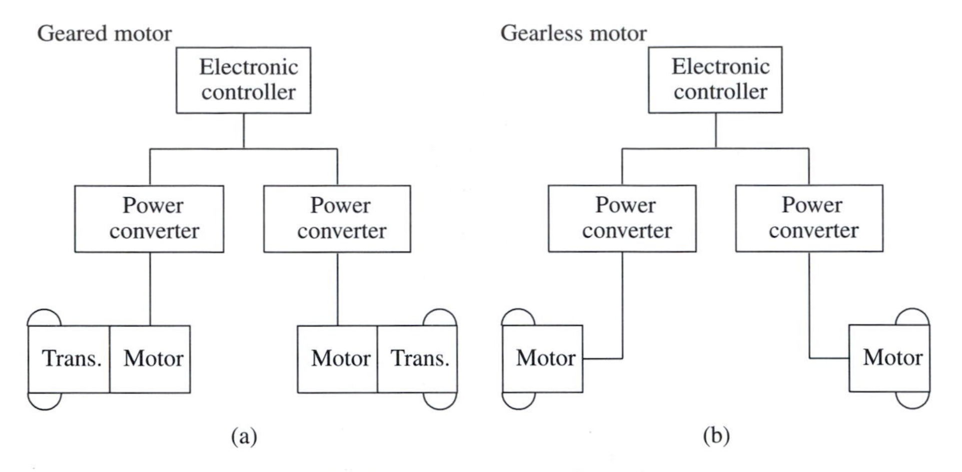

Fig. 5.3. In-wheel motor configurations.

reduction of interior and luggage spaces, the increase in vehicle weight and cost, as well as the degradation of vehicle performances. Since different EV types adopt different system voltage levels, the design of EV motors needs to cater for different EVs. Roughly, the system voltage is governed by the battery weight which is about 30% of the total vehicle weight. In practice, higher power motors adopt higher voltage levels. For examples, the GM EV1 adopts the 312-V voltage level for its 102-kW motor, whereas the Reva EV adopts the 48-V voltage level for its 13-kW motor.

(5) Integration—the integration of the motor with the converter, controller, transmission and energy source is prime important consideration. The EV motor designer should fully understand the characters of these components, thus to design the motor under this given environment. It is quite different with the normal standard motors under standard power source for normal industrial drives.

Classification of EV motors

Electric motors have been available for over a century. The evolution of motors, unlike that of electronics and computer science, has been long and relatively slow. Nevertheless, the development of motors is continually fuelled by new materials, sophisticated topologies, powerful computer-aided design (CAD) as well as modern power electronics and microelectronics. As illustrated in Fig. 5.4, those motors applicable to electric propulsion can be classified as two main groups, namely the commutator motors and commutatorless motors. The former simply denote that they generally consist of the commutator, while the latter have no commutator. Moreover, the shaded motor types indicate that they have ever been adopted by recent EVs. Table 5.3 also illustrates their recent applications to flagship EVs.

To keep up with the stringent requirement and fast changing motor topologies, the design of motors have turned to CAD. Basically, there are two major design

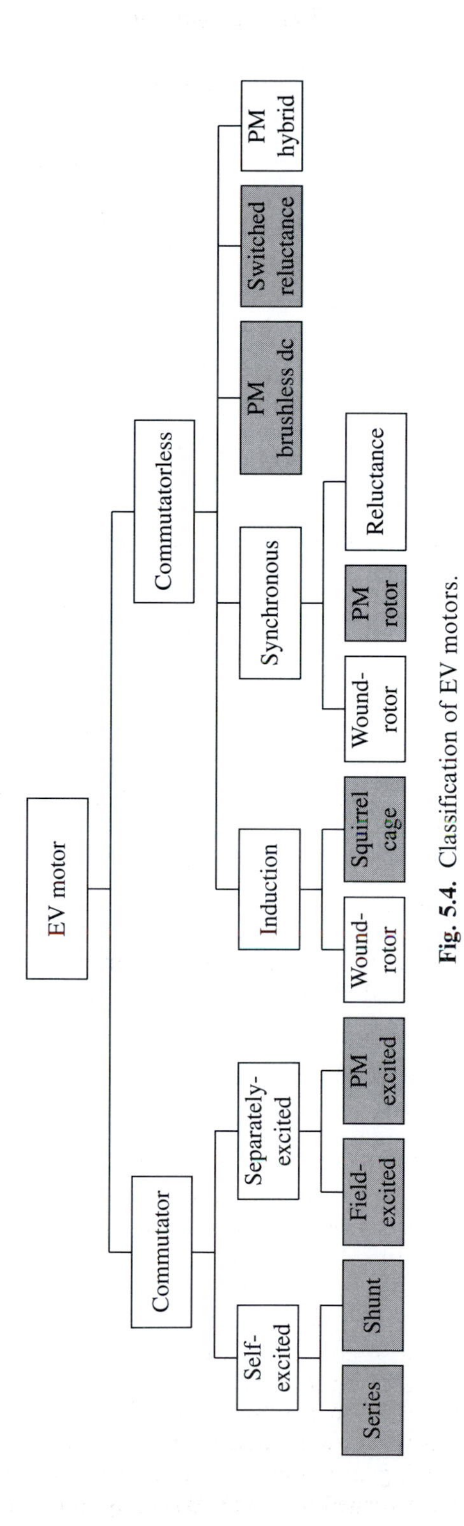

Fig. 5.4. Classification of EV motors.

Table 5.3 Applications of EV motors

EV models	EV motors
Fiat Panda Elettra	Series dc motor
Mazda Bongo	Shunt dc motor
Conceptor G-Van	Separately excited dc motor
Suzuki Senior Tricycle	PM dc motor
Fiat Seicento Elettra	Induction motor
Ford Th!nk City	Induction motor
GM EV1	Induction motor
Honda EV Plus	PM synchronous motor
Nissan Altra	PM synchronous motor
Toyota RAV4	PM synchronous motor
Chloride Lucas	Switched reluctance motor

approaches—circuit and field. In essence, the circuit approach is based on equivalent circuit analysis while the field approach depends on electromagnetic field analysis. The field approach takes the advantages of more accurate results, greater knowledge of the critical areas as well as capabilities of handling complicated machine geometry and nonlinear materials. Recently, the finite element method (FEM) has been regarded as one of the most powerful tools for electromagnetic field analysis of EV motors. The FEM outranks other numerical methods because of its flexibility and applicability in stress and thermal field analyses. Moreover, with the use of computer graphics, the analysis can be carried out visually and interactively.

The basic consideration of motor design includes magnetic loading—the peak of fundamental component of radial flux density in the air-gap of the motor, electric loading—the total r.m.s. current per unit length of periphery of the motor or ampere-turns per unit periphery, power per unit volume and weight, torque per unit volume and weight, flux density at each part of the magnetic circuit, speed, torque and power, losses and efficiency, and thermal design and cooling.

The corresponding key issues are better utilization of steel, magnet and copper, better electromagnetic coupling, better geometry and topology, better thermal design and cooling, understanding the limits on the motor performance, and understanding the relationship among geometry, dimensions, parameters and performance, thus to achieve higher power per unit weight, higher torque per unit weight and better performance.

Recent EV motors

Traditionally, dc commutator motors have been loosely named as dc motors. Their control principle is simple because of the orthogonal disposition of field and armature mmfs. By replacing the field winding of dc motors with PMs, PM dc motors permit a considerable reduction in stator diameter due to the efficient use of radial space. Owing to the low permeability of PMs, armature reaction is usually reduced and commutation is improved. However, the principle problem of dc motors, due to their commutators and brushes, makes them less reliable and unsuitable for maintenance-free operation. Nevertheless, because of mature technology and simple control, dc motors have ever been prominent in electric propulsion. Actually, various types of dc motors, including series, shunt, separately excited and PM excited, have ever been adopted by recent EVs.

Recently, technological developments have pushed commutatorless motors to a new era, leading to take the advantages of higher efficiency, higher power density, lower operating cost, more reliable and maintenance-free over dc commutator motors. As high reliability and maintenance-free operation are prime considerations for electric propulsion in EVs, commutatorless motors are becoming attractive.

Induction motors are a widely accepted commutatorless motor type for EV propulsion because of their low cost, high reliability and free from maintenance. However, conventional control of induction motors such as variable-voltage variable-frequency (VVVF) cannot provide the desired performance. One major reason is due to the nonlinearity of their dynamic model. With the advent of microcomputer era, the principle of field-oriented control (FOC) of induction motors has been accepted to overcome their control complexity due to the nonlinearity. Notice that FOC is also known as vector control or decoupling control. Nevertheless, these EV induction motors employing FOC still suffer from low efficiency at light loads and limited constant-power operating region. Recently, an on-line efficiency-optimizing control scheme has been developed for these EV induction motors (Yamada *et al.*, 1996), which can reduce the consumed energy by about 10% and increase the regenerative energy by about 4%, leading to extend the driving range of EVs by more than 14%. On the other hand, an electrically pole changing scheme has been developed for EV induction motors (Mori *et al.*, 1996), which can significantly extend the constant-power operating region to over four times the base speed.

By replacing the field winding of conventional synchronous motors with PMs, PM synchronous motors can eliminate conventional brushes, slip-rings and field copper losses. Actually, these PM synchronous motors are also called as PM brushless ac motors or sinusoidal-fed PM brushless motors because of their sinusoidal ac current and brushless configuration. As these motors are essentially synchronous motors, they can run from a sinusoidal or PWM supply without electronic commutation. When PMs are mounted on the rotor surface, they behave as non-salient synchronous motors because the permeability of PMs is similar to that of air. By burying those PMs inside the magnetic circuit of the rotor, the saliency causes an additional reluctance torque which leads to facilitate a wider speed range at constant-power operation. On the other hand, by abandoning the field winding or PMs while purposely making use of the rotor saliency, synchronous reluctance motors are generated. These motors are generally simple and cheap, but with relatively low output power. Similar to induction motors, those PM synchronous motors usually employ FOC for high-performance applications. Because of their inherent high power density and high efficiency, they have been accepted to have great potential to compete with induction motors for EV applications. Recently, a self-tuning control has been developed for PM synchronous motors (Chan and Chau, 1996), which can enable them to achieve optimal efficiency throughout the operating region.

By virtually inverting the stator and rotor of PM dc motors, PM brushless dc motors are generated. Notice that the name containing the 'dc' term may be misleading, since it does not refer to a dc current motor. Actually, these motors are fed by rectangular ac current, hence also called as rectangular-fed PM brushless motors. The most obvious advantage of these motors is the removal of brushes, leading to eliminate many problems associated with brushes. Another advantage is the ability to produce a larger torque because of the rectangular interaction between current and flux. Moreover, the brushless configuration allows more cross-sectional area for the armature winding. Since the conduction of heat through the frame is improved, an increase in electric loading causes higher power density. Different from PM synchronous motors, these PM brushless dc motors generally operate with shaft position sensors. Recently, a phase-decoupling PM brushless dc motor has been developed for EVs, which offers the merits of outstanding power density, no cogging torque and excellent dynamic performance (Chan *et al.*, 1993; Chan *et al.*, 1994). Also, it can adopt advanced conduction angle control to greatly extend the constant-power operating range (Chan, Jiang, Xia and Chau, 1995).

SR motors have been recognized to have considerable potential for EV applications. Basically, they are direct derivatives of single-stack variable-reluctance stepping motors. SR motors have the definite advantages of simple construction, low manufacturing cost and outstanding torque-speed characteristics for EV propulsion. Although they possess the simplicity in construction, it does not imply any simplicity of their design and control. Because of the heavy saturation of pole tips and the fringe effect of poles and slots, their design and control are difficult and subtle. Also, they usually exhibit acoustic noise problems. Recently, an optimum design approach to SR motors has been developed (Chan, Jiang and Zhou, 1995), which employs finite element analysis to minimize the total motor losses while taking into account the constraints of pole arc, height and maximum flux density. Also, fuzzy sliding mode control has been developed for those EV SR motors so as to handle the motor non-linearities and minimize the control chattering (Chan, Jiang *et al.*, 1996; Zhan *et al.*, 1999).

Recently, a new research direction has been identified on the development of PM hybrid motors for EV applications. In principle, there are many PM hybrids in which three of them have been actively investigated, namely the PM and reluctance hybrid, the PM and hysteresis hybrid, and the PM and field-winding hybrid. Firstly, by burying PMs inside the magnetic circuit of rotor, the PM synchronous motor can easily incorporate both PM torque and synchronous

Table 5.4 Evaluation of EV motors

	Dc motor	Induction motor	PM brushless motor	SR motor	PM hybrid motor
Power density	2.5	3.5	5	3.5	4
Efficiency	2.5	3.5	5	3.5	5
Controllability	5	4	4	3	4
Reliability	3	5	4	5	4
Maturity	5	5	4	4	3
Cost	4	5	3	4	3
Total	22	26	25	23	23

reluctance torque. On the other hand, by incorporating PMs into the SR structure, another PM and reluctance hybrid is generated which is so-called the doubly salient PM (DSPM) motor (Liao *et al.*, 1995; Cheng, Chau, Chan and Zhou, 2000). Recent development of this DSPM motor has shown that it is of high efficiency, high power density and wide speed range (Chau *et al.*, 1999; Cheng, Chau, Chan, Zhou and Huang, 2000). Secondly, a new PM hybrid motor, incorporating both PM torque and hysteresis torque, has been introduced (Rahman and Qin, 1997). By inserting PMs into the slots at the inner surface of the hysteresis ring, this PM hysteresis hybrid motor can offer unique advantages such as high starting torque as well as smooth and quiet operation for EV applications. Thirdly, another new PM hybrid motor has recently been developed for EVs, which comprises of both PMs in the rotor and a dc field winding in the inner stator (Chan, Zhang *et al.*, 1996). By controlling the direction and magnitude of the dc field current, the air-gap flux of this motor can be flexibly adjusted, hence the torque-speed characteristics can be easily shaped to meet the special requirements for EV propulsion.

In order to evaluate the aforementioned EV motor types, a point grading system is adopted. The grading system consists of six major characteristics and each of them is graded from 1 to 5 points. As listed in Table 5.4, this evaluation indicates that induction motors are relatively most acceptable. When the cost and maturity of PM brushless (including ac or dc) motors have significant improvements, these motors will be most attractive. Conventional dc motors seem to be losing their competitive edges, whereas both SR and PM hybrid motors have increasing potentials for EV propulsion.

5.1.2 POWER ELECTRONICS

EV power devices

In the past decades, power semiconductor device technology has made tremendous progress. These power devices have grown in power rating and performance by an evolutionary process. Among existing power devices, the power diode behaves as an uncontrolled switch, whereas the others, including the thyristor, gate turn-off thyristor (GTO), power bipolar-junction transistor (BJT), power metal-oxide field-effect transistor (MOSFET), insulated-gate bipolar transistor (IGBT), static-induction transistor (SIT), static-induction thyristor (SITH) and MOS-controlled thyristor (MCT), are externally controllable. Active research is still being pursued on the development of high performance power devices.

Before selecting the preferred power devices for electric propulsion, the following requirements have to be considered:

- Ratings—the voltage rating is based on the battery nominal voltage, maximum voltage during charging, and maximum voltage during regenerative braking. On the other hand, the current rating depends on the motor peak power rating and number of power devices connected in parallel. When paralleling these devices, on-state and switching characteristics have to be matched.
- Switching frequency—switching at higher frequencies can bring down the filter size and help to meet the electromagnetic interference (EMI) limitation requirements. Over the switching frequency of 20 kHz, there is no acoustic noise problem.
- Power losses—the on-state conduction drop or loss should be minimum while the switching loss should be as low as possible. Since higher switching frequencies increase the switching loss, switching the device at about 10 kHz seems to be an optimum for efficiency, power density, acoustic noise and EMI considerations. The leakage current should also be less than 1 mA to minimize the off-state loss.
- Base/gate driverability—the device should allow for simple and secure base/gate driving. The corresponding driving signal may be either triggering voltage/current or linear voltage/current. The voltage-mode driving involves very little energy and is generally preferable.
- Dynamic characteristics—the dynamic characteristics of the device should be good enough to allow for high dv/dt capability, high di/dt capability and easy paralleling. The internal anti-parallel diode should have similar dynamic characteristics as the main device.
- Ruggedness—the device should be rugged to withstand a specific amount of avalanche energy during over-voltage and be protected by fast semiconductor fuses during over-current. It should operate with no or minimal use of snubber circuits. Since EVs are frequently accelerated and decelerated, the device is subjected to thermal cycling at frequent intervals. It should reliably work under these conditions of thermal stress.
- Maturity and cost—since the cost of power devices is one of the major parts in the total cost of electric propulsion systems, these devices should be economical. Some recent power devices such as the high-power MCT are not yet mature for EV applications.

Taking into account the above requirements, the GTO, power BJT, power MOSFET, IGBT and MCT are considered for electric propulsion. The thyristor is not considered because it requires additional commutating components to

Table 5.5 Evaluation of EV power devices

	GTO	BJT	MOSFET	IGBT	MCT
Ratings	5	4	2	5	3
Switching frequency	1	2	4	4	4
Power losses	2	3	4	4	4
Base/gate driveability	2	3	5	5	5
Dynamic characteristics	2	3	5	5	5
Ruggedness	3	3	5	5	5
Maturity	5	5	4	4	2
Cost	4	4	4	4	2
Total	24	27	33	36	30

turnoff and its switching frequency is limited to 400 Hz. The SIT and SITH are also excluded because of their normally turn-on property and limited availability. In order to evaluate their suitability, a point grading system is adopted, which consists of eight major characteristics and each of them is graded from 1 to 5 points. From Table 5.5, the power MOSFET, IGBT and MCT score high points which indicate that they are particularly suitable for EV propulsion. Due to its highest score, the IGBT is almost exclusively used for modern EVs. Nevertheless, the power MOSFET has also been accepted for those relatively low-power electric tricycles and bikes.

EV power converters

The evolution of power converter topologies normally follows that of power devices, aiming to achieve high power density, high efficiency, high controllability and high reliability (Bose, 1992). Power converters may be ac-dc, ac-ac at the same frequency, ac-ac at different frequencies, dc–dc or dc–ac. Loosely, dc-dc converters are known as dc choppers while dc-ac converters are known as inverters, which are respectively used for dc and ac motors for electric propulsion.

Initially, dc choppers were introduced in the early 1960s using force-commutated thyristors that were constrained to operate at low switching frequency. Due to the advent of fast-switching power devices, they can now be operated at tens or hundreds of kilohertz. In electric propulsion applications, two-quadrant dc choppers are desirable because they convert battery dc voltage to variable dc voltage during the motoring mode and revert the power flow during regenerative braking. Furthermore, four-quadrant dc choppers are employed for reversible and regenerative speed control of dc motors. A four-quadrant dc chopper is shown in Fig. 5.5.

Inverters are generally classified into voltage-fed and current-fed types. Because of the need of a large series inductance to emulate a current source, current-fed inverters are seldom used for electric propulsion. In fact, voltage-fed inverters are almost exclusively used because they are very simple and can have power flow in either direction. A typical three-phase full-bridge voltage-fed inverter is shown

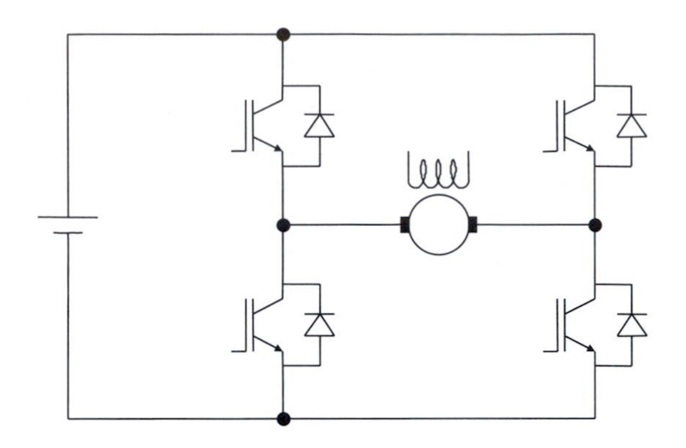

Fig. 5.5. Four-quadrant dc chopper.

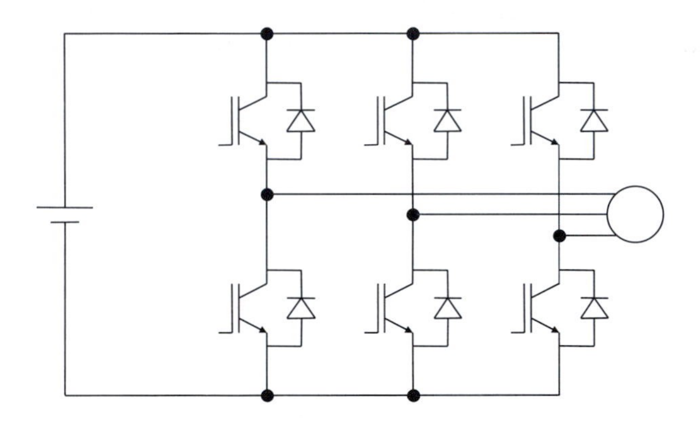

Fig. 5.6. Three-phase full-bridge voltage-fed inverter.

in Fig. 5.6. Its output waveform may be rectangular, six-step or PWM, depending on the switching strategy for different applications. For example, a rectangular output waveform is produced for a PM brushless dc motor, while a six-step or PWM output waveform is for an induction motor. It should be noted that the six-step output is becoming obsolete because its amplitude cannot be directly controlled and its harmonics are rich. On the other hand, the PWM waveform is harmonically optimal and its fundamental magnitude and frequency can be smoothly varied for speed control.

Starting from the last decade, numerous PWM switching schemes have been developed for voltage-fed inverters, focusing on the harmonic suppression, better utilization of dc voltage, tolerance of dc voltage fluctuation as well as suitability for real-time and microcontroller-based implementation (Bose, 1992). These schemes can be classified as voltage-controlled and current-controlled PWM. The state-of-the-art voltage-controlled PWM schemes are natural or sinusoidal PWM, regular or uniform PWM, harmonic elimination or optimal PWM, delta PWM, carrierless or random PWM, and equal-area PWM. On the other hand, the use of current control for voltage-fed inverters is particularly attractive for high-performance motor drives because the motor torque and flux are directly related to the controlled current. The state-of-the-art current-controlled PWM schemes are hysteresis-band or band-band PWM, instantaneous current control with voltage PWM, and space vector PWM.

Soft-switching EV converters

Instead of using hard or stressed switching, power converters can adopt soft or relaxed switching. The key of soft switching is to employ a resonant circuit to shape the current or voltage waveform such that the power device switches at zero-current or zero-voltage condition. In general, the use of soft-switching converters possess the following advantages:

- Due to zero-current or zero-voltage switching condition, the device switching loss is practically zero, thus giving high efficiency.
- Because of low heat sinking requirement and snubberless operation, the converter size and weight are reduced, thus giving high power density.
- The device reliability is improved because of minimum switching stress during soft switching.
- The EMI problem is less severe and the machine insulation is less stressed because of lower dv/dt resonant voltage pulses.
- The acoustic noise is very small because of high frequency operation.

On the other hand, their key drawbacks are the additional cost of the resonant circuit and the increased complexity. Although soft-switching dc–dc converters have been widely accepted by switched-mode power supplies, the corresponding development for EV propulsion is much slower. As the pursuit of power converters having high efficiency and high power density for EV propulsion is highly desirable,

Table 5.6 Comparison of hard switching and soft switching for EV converters

	Hard switching	Soft switching
Switching loss	Severe	Almost zero
Overall efficiency	Norm	Possibly higher
Heat-sinking requirement	Norm	Possibly lower
Hardware count	Norm	More
Overall power density	Norm	Possibly higher
EMI problem	Severe	Low
Dv/dt problem	Severe	Low
Modulation scheme	Versatile	Limited
Maturity	Mature	Developing
Cost	Norm	Higher

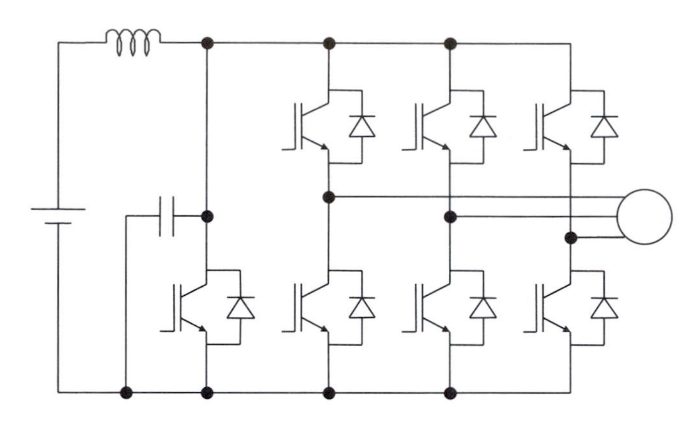

Fig. 5.7. Three-phase voltage-fed resonant dc link inverter.

the development of EV soft-switching power converters is in progress (Chan, Chau, Chan and Yao, 1997; Lai, 1997; Murai *et al.*, 1997). Table 5.6 gives a comparison between hard-switching and soft-switching converters for EV propulsion.

Although there have been many soft-switching dc–dc converters developed for switched-mode power supplies, these converters cannot be directly applied to dc motors for EV propulsion. Apart from suffering excessive voltage and current stresses, they cannot handle backward power flow during regenerative braking. It should be noted that the capability of regenerative braking is very essential for EVs as it can extend the vehicle driving range by up to 25%. Recently, a new soft-switching dc–dc converter, having the capability of bidirectional power flow for motoring and regenerative braking as well as the minimum hardware count, has been developed for EV dc motors (Chau *et al.*, 1997).

The development of soft-switching inverters for ac motors (including induction motors, PM brushless motors and PM hybrid motors) has become a research direction in power electronics (Chen and Lipo, 1996; Li *et al.*, 1996). Figure 5.7 shows a milestone of soft-switching inverters, namely the three-phase voltage-fed resonant dc link inverter developed in 1986 (Divan, 1986). Consequently, many improved soft-switching topologies have been proposed, such as the quasi-resonant dc link, series resonant dc link, parallel resonant dc link, synchronized resonant dc link, resonant transition, auxiliary resonant commutated pole and auxiliary resonant snubber inverters. A number of development goals of soft-switching inverters for EV propulsion have been identified, namely efficiency over 95%, power density over $3.5\,\mathrm{W/cm^3}$, switching frequency over 10–20 kHz, dv/dt below $1000\,\mathrm{V}/\mu\mathrm{s}$, zero EMI, zero failure before the end of the vehicle life, and redundant with 'limp-home' mode. Recently, the delta-configured auxiliary resonant snubber version has satisfied most of these goals, and has been demonstrated to achieve an output power of 100 kW.

Compared with the development of soft-switching inverters for ac motors, the development for SR motors has been very little (Cho *et al.*, 1997). Recently, a new soft-switching converter, so-called the zero-voltage-transition version, has been particularly developed for SR motors (Ching *et al.*, 1998). This new converter possesses the advantages that all main switches and diodes can operate at zero-voltage condition, unity device voltage and current stresses, as well as wide operating range. Moreover, it offers simple circuit topology, minimum hardware count and low cost, leading to achieve high switching frequency, high power density and high efficiency.

5.1.3 MICROELECTRONICS

Since the introduction of microcomputers in 1970, the microelectronics technology has gone through an intense evolution in last three decades. Modern microelectronic devices can generally be classified as microprocessors, microcontrollers, digital signal processors (DSPs), and transputers.

Microprocessor technology has been used to recognize the milestone of the development of microelectronics, such as the 8086, 80186, 80286, 80386, 80486, Pentium, Pentium II and Pentium III. Microprocessors are the CPU of microcomputer systems, which decode instructions, control activities as well as perform all arithmetic and logical computations. Unlike microprocessors, microcontrollers, such as the 8096, 80196 and 80960, include all resources (CPU, ROM or EPROM, RAM, DMA, timers, interrupt sources, A/D and D/A converters and I/O ports) to serve as stand-alone single-chip controllers. Thus, microcontroller-based electric propulsion systems possess definite advantages of minimum hardware and compact software. Digital signal processors (DSPs), such as the TMS320C30, TMS320C40 and i860, have the capability of high-speed floating-point computation to implement sophisticated control algorithms for high-performance motors for electric propulsion. On the other hand, transputers, such as the T400, T800 and T9000, are particularly designed for parallel processing applications. By employing multiple chips of transputers, any sophisticated control algorithms can be implemented.

By integrating microelectronic devices and power devices on the same chip (like the integration of brain and muscle), power ICs (PICs), loosely named as 'smart power', aim to further reduce the cost, minimize the size and improve the reliability. The PIC may include the power module, control, protection, communication and cooling. The main problems in PIC synthesis are the isolation between high-voltage and low-voltage devices as well as cooling. Nevertheless, this technology has promising applications to electric propulsion in near future. The key is the integrating and packaging.

5.1.4 CONTROL STRATEGIES

Conventional linear control such as PID can no longer satisfy the stringent requirement placed on high-performance motor drives. In recent years, many modern control strategies have been proposed. The state-of-the-art control strategies that have been proposed for motor drives are adaptive control, variable structure control, fuzzy control and neural network control.

Adaptive control includes self-tuning control (STC) and model-referencing adaptive control (MRAC). Using STC, the controller parameters are tuned to adapt to system parameter variations. The key is to employ an identification block to track changes in system parameters and to update the controller parameters through controller adaptation in such a way that a desired closed-loop performance can be obtained. Using MRAC, the output response is forced to track the response of a reference model irrespective of system parameter variations. Based on an adaptation algorithm that utilizes the difference between the reference model and system outputs, the controller parameters are adjusted to give a desired closed-loop performance. Recently, both MRAC and STC have been applied to commutatorless motor drives for electric propulsion in EVs.

Variable structure control (VSC) has recently been applied for motor drives to compete with adaptive control. Using VSC, the system can be designed to provide parameter-insensitive features, prescribed error dynamics and simplicity in implementation. Based on a set of switching control laws, the system is forced to follow a predefined trajectory in the phase plane irrespective of system parameter variations.

Emerging technologies such as fuzzy logic and neural networks have recently been introduced into the field of motor drives. Fuzzy control is essentially a linguistic process which is based on the prior experience and heuristic rules used by human operators. Making use of neural network control (NNC), the controller can possibly interpret the behaviour of system dynamics, then self-learn and self-adjust accordingly. Furthermore, these state-of-the-art control strategies can incorporate one another such as adaptive fuzzy control, fuzzy NNC and fuzzy VSC. In near future, controllers incorporating artificial intelligence (AI) can permit diagnosis of systems and correction of faults to supplant the need of human intervention.

5.2 Dc motor drives

Dc motor drives are classified as wound-field and PM dc types. The former has the field winding and the field can be controlled by the dc current, whereas the latter has no field winding and the PM field is uncontrollable. Various dc motor drives have ever been widely applied to different EVs because of their technological maturity and control simplicity.

5.2.1 SYSTEM CONFIGURATIONS

Dc motor drives have been widely used for electric propulsion. The earliest system configuration of dc motor drives consisted of a string of resistors connected in series and/or in parallel with the dc motor (Unnewehr, 1982). Motor voltage is equal to the battery voltage minus the voltage drop across the resistors and can be increased by operating contactors to short out a portion of the resistance. A basic configuration of

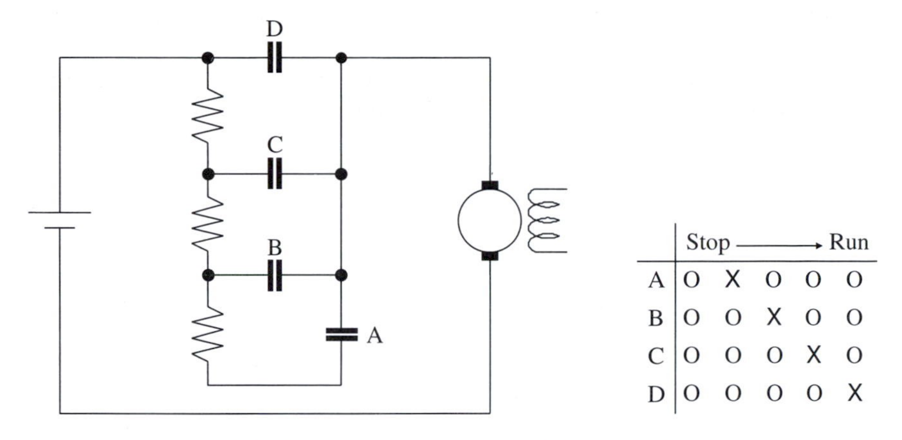

Fig. 5.8. Basic resistance control of dc motor drives.

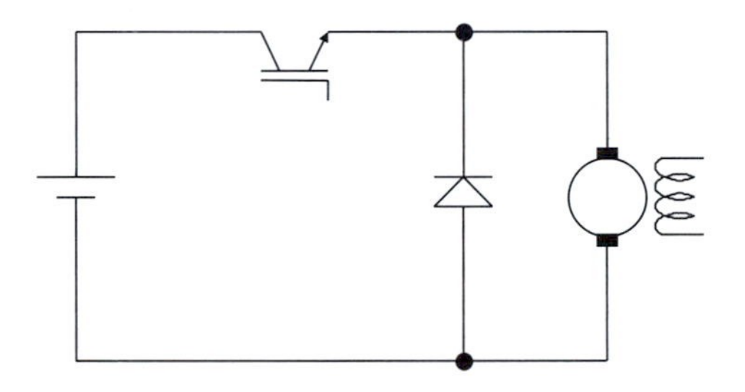

Fig. 5.9. Basic one-quadrant dc chopper control.

resistance control of dc motor drives is shown in Fig. 5.8, where A, B, C and D are externally controlled contactors. This system is satisfactory for EVs which operate almost exclusively at rated speed and only require to provide smooth acceleration at start up. Although this resistance control is simple and low cost, the main drawback is its poor efficiency because considerable energy is lost as heat in the resistors. In addition, it cannot offer smooth control and results in jerkiness. With the rapid development of power electronics, resistance control is obsolete while dc chopper control has been used extensively for electric propulsion because of the advantages of small size, lightweight, high efficiency and high controllability, especially offering smooth acceleration at the selected rate to the desired speed. Figure 5.9 shows a basic one-quadrant dc chopper for speed control of dc motor drives.

5.2.2 DC MOTORS

The name applied to wound-field dc motors is usually determined by the mutual interconnection between the field and armature windings. As shown in Fig. 5.10, common types of wound-field dc motors are separately excited, series, shunt, and cumulative compound (Dubey, 1989). Without external control, their torque-

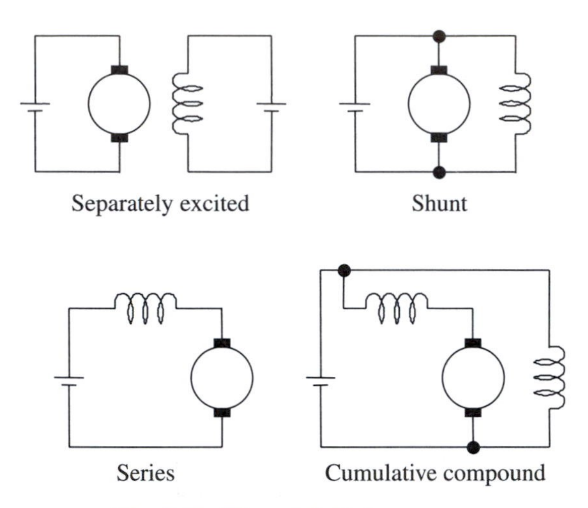

Fig. 5.10. Wound-field dc motors.

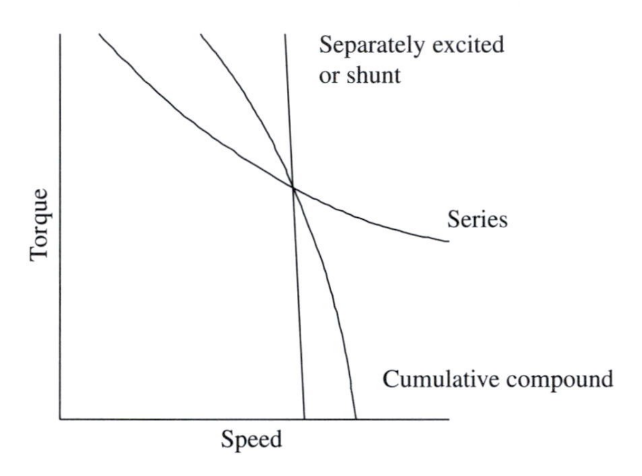

Fig. 5.11. Torque-speed characteristics of wound-field dc motors.

speed characteristics at the rated voltage are shown in Fig. 5.11. In the case of separately excited dc motors, the field and armature voltages can be controlled independent of each other. The torque-speed characteristic is linearly related that speed decreases as torque increases and speed regulation depends on the armature circuit resistance. In series dc motors, the field current is the same as the armature current. An increase in torque is accompanied by an increase in the armature current and hence an increase in flux. The speed must drop to maintain the balance between the supply and induced voltages. The torque-speed characteristic has an inverse relationship. In shunt dc motors, the field and armature are connected to a common voltage source. The corresponding characteristic is similar to that of separately excited dc motors. In cumulative compound dc motors, the mmf of the series field is in the same direction as the mmf of the shunt field. The characteristic lies between those of series and shunt dc motors, depending on the relative strength of the series and shunt fields. It should be noted that series dc motors have been widely accepted for conventional EV propulsion because of their very high torque at low speeds. They are considered to possess the highest torque-per-ampere ratio than any other dc motor types. This feature can greatly reduce the battery drain during vehicle acceleration and hill climbing.

By replacing the field winding and pole structure with PMs, PM dc motors can readily be generated from those wound-field dc motors. Compared with wound-field dc motors, PM dc motors have relatively higher power density and higher efficiency because of the space-saving benefit by PMs and the absence of field losses. Owing to the low permeability of PMs, similar to that of air, armature reaction is usually reduced and commutation is improved. However, since the field excitation in PM dc motors is uncontrollable, they cannot readily attain the operating characteristics similar to those of wound-field dc motor drives.

Both wound-field and PM dc motors suffer from the same problem due to the use of commutators and brushes. Commutators cause torque ripples and limit the motor speed, while brushes are responsible for friction and radio-frequency interference (RFI). Moreover, due to the wear and tear, periodic maintenance of commutators and brushes is always required. These drawbacks make them less reliable and unsuitable for maintenance-free operation, and limit them to be widely applied for modern EV propulsion.

As mentioned before, the major advantages of dc motors are their maturity and simplicity. The simplicity is mainly due to their simple control because the air-gap flux Φ and the armature current I_a, hence the motor speed ω_m and torque T, can be independently controlled. No matter wound-field or PM dc motors, they are governed by the following basic equations:

$$E = K_e \Phi \omega_m$$
$$V_a = E + R_a I_a$$
$$T = K_e \Phi I_a,$$

where E is the back emf, V_a is the armature voltage, R_a is the armature resistance, and K_e is named as the back emf constant or torque constant. For those wound-field dc motors, Φ is linearly related to the field current I_f which may be independently controlled, dependent on I_a, dependent on V_a, or dependent on both I_a and V_a, respectively for those separately excited, series, shunt, or cumulative compound types. In contrast, Φ is essentially uncontrollable for those PM dc motors.

The basic consideration of dc motor design includes main dimensions—namely armature outer-diameter and core length, optimization of the ratio of armature outer-diameter to core length, air-gap length, number of poles, number of armature slots, armature tooth width and slot depth, number of turns per coil, slot fill factor, number of commutator bars, commutation, flux density at each part of the magnetic circuit, magnetizing current, thermal resistance at each part of the thermal circuit, speed, torque and efficiency, torque per unit weight, and weight of copper and magnetic iron core used (West and Jack, 2000).

5.2.3 DC–DC CONVERTERS

When dc–dc converters adopt the chopping mode of operation, they are usually termed as dc choppers and are extensively used for voltage control of dc motor drives. These dc choppers are classified as first-, second-, two- and four-quadrant versions. The first-quadrant dc chopper is suitable for motoring and the power flow is from the source to the load, whereas the second-quadrant one is for regenerative braking and the power flow is out from the load into the source. As regenerative braking is very essential for EVs which can extend the vehicle driving range by up to 25%, the two-quadrant dc chopper shown in Fig. 5.12 is preferred as it is suitable to both motoring and regenerative braking for EV propulsion. Moreover, instead of using mechanical contactors to achieve reversible operation, the four-quadrant dc chopper can be employed so that motoring and regenerative braking in both forward and reversible operations are controlled electronically.

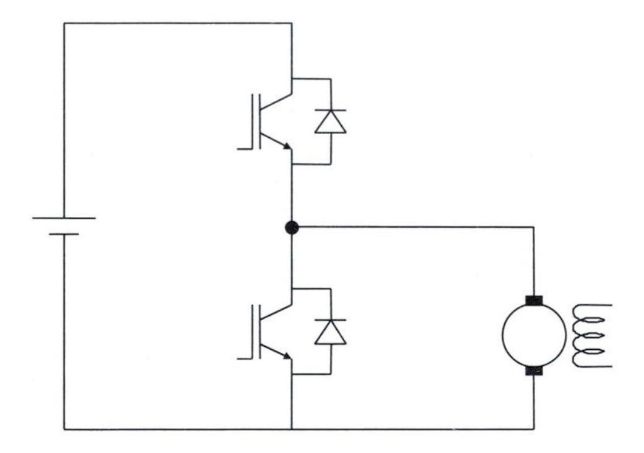

Fig. 5.12. Two-quadrant dc choppers for EV propulsion.

As shown in Fig. 5.13, there are three ways in which the chopper output voltage can be varied, namely pulse-width-modulated (PWM) control, frequency-modulated control, and current-limit control (Dubey, 1989). In the first method, the chopper frequency is kept constant and the pulse width is varied. The second method has a constant pulse width and a variable chopping frequency. In the third method, both the pulse width and frequency are varied to control the load current between certain specified maximum and minimum limits. For conventional dc motor drives for EV propulsion, PWM control of two-quadrant dc chopper is generally adopted. The corresponding control is based on the variation of duty cycle δ:

$$V_a = \delta V_s$$

$$I_a = \frac{V_a - E}{R_a},$$

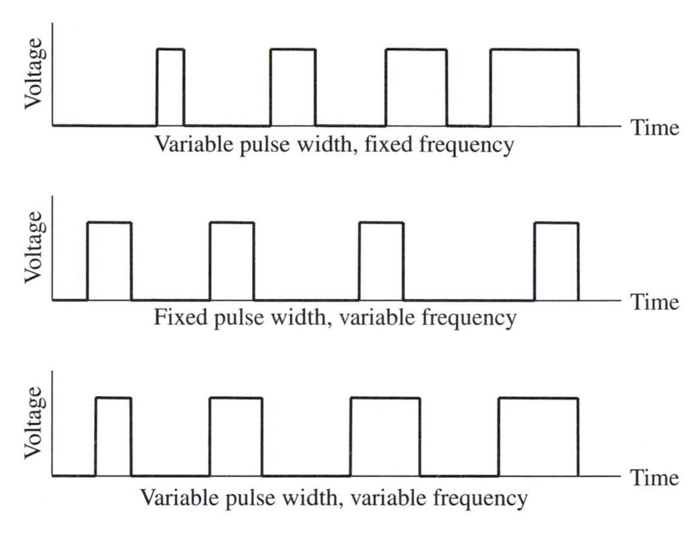

Fig. 5.13. Dc chopper controlled output voltages.

where V_s is the dc supply voltage. Hence, the motoring operation ($I_a > 0$) occurs when $\delta > (E/V_s)$, and that regenerative braking ($I_a < 0$) occurs when $\delta < (E/V_s)$. The no-load operation is obtained when $\delta = (E/V_s)$. As the current is flowing all the time, the discontinuous conduction mode does not occur.

Soft-switching dc–dc converters are seldom used for voltage control of dc motor drives. The major reason is due to the fact that the corresponding development has been much less than that for switched-mode power supplies. Also, those available soft-switching dc–dc converters cannot handle backward power flow during regenerative braking. Until recently, we have purposely developed a two-quadrant soft-switching dc–dc converter, namely the two-quadrant zero-voltage transition (2Q-ZVT) converter, for EV dc motor drives, which possesses the advantages of high efficiency for both motoring and regenerative braking, as well as minimum voltage and current stresses. Figure 5.14 shows this 2Q-ZVT converter-fed dc motor drive. Its equivalent circuits and operating waveforms in both the motoring and regenerating modes are shown in Figs. 5.15 and 5.16, respectively. The output voltage V_0 of this converter is governed by the voltage-conversion ratio $\mu_m = V_0/V_i$ in the motoring mode of operation, which is related to a new controllable duty ratio δ_m as given by:

$$\mu_m = \delta_m + \frac{f}{4\pi} + \left(\pi - 2 + \frac{1}{\lambda_m}\right),$$

where δ_m is defined as the normalized total duration of S4, S5 and S6, f is the normalised switching frequency, and λ_m is the normalized load current. Hence, when f is selected as 0.04, the operating characteristics of μ_m vs. λ_m and μ_m vs. δ_m

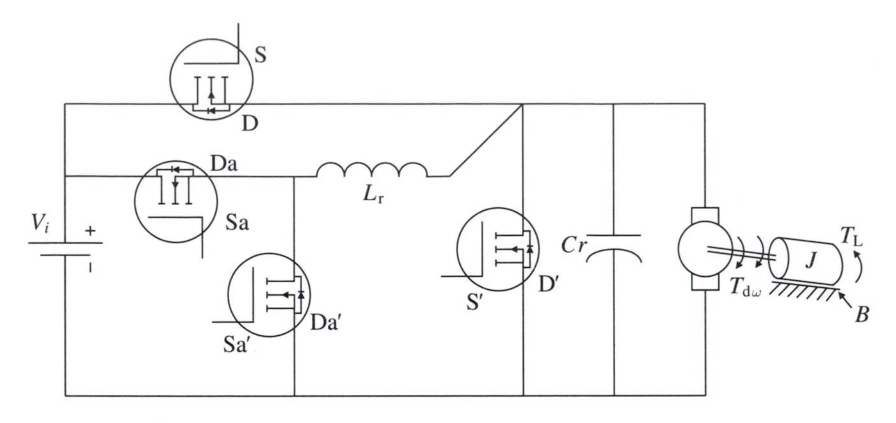

Fig. 5.14. Two-quadrant zero-voltage transition converter-fed dc motor drive.

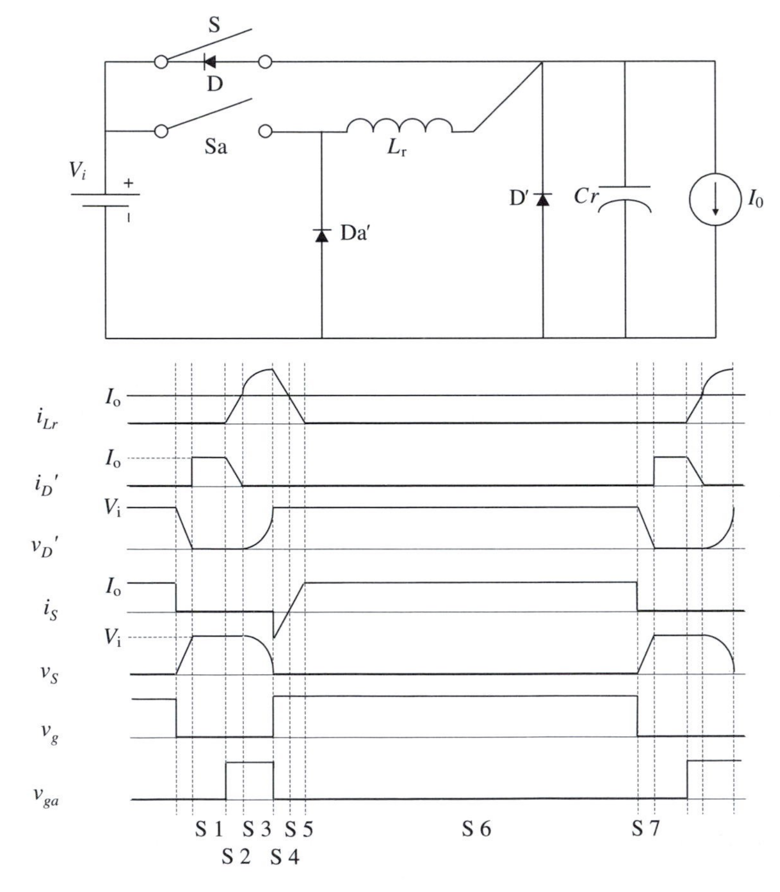

Fig. 5.15. Equivalent circuits and operating waveforms in motoring mode.

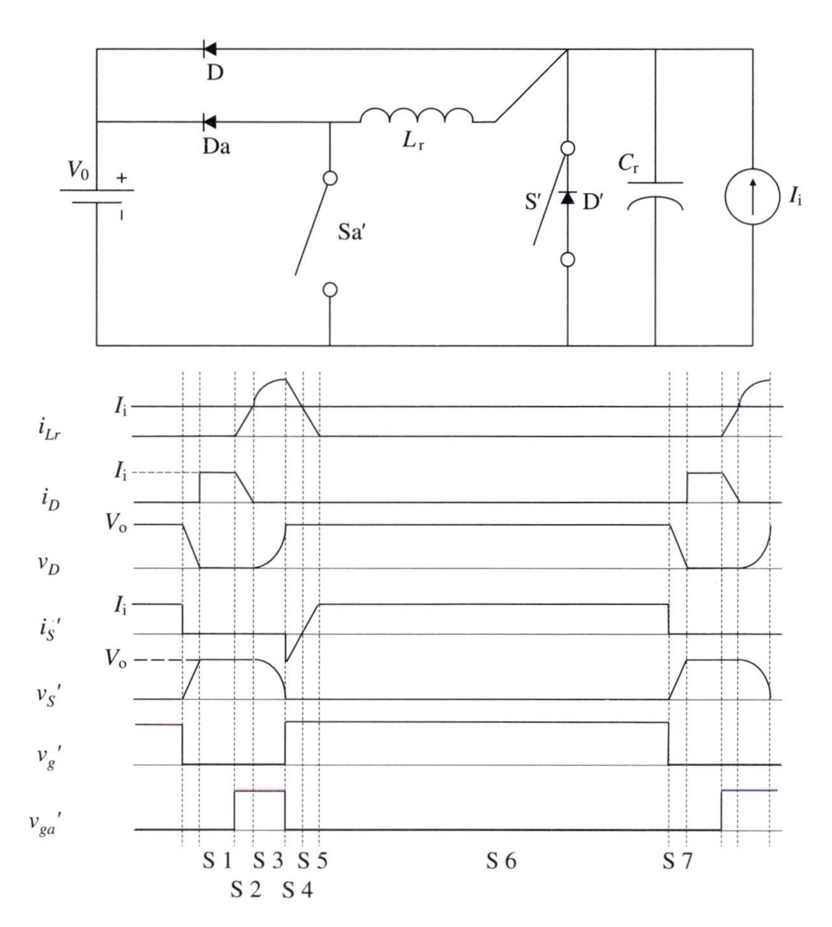

Fig. 5.16. Equivalent circuits and operating waveforms in regenerating mode.

are illustrated in Fig. 5.17. Similarly, the voltage-conversion ratio μ_r in the regenerating mode can be determined as:

$$\mu_r = \frac{1}{1 - \left[\delta_r + \frac{f}{4\pi} + \left(\pi - 2 + \frac{1}{\lambda_r}\right)\right]},$$

where the duty ratio δ_r is also defined as the normalized total duration of S4, S5 and S6, and λ_r is the normalized regenerative load current. When f is selected as 0.04, the corresponding characteristics, namely μ_r vs. λ_r and μ_r vs. δ_r, are shown in Fig. 5.18.

5.2.4 SPEED CONTROL

In general, speed control of dc motor drives can be accomplished by two methods, namely armature control and field control. When the armature voltage of the dc motor is reduced, the armature current and hence the motor torque decrease,

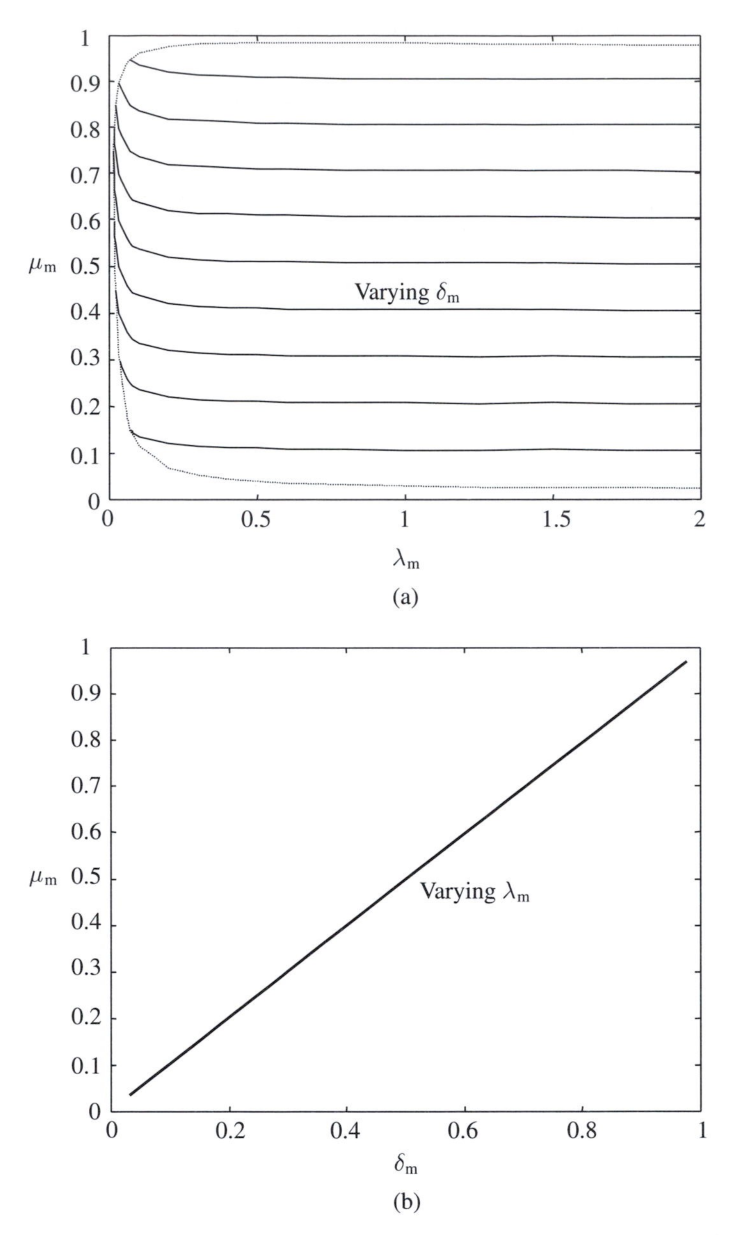

Fig. 5.17. Operating characteristics in motoring mode.

causing the motor speed to decrease. In contrast, when the armature voltage is increased, the motor torque increases, causing the motor speed to increase. Since the maximum allowable armature current remains constant and the field is fixed,

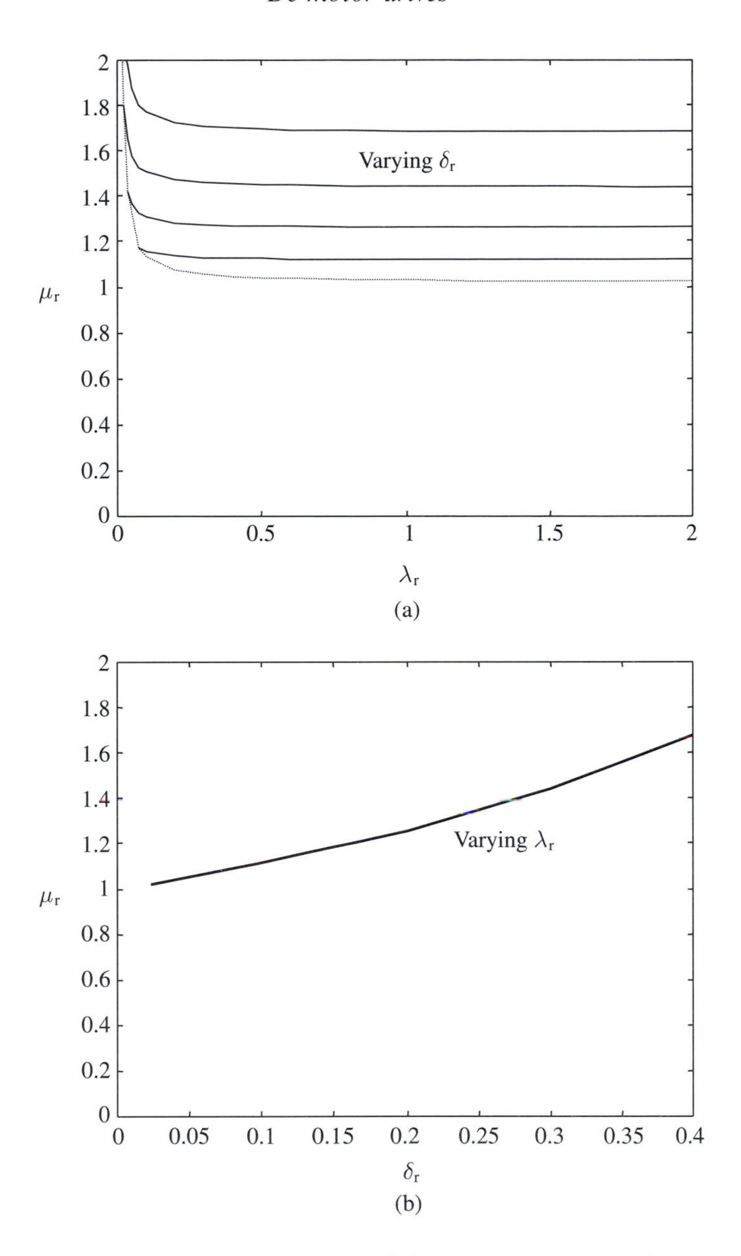

Fig. 5.18. Operating characteristics in regenerating mode.

this armature voltage control has the advantage of retaining the maximum torque capability at all speeds. However, since the armature voltage cannot be further increased beyond the rated value, this control is used only when the dc motor drive operates below its base speed. On the other hand, when the field voltage of the dc

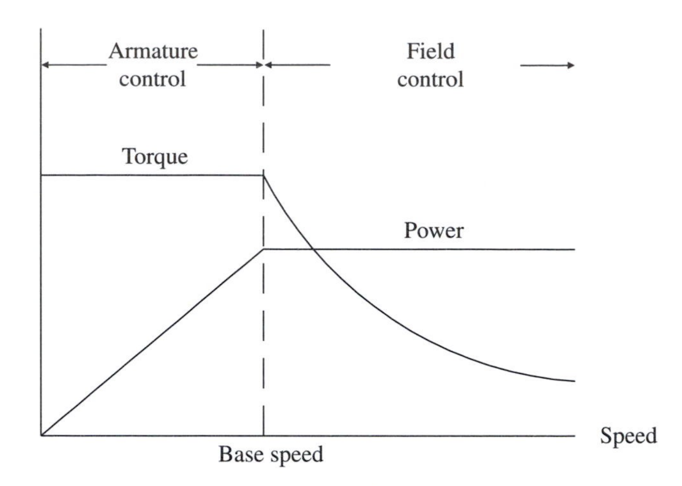

Fig. 5.19. Combined armature and field control of dc motor drive.

motor is weakened while the armature voltage is fixed, the motor induced emf decreases. Because of low armature resistance, the armature current will increase by an amount much larger than the decrease in the field. Thus, the motor torque is increased, causing the motor speed to increase. Since the maximum allowable armature current is constant, the induced emf remains constant for all speeds when the armature voltage is fixed. Hence, the maximum allowable motor power becomes constant so that the maximum allowable torque varies inversely with the motor speed. Therefore, in order to achieve wide-range speed control of dc motor drives for EVs, armature control has to be combined with field control. By maintaining the field constant at the rated value, armature control is employed for speeds from standstill to the base speed. Then, by keeping the armature voltage at the rated value, field control is used for speeds beyond the base speed. The corresponding maximum allowable torque and power in the combined armature and field control are shown in Fig. 5.19. Moreover, Fig. 5.20 shows the corresponding torque-speed characteristics of a separately excited dc motor drive during motoring and regenerative braking.

5.3 Induction motor drives

As mentioned before, commutatorless motor drives offer a number of advantages over conventional dc commutator motor drives for electric propulsion. At present, induction motor drives is the most mature technology among various commutatorless motor drives. There are two types of induction motors, namely wound-rotor and squirrel-cage. Because of high cost, need of maintenance and lack of sturdiness, wound-rotor induction motors are less attractive than squirrel-cage counterparts, especially for electric propulsion in EVs. Hence, squirrel-cage

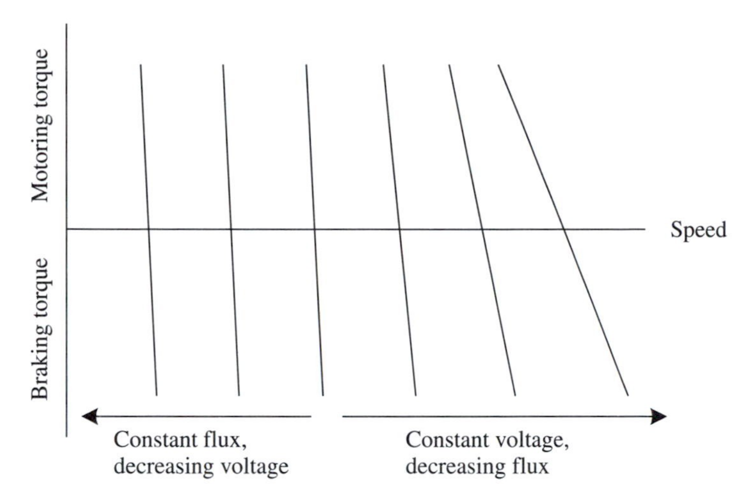

Fig. 5.20. Characteristics of a separately excited dc motor drive.

induction motors are loosely named as induction motors. Apart from the common advantages of commutatorless motor drives, induction motor drives possess the definite advantages of low cost and ruggedness. These advantages can generally outweigh their major disadvantage of control complexity, and facilitate them to be widely accepted for EV propulsion.

5.3.1 SYSTEM CONFIGURATIONS

The system configuration of induction motor drives for EV propulsion can be categorized into single- and multiple-motor types. The single-motor system configuration shown in Fig. 5.21 consists of a three-phase squirrel-cage induction motor, a three-phase voltage-fed PWM inverter, an electronic controller as well as reduction and differential gears. As discussed before, the multiple-motor system configuration consists of multiple motors, multiple inverters, centralized or distributed controllers and optional reduction gears. Both of them have their

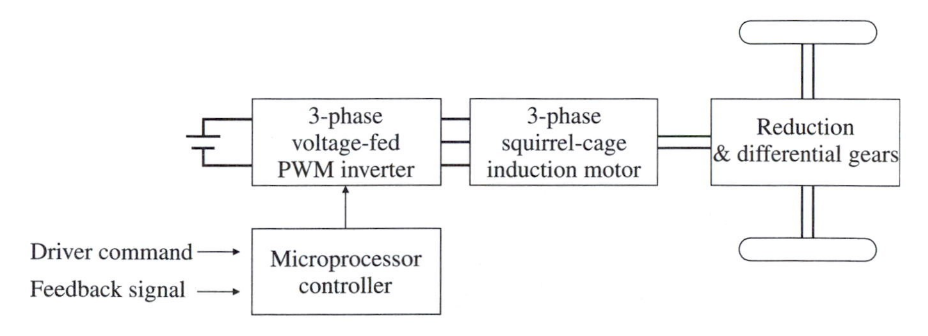

Fig. 5.21. Basic EV induction motor drive configuration.

individual merits and have been employed in modern EVs. Nevertheless, both of these system configurations should have the features that the three-phase induction motor is specially designed and closely integrated with the transaxle, the three-phase PWM inverter is inherently regenerative and has low harmonic distortion, the electronic controller can perform all aspects of motor drive control, and the reduction gear is usually of fixed gearing to provide both high torque at low speeds for hill climbing and low torque at high speeds for cruising.

5.3.2 INDUCTION MOTORS

Induction motors used for electric propulsion are principally similar to that for industrial applications. Nevertheless, these induction motors need to be specially designed. Laminated thin silicon cores should be used for the rotor and stator to reduce the iron loss, while copper bars should be adopted for the squirrel-cage to reduce the winding loss. Stator coils should adopt the Class C insulation, and be directly cooled by oil with low viscosity. All housings should be made of cast aluminium to reduce the total motor weight. Reasonable high-voltage low-current motor design should be employed to reduce the cost and size of the power inverter, although the voltage level of the motor is limited by the number, weight and type of EV batteries. High-speed operation should be adopted to minimize the motor size and weight, although the maximum speed of the motor is limited by the bearing friction and windage losses as well as the transaxle tolerance. Low stray reactance is also necessary to favour flux-weakening operation. Concerning on motor performances for EV operation, high torque at low speeds, low torque at high speeds and instantaneous overloading capability are desired for hill climbing, highway cruising and vehicle overtaking, respectively.

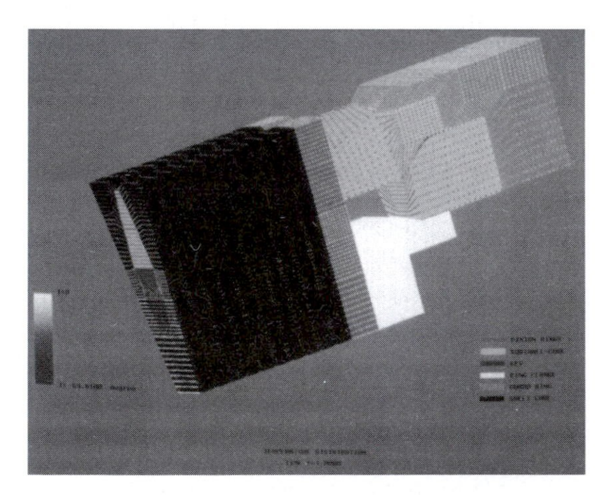

Fig. 5.22. EV induction motor thermal field analysis.

In order to optimize the motor geometries and parameters, CAD techniques are usually employed. In general, the two-dimensional finite element method (FEM) is used to carry out both steady-state and dynamic electromagnetic field analyses. Moreover, it is becoming interested in three-dimensional FEM-based thermal field analysis of induction motors. The reason is due to the fact that the skin effect of the motor generally causes a considerable variation of the loss density distribution with respect to time during starting, hence resulting in a serious transient thermal stress on both rotor bars and end-rings. Figure 5.22 shows the transient temperature distribution of one rotor slot pitch of an EV induction motor by using FEM-based thermal field analysis.

The basic consideration of induction motor design includes stator core outer- and inner-diameters, core length, air-gap length, number of poles, number of stator slots, number of rotor slots, stator tooth width and slot depth, rotor tooth width and slot depth, number of turns per phase, slot fill factor, flux density at each part of the magnetic circuit, magnetizing current, thermal resistance at each part of the thermal circuit, speed, torque and efficiency, torque per unit weight, and weight of copper and magnetic iron core used (West and Jack, 2000).

5.3.3 INVERTERS

For EV propulsion, three-phase voltage-fed PWM inverters are almost exclusively used for induction motor drives. The inverter design highly depends on the technology of power devices. At present, the IGBT based inverter is most attractive, and has been accepted by many modern EVs. Since the hard-switching inverter topology is almost fixed, the inverter design generally depends on the selection of power devices and PWM switching schemes. The selection of power devices is based on the criteria that the voltage rating is at least twice the nominal battery voltage because of the voltage surge during switching, the current rating is large enough so that there is no need to connect several power devices in parallel, and the switching speed is sufficiently high to suppress motor harmonics and acoustic noise levels. The module is a two-in-one type, namely two devices in series with an anti-parallel diode across each device, to minimize wiring and stray impedance. On the other hand, the selection of PWM switching schemes is based on the criteria that the magnitude and frequency of the fundamental component of the output waveform can be smoothly varied, the harmonic distortion of the output waveform is minimum, the switching algorithm can be real-time implemented with minimum hardware and compact software, and the large fluctuation of battery voltage (possibly from −35% to +25% of the nominal value) can be handled. There are numerous PWM schemes that have been available, such as the sinusoidal PWM, uniform PWM, optimal PWM, delta PWM, random PWM, equal-area PWM, hysteresis-band PWM, and space vector PWM. Both the current-controlled hysteresis-band PWM and space vector PWM have been widely used for induction motor drives in EVs. Nevertheless, the voltage-controlled equal-area

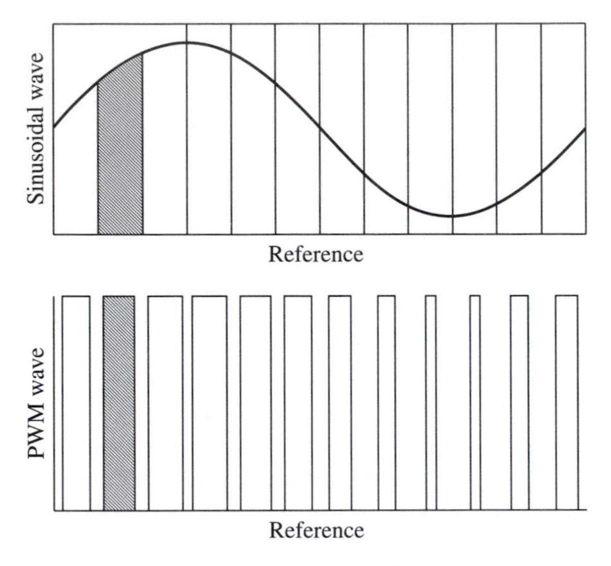

Fig. 5.23. Equal-area PWM.

PWM scheme is particularly designed for battery-powered induction motor drives in EVs (Chan and Chau, 1991).

The definite merits of the equal-area PWM are low harmonic distortion, minimum hardware, compact software, real-time implementary and tolerance of dc voltage fluctuation. The principle is to divide one period of sinusoidal wave into a number of segments, and to evaluate the pulse widths so that each pulse area is equal to the area of the related segment of the sinusoidal wave with respect to the same reference. As illustrated in Fig. 5.23, the required output voltage is represented by a sinusoidal wave:

$$V_0 = V_\mathrm{d} + V_\mathrm{a} \sin \omega t,$$

where V_d is the dc supply voltage, V_a is the amplitude of the required output voltage, and ω is the required angular output frequency. Hence, the shaded area under this sinusoidal wave is given by:

$$A_j = \int_{t_j}^{t_{j+1}} (V_\mathrm{d} + V_\mathrm{a} \sin \omega t) dt,$$

where t_j and t_{j+1} are the time intervals of the jth segment. On the other hand, the corresponding shaded pulse area under the PWM wave is given by:

$$B_j = 2V_\mathrm{d} P_\mathrm{j},$$

where P_j is the pulse width for the jth segment. Hence, by equating these two areas, P_j can be obtained as:

$$P_j = \frac{1}{\omega}\left[\frac{\theta_{j+1} - \theta_j}{2} + \frac{V_\mathrm{a}}{V_\mathrm{d}} \frac{\cos \theta_j - \cos \theta_{j+1}}{2}\right],$$

where $\theta_j = \omega t_j$ and $\theta_{j+1} = \omega t_{j+1}$ are the corresponding angle intervals of the jth segment. Similarly, the corresponding notch width N_j can also be deduced as:

$$N_j = \frac{1}{\omega}\left[\frac{\theta_{j+1} - \theta_j}{2} - \frac{V_\mathrm{a}}{V_\mathrm{d}}\frac{\cos\theta_j - \cos\theta_{j+1}}{2}\right].$$

The calculated pulse width are located at the centre of each segment, which gives output symmetry at $\pi/2$ and $3\pi/2$, hence reducing the required calculation and harmonic distortion. Since the expressions of the pulse width and the notch width are nearly the same except the sign of the second term, the required calculation can be further reduced which is highly desirable for microcontroller implementation. Increasingly, since all pulse widths are expressed as an algebraic equation and directly depend on the instantaneous dc voltage level, they can be computed in real time with minimum hardware and compact software as well as be adjusted automatically for a changing battery voltage.

The development of soft-switching inverters for induction motor drives has been conducted since the advent of the well-known resonant dc link inverter in 1986. Subsequently, many improved soft-switching topologies have been proposed, such as the quasi-resonant dc link, series resonant dc link, parallel resonant dc link, synchronized resonant dc link, resonant transition, auxiliary resonant commutated pole, and auxiliary resonant snubber inverters. Among them, the auxiliary resonant snubber (ARS) inverter type is being actively developed for EV propulsion.

By using auxiliary switches and resonant inductors along with resonant snubber capacitors to achieve the soft-switching condition, two three-phase topologies of the ARS inverter are shown in Fig. 5.24. These inverter topologies offer the merits that all main power devices can operate at the zero-voltage switching (ZVS) condition, while all auxiliary power switches also operate at the zero-current switching (ZCS) condition. Moreover, the parasitic inductance and stray capacitance of these topologies are utilized as a part of the resonant components, while there are less over-voltage or over-current penalty in the main power switches. Thus, these ARS inverter topologies have promising applications to EV propulsion.

Between the star-configured (Y-ARS) and the delta-configured (Δ-ARS) inverter topologies, the Δ-ARS inverter is relatively more attractive for EV propulsion because of the advantages of higher power capability, no floating-voltage or over-voltage penalty on the auxiliary power switches, no need of using additional voltage or current sensors, and no need of using anti-paralleled fast reverse recovery diodes across the resonant switches. The principle of operation can be easily described by its single-phase topology shown in Fig. 5.25 because the three-phase topology actually works as a segmented single-phase one. As illustrated in Fig. 5.26, there are 10 operating modes within a complete cycle. The Mode 0 ($t < t_0$) is the initial stage. At the Mode 1 ($t_0 \sim t_1$), $\mathrm{S_{r1}}$ is turned on and the resonant inductor current begins to increase linearly. At the Mode 2 ($t_1 \sim t_2$), the turnoff of $\mathrm{S_1}$ and $\mathrm{S_4}$ starts the resonance between the resonant inductors and snubber capacitors. When $\mathrm{C_1}$ and $\mathrm{C_4}$ are charged to the dc link voltage while $\mathrm{C_2}$

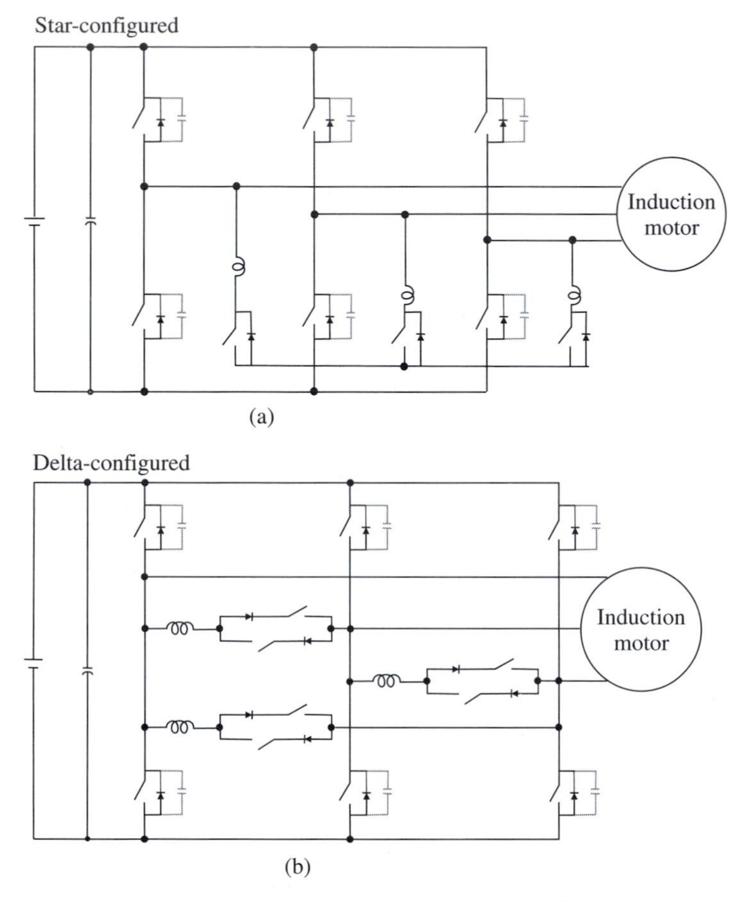

Fig. 5.24. Three-phase ARS inverter topologies.

and C_3 are discharged to zero, it provides the ZVS condition for both S_2 and S_3. At the Mode 3 ($t_2 \sim t_3$), S_2 and S_3 are turned on with ZVS. The resonant inductor current begins to decrease linearly, and both D_2 and D_3 start freewheeling. At the Mode 4 ($t_3 \sim t_4$), the resonant inductor current is zero so that S_{r1} can be turned off with ZCS. The load current continues freewheeling via D_2 and D_3. At the Mode 5 ($t_4 \sim t_5$), S_{r2} is turned on. The resonant inductor current begins to increase linearly. At the Mode 6 ($t_5 \sim t_6$), the current in both S_2 and S_3 changes the direction. Thus, the resonant inductor current continues to increase, and exceeds the load current. When this inductor current becomes sufficiently high that the stored energy in the resonant inductors can charge and discharge the snubber capacitors, both S_2 and S_3 are turned off. At the Mode 7 ($t_6 \sim t_7$), resonance between the resonant inductors and snubber capacitors begins to take place. When C_2 and C_3 are charged to full voltage while C_1 and C_4 are discharged to zero, it provides the desired ZVS condition for turning on S_1 and S_4. At the Mode 8

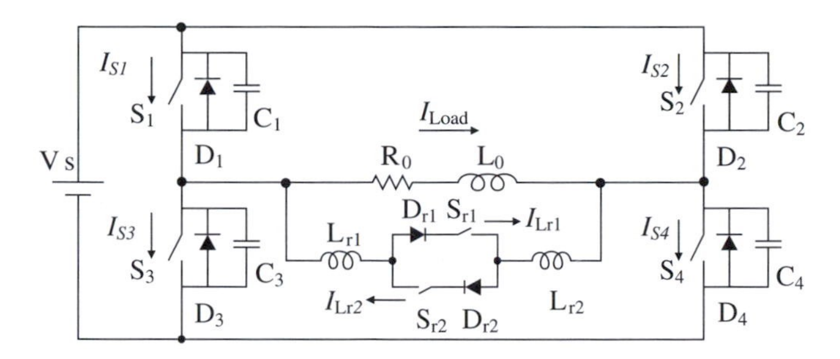

Fig. 5.25. Single-phase ARS inverter topology.

($t_7 \sim t_8$), S_1 and S_4 are turned on with ZVS, and the resonant inductor current starts to decrease while D_1 and D_4 become conducting. At the Mode 9 ($t_8 \sim t_9$), the resonant current continues to decrease linearly. When the resonant inductor current decreases to zero, S_{r2} can be turned off with ZCS. The next operating stage is the Mode 0, and the whole operating process repeats cyclically. The corresponding waveforms are shown in Fig. 5.27 (Chan, Chau and Yao, 1997).

Although this ARS inverter has promising applications to EV propulsion, it still needs continual improvement before practically applying to EVs. Particularly, the corresponding control complexity should be alleviated, while the corresponding PWM switching scheme needs to be modified to enable variable speed control of induction motor drives. In fact, commercialized EVs have not yet adopted soft-switching inverters for electric propulsion.

5.3.4 SPEED CONTROL

Speed control of induction motors is considerably more complex than that of dc motors because of the non-linearity of the dynamic model with coupling between direct and quadrature axes. A number of control strategies have been developed to allow induction motor drives to be applicable for electric propulsion. There are three state-of-the-art control strategies, namely variable-voltage variable-frequency (VVVF) control, field-oriented control (FOC) which is also called as vector control or decoupling control, and pole-changing control. The basic equation of induction motor speed control is governed by:

$$N = N_s(1 - s) = \frac{60f}{p}(1 - s),$$

where N is the motor speed, N_s is the rotating-field synchronous speed, s is the slip, p is the number of pole pairs, and f is the supply frequency. Thus, the motor speed can be controlled by variations in f, p and/or s. In general, more than one of these control variables are adopted. On top of these control strategies, sophisticated control algorithms such as adaptive control, variable-structure control and optimal control have also been employed to achieve faster response, higher efficiency and wider operating ranges.

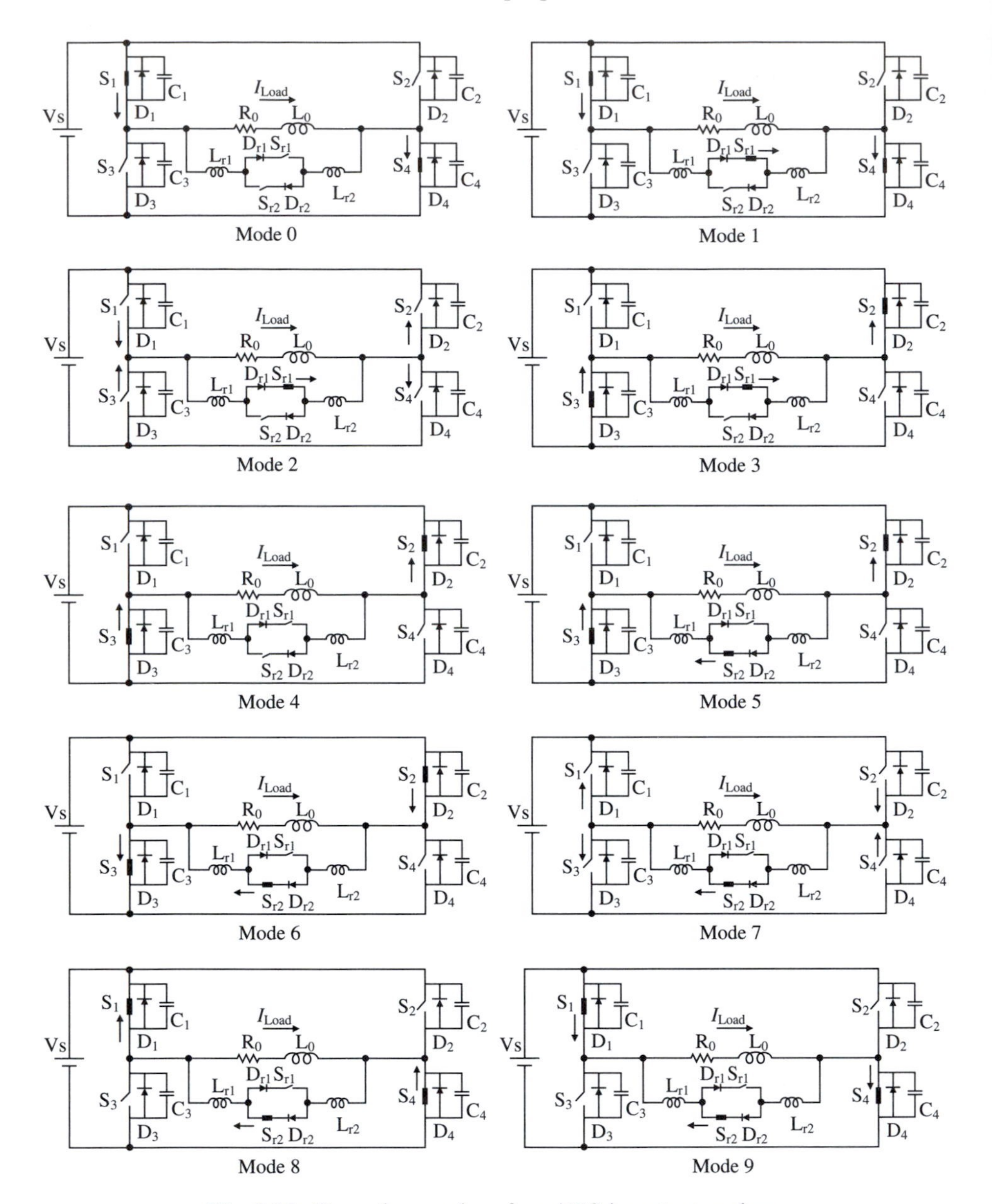

Fig. 5.26. Operating modes of an ARS inverter topology.

Variable-voltage variable-frequency control

Figure 5.28 shows the functional block diagram of VVVF control of induction motor drives. This strategy is based on constant volts/hertz control for frequencies below the motor rated frequency, whereas variable-frequency control with

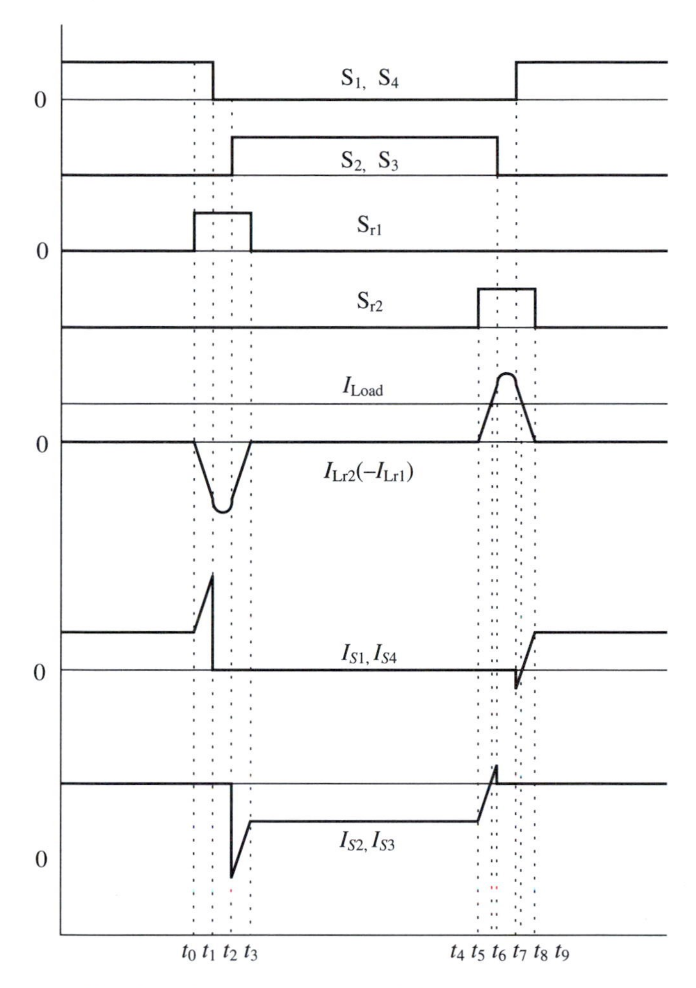

Fig. 5.27. Waveforms of an ARS inverter topology.

constant rated voltage for frequencies beyond the rated frequency. For very low frequencies, voltage boosting is applied to compensate the difference between the applied voltage and induced emf due to the stator resistance drop. As shown in Fig. 5.29, the induction motor drive characteristics can be divided into three operating regions. The first region is called the constant-torque region in which the motor can deliver its rated torque for frequencies below the rated frequency. In the second region, called the constant-power region, the slip is increased to the maximum value in a pre-programmed manner so that the stator current remains constant and the motor can maintain its rated power capability. In the high speed region, the slip remains constant while the stator current decreases. Thus, the torque capability declines with the square of speed. Because of the disadvantages

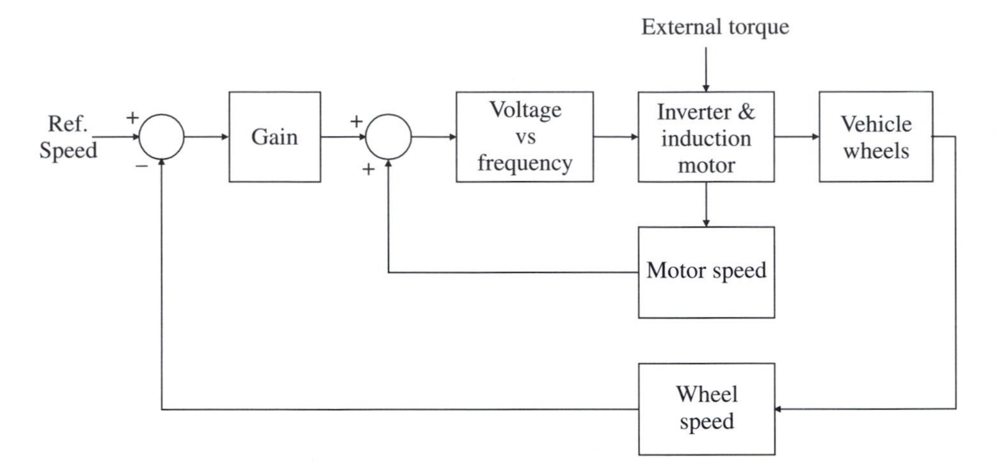

Fig. 5.28. VVVF control of induction motor drives.

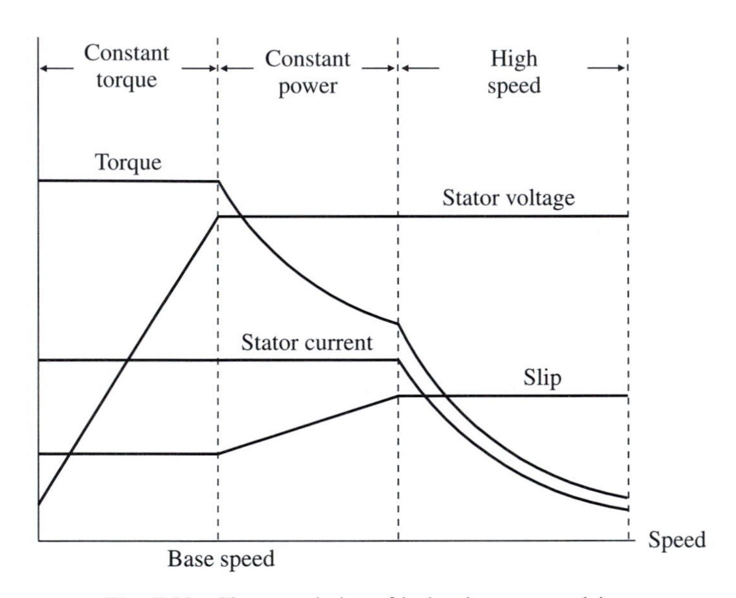

Fig. 5.29. Characteristics of induction motor drives.

of air-gap flux drifting and sluggish response, the VVVF control strategy is becoming less attractive for high-performance EV induction motor drives.

Field-oriented control

In order to improve the dynamic performance of induction motor drives for EV propulsion, FOC is preferred to VVVF control. By using FOC, the mathematical

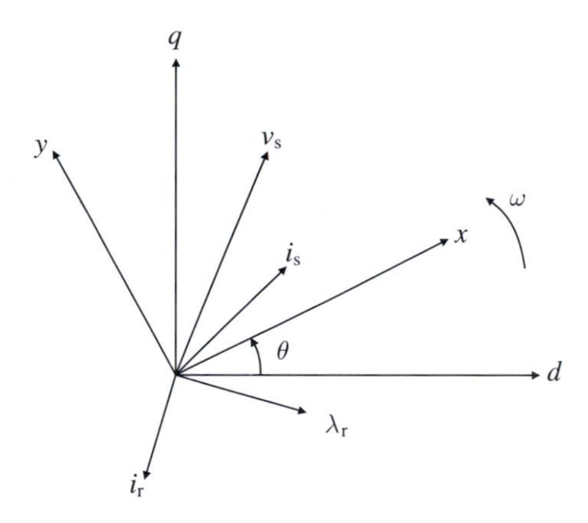

Fig. 5.30. x–y frame rotating synchronously in general.

model of induction motors is transformed from the stationary reference frame (d–q frame) to the general synchronously rotating frame (x–y frame) as shown in Fig. 5.30. Thus, at steady state, all the motor variables such as supply voltage v_s, stator current i_s, rotor current i_r and rotor flux linkage λ_r can be represented by dc quantities. When the x-axis is purposely selected to be coincident with the rotor flux linkage vector, the reference frame (α–β frame) becomes rotating synchronously with the rotor flux as shown in Fig. 5.31 where $i_{s\alpha}$ and $i_{s\beta}$ are the α-axis component and β-axis component of stator current, respectively. Hence, the motor torque T can be obtained as:

$$T = \frac{3}{2} p \frac{M}{L_r} \lambda_r i_{s\beta},$$

where M is the mutual inductance per phase and L_r is the rotor inductance per phase. Since λ_r can be written as $M i_{s\alpha}$, the torque equation can be rewritten as:

$$T = \frac{3}{2} p \frac{M^2}{L_r} i_{s\alpha} i_{s\beta}.$$

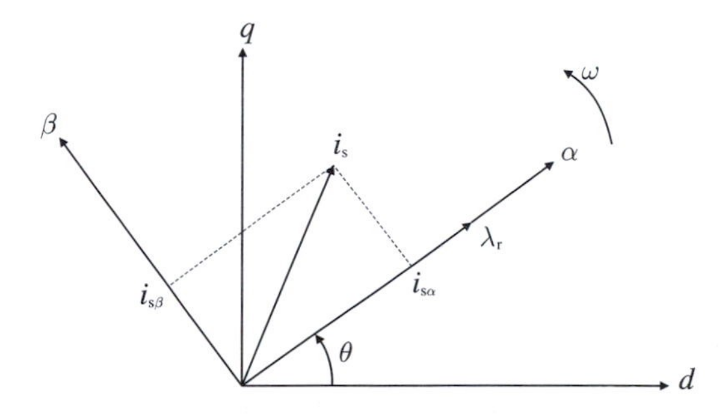

Fig. 5.31. α–β frame rotating synchronously with rotor flux.

This torque equation is very similar to that of separately excited dc motors. Namely, $i_{s\alpha}$ resembles to the field current I_f while $i_{s\beta}$ resembles to the armature current I_a. Thus, $i_{s\alpha}$ can be considered to be the field component of i_s which is responsible for establishing the air-gap flux. On the other hand, $i_{s\beta}$ can be considered to be the torque component of i_s which produces the desired motor torque. Therefore, by means of this FOC, the motor torque can be effectively controlled by adjusting the torque component as long as the field component remains constant. Hence, induction motor drives can offer the desired fast transient response similar to that of separately excited dc motor drives. The corresponding block diagram of induction motor drives using FOC is shown in the Fig. 5.32.

It should be noted that in order to attain the above FOC, the rotor flux linkage vector is always aligned with the α-axis. This criterion, so-called the decoupling condition, can be attained through slip frequency ω_{slip} control as given by:

$$\omega_{slip} = \frac{R_r i_{s\beta}}{L_r i_{s\alpha}},$$

where R_r is the rotor resistance per phase.

Since the advent of FOC, a number of methods have been proposed for implementation. Basically, these methods can be classified into two groups, namely direct FOC and indirect FOC. The direct FOC requires the direct measurement of the rotor flux, which not only increases the complexity in implementation but also suffers from the unreliability in measurement at low speeds.

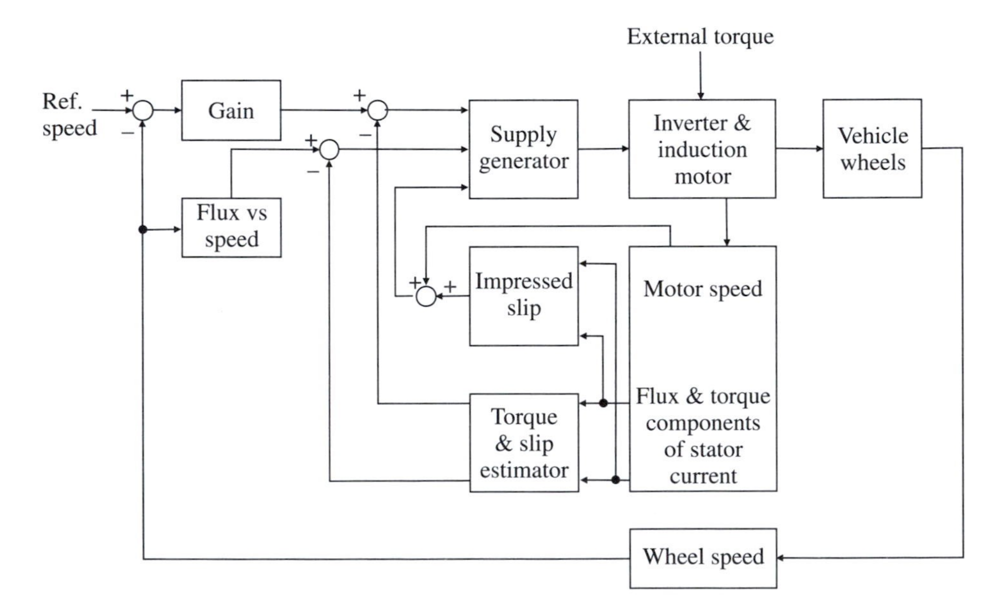

Fig. 5.32. FOC of induction motor drives.

Thus, it is seldom used for EV propulsion. The indirect FOC determines the rotor flux by calculation, instead of direct measurement. This method takes the definite advantage of easier implementation than the direct FOC. Therefore, the indirect FOC is attractive for application to high-performance EV propulsion.

Adaptive control

Although the indirect FOC has been widely used for high-performance induction motor drives, it still suffers from some drawbacks for application to EVs. Particularly, the rotor time constant L_r/R_r (which has a dominant effect on the decoupling condition) changes severely with the operating temperature and magnetic saturation, leading to deteriorate the desired FOC. In general, there are two ways to solve this problem. One way is to perform on-line identification of the rotor time constant and accordingly to update the parameters used in the FOC controller. The other way is to adopt a sophisticated control algorithm to enable the FOC controller insensitive to motor parameter variations.

Recently, a model-reference adaptive control (MRAC) algorithm has been employed for FOC control of an EV induction motor drive (Chan *et al.*, 1990). Firstly, a reference model is made to satisfy the desired dynamic performance of the motor drive. This reference model is designed to be an optimal system in general. Then, an adaptive mechanism is adopted which aims to force the motor drive following the reference model even after variations of system parameters

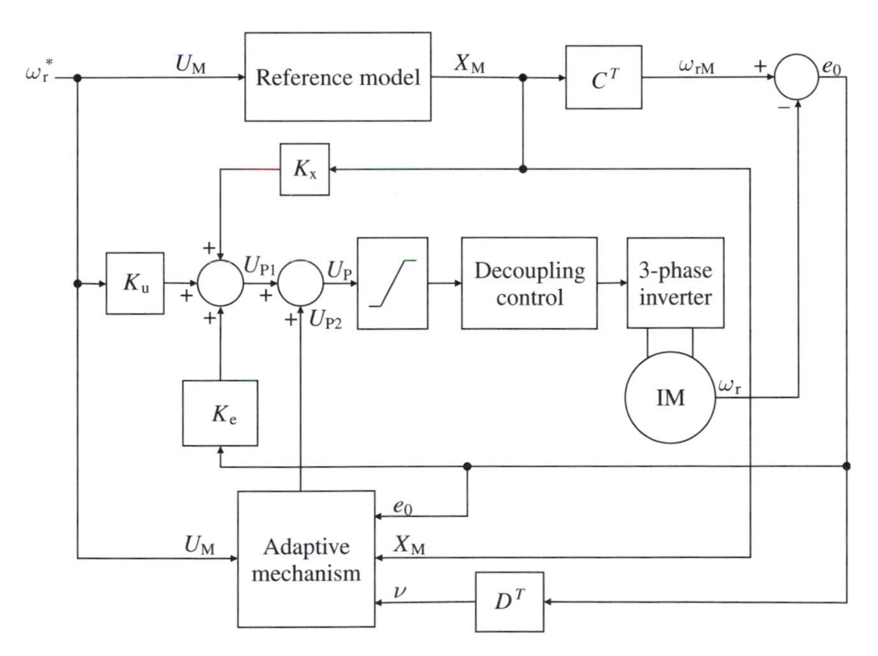

Fig. 5.33. MRAC of an EV induction motor drive.

such as the change of L_r/R_r due to prolonged operation. The main criterion of the adaptive mechanism is to assure robustness with asymptotically zero error between the outputs of the reference model and the motor drive. The definite advantage of this MRAC scheme is that there is no need to carry out explicit parameter identification or estimation in the synthesis of the motor drive control input. In fact, only the command input, the controlled motor drive output and the reference model are required to establish this control scheme. Figure 5.33 shows the corresponding functional block diagram in which U_M is the model input, U_P is the motor drive control input, X_M is the state vector of the reference model, ω_r^* is the desired motor speed, ω_r is the actual motor speed, ω_{rM} is the reference model motor speed, e_o is the feedback error between ω_{rM} and ω_r, as well as K_e, K_u and K_x are the constant gains. The motor drive control input is constituted by two parts, namely the regular input U_{P1} generated by the linear model following control and the adaptation input U_{P2} generated by the adaptive mechanism, as given by:

$$\begin{aligned} U_P &= U_{P1} + U_{P2} \\ U_{P1} &= K_e e_0 + K_u U_M + K_x X_M \\ U_{P2} &= \Delta K_e(\varepsilon, t) e_0 + \Delta K_u(\varepsilon, t) U_M + \Delta K_x(\varepsilon, t) X_M \end{aligned},$$

where ΔK_e, ΔK_u and ΔK_x are the adaptive gains which are functions of the state error vector ε. In order to ensure that ε can asymptotically reduce to zero for all initial conditions, the following PI adaptation laws can be adopted:

$$\begin{aligned} \Delta K_e &= \int_0^t M_1 \nu (R_1 e_0)^T d\tau + M_2 \nu (R_2 e_0)^T \\ \Delta K_u &= \int_0^t N_1 \nu (S_1 U_M)^T d\tau + N_2 \nu (S_2 U_M)^T, \\ \Delta K_x &= \int_0^t L_1 \nu (Q_1 X_M)^T d\tau + L_2 \nu (Q_2 X_M)^T \end{aligned}$$

where ν is the compensated error feedback, and M_1, M_2, N_1, N_2, L_1, L_2, R_1, R_2, S_1, S_2, Q_1, Q_2 are chosen to be positive constants.

Sliding-mode control

Instead of using the MRAC technique to guarantee the FOC of EV induction motor drives, the sliding-mode control has also been employed. This sliding-mode control approach takes the merits of fast response, insensitivity to plant parameter variations and simplicity in design and implementation. Several methods of applying sliding-mode control to induction motor drives have been implemented. All of these methods have a common feature that the analysis and design of the sliding-mode controller are based on the mathematical model of FOC of induction motor drives. However, when there are significant change in system parameters such as the change of L_r/R_r due to prolonged operation, the decoupling condition may be violated. Hence, the motor can no longer be represented by a second-order model, but is only described by a nonlinear fifth-order model. This implies that the structure of the motor drive has changed. Although sliding-mode control is insensitive to the motor drive parameter variations, its performance can be greatly affected by structural changes in the motor drive. Therefore, these methods cannot work well under a wide range of motor parameter variations such as the severe changes in L_r/R_r due to high operating temperatures and high magnetic saturation in EV induction motors.

Recently, a new sliding-mode control method has been proposed for an EV induction motor drive operating at high temperatures and/or high saturation (Chan and Wang, 1996). Different from the conventional FOC that the frame is rotating synchronously with the rotor flux linkage vector, it is selected to be rotating synchronously with the stator current vector as shown in Fig. 5.34. Thus, based on this $\varepsilon-\eta$ frame, the torque generated by the motor can be expressed as:

$$T = \frac{3}{2} p \frac{M}{L_r} i_s \lambda_{r\eta},$$

where i_s is the stator current and $\lambda_r\eta$ is the η-axis component of rotor flux. By defining s_1 and s_2 as:

$$s_1 = i_s - i_s^*$$
$$s_2 = \lambda_{r\eta} - \lambda_{r\eta}^*,$$

where the asterisk represents the reference value of the corresponding quantities, the system motion can be brought into the intersection of the hyperplanes ($s_1 = 0$

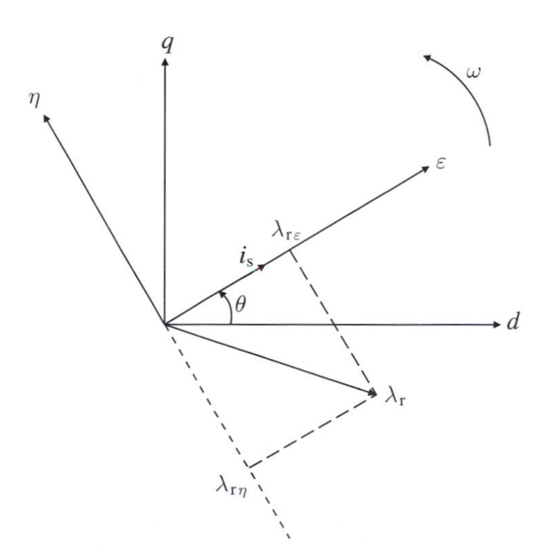

Fig. 5.34. ε–η frame rotating synchronously with stator current.

and $s_2 = 0$) and is kept sliding along the intersection. The corresponding control law, so-called the switching law, that is able to obtain a stable sliding-mode motion along the intersection is to switch the ε-axis component of stator voltage vector $u_{s\varepsilon}$ and the slip frequency ω_{slip} on two hyperplanes as:

$$\text{if } s_1 > 0,\ u_{s\varepsilon} - \left(c_1 i_s + c_2\lambda_{r\varepsilon} + c_3\omega_r\lambda_{r\eta}\right) - \delta$$

$$\text{if } s_1 < 0,\ u_{s\varepsilon} > -\left(c_1 i_s + c_2\lambda_{r\varepsilon} + c_3\omega_r\lambda_{r\eta}\right) + \delta$$

$$\text{if } \lambda_{r\varepsilon}s_2 > 0,\ \omega_{slip} > -\frac{R_r}{L_r}\frac{\lambda_{r\eta}}{\lambda_{r\varepsilon}} + \delta$$

$$\text{if} \lambda_{r\varepsilon}s_2 < 0,\ \omega_{slip} < -\frac{R_r}{L_r}\frac{\lambda_{r\eta}}{\lambda_{r\varepsilon}} - \delta,$$

where δ is any small positive number and the constants c_1, c_2 and c_3 are given by:

$$c_1 = -R_s - \frac{R_r M^2}{L_r^2}$$

$$c_2 = \frac{MR_r}{L_r^2}$$

$$c_3 = \frac{M}{L_r}.$$

Once $u_{s\varepsilon}$ and ω_{slip} are determined, the η-axis component of stator voltage vector $u_{s\eta}$ can be easily deduced by using the system equation. The above control law possesses two important features. Firstly, the sliding-mode motion along the intersection of the two switching hyperplanes ($s_1 = 0$ and $s_2 = 0$) is achieved by two independent sliding-mode motions on the two hyperplanes. It follows that the two components of torque, namely i_s and $\lambda_{r\eta}$, are decoupled. Secondly, since the positions of the coordinate axes of the frame rotating synchronously with the stator current vector can easily be determined by measurement, the motor model remains structurally unchanged when there are variations in motor parameters. Moreover, because the sliding-mode control law is inherently based on inequalities instead of equalities, the decoupling condition will not be destroyed even when the motor parameters severely change with the operating temperature and magnetic saturation. Therefore, the distinct merit of sliding-mode control, namely insensitivity to system parameter variations, can be fully realized. The block diagram of this sliding-mode control for an EV induction motor drive is shown in Fig. 5.35.

Efficiency-optimizing control

Using the conventional FOC in induction motor drives, the field component current is usually controlled to be approximately constant for various loads in the constant torque region. For most operating conditions, the corresponding core

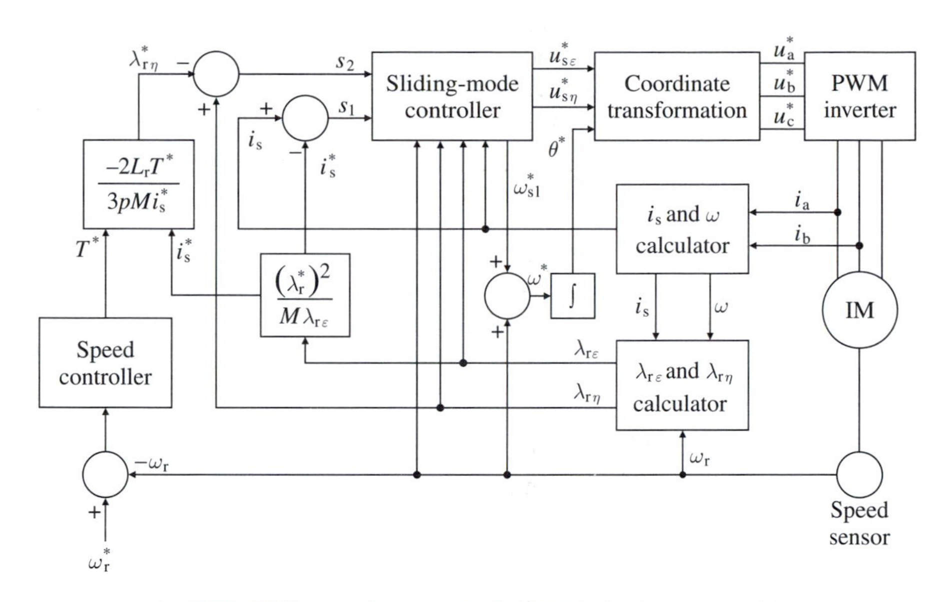

Fig. 5.35. Sliding-mode control of an EV induction motor drive.

loss is insignificant compared with the copper loss, hence the conventional FOC can offer the merit of maximum torque per ampere. However, for light-load operation, the core loss becomes comparable with the copper loss, causing

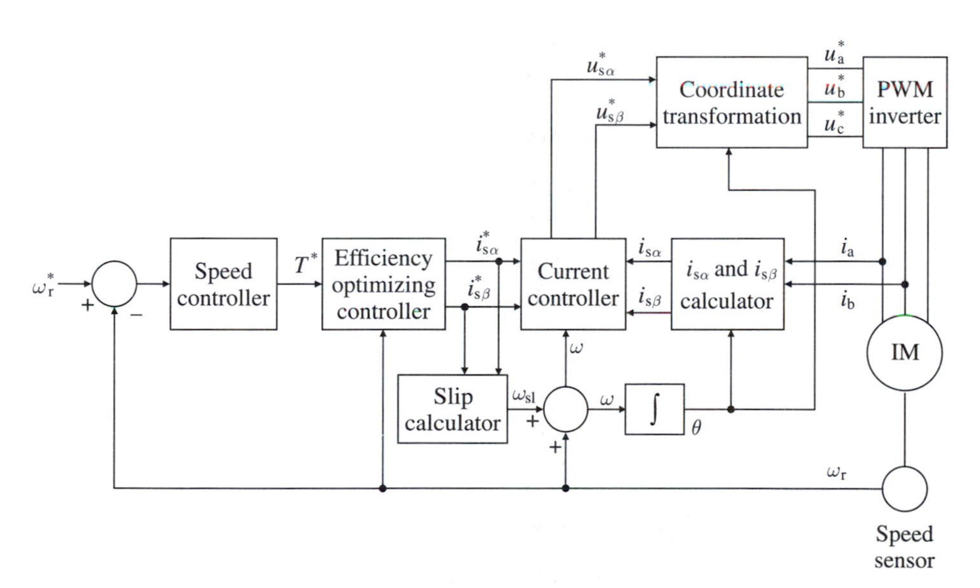

Fig. 5.36. Efficiency-optimizing FOC of an EV induction motor drive.

a problem of low efficiency. Due to the limited on-board energy storage in EVs as well as the wide ranges of load and speed for EV operation, the efficiency of EV motor drives should be optimized throughout the whole operating region.

Figure 5.36 shows the efficiency-optimizing FOC of an EV induction motor drive (Yamada *et al.*, 1996). Based on the equivalent circuit of the FOC-based induction motor, the total motor loss P_{loss} can be expressed as:

$$P_{\text{loss}} = (R_s + R'_r)\frac{TA}{pM'} + (R_s + R'_m)\frac{T}{pM'A} + \frac{2R_m}{\omega M}\frac{R_s T}{pM'} + P_m,$$

where p is the number of pole-pairs, R_s is the stator resistance, R_r is the rotor resistance, R_m is the equivalent core loss resistance, M is the mutual inductance, L_s is the stator inductance, L_r is the rotor inductance, ω is the supply angular frequency, T is the motor torque, P_m is the mechanical loss, $R'_r = \alpha^2 R_r$, $R'_m = \alpha R_m$, and A is the ratio of the torque-component stator current $i_{s\beta}$ to the field-component stator current $i_{s\alpha}$. To minimize this loss at given T and ω, P_{loss} is differentiated with respect to A. By setting $dP_{\text{loss}}/dA = 0$, the condition for achieving optimal efficiency can be derived as:

$$A = \sqrt{\frac{R_s + R'_m}{R_s + R'_r}}.$$

Thus, the command values of $i_{s\beta}$ and $i_{s\alpha}$ for efficiency-optimizing FOC can be obtained as:

$$i^*_{s\alpha} = \sqrt{T^*/(3pM'A)}$$
$$i^*_{s\beta} = T^*/(3PM'i^*_{s\alpha}).$$

It should be noted that R_s varies with the temperature, R_m is affected by the frequency and α changes with the level of magnetic saturation. So, the optimal ratio A needs to be updated continually.

Comparing the applications of FOC and efficiency-optimizing FOC to EVs, the drive system efficiency is improved by about 17% and the driving range is extended by over 26% for constant-speed operation at 40 km/h. Based on the 10.15 Mode driving cycle, which is derived from the actual driving profile in Japan, the EV driving range can be improved by over 14%.

Pole-changing control

It is well known that a change of the number of pole pairs of induction motor drives can adjust the rotating-field synchronous speed. The squirrel-cage type takes a definite advantage over the wound-rotor type that it is able to automatically adapt the pole number of the rotor to that of the stator. In early time, this pole-changing control was implemented by using mechanical contactors, and only two or three discrete speeds were achieved. With the advancement of power electronics and control technologies, the pole-changing control can be implemented electronically. Its basic principle can be illustrated by Fig. 5.37. Each stator winding consists of two coil groups, and the change in the current direction in these coil groups causes the change of pole numbers. Figure 5.38 shows a newly proposed dual-inverter six-phase pole-changing EV induction motor drive that can offer both 4-pole and 8-pole operations (Chau *et al.*, 2000). The corresponding maximum torque characteristics are shown in Fig. 5.39. Hence, the high-speed constant-power capability can be remarkably extended, which is particularly desirable for EV cruising.

Figure 5.40 shows a 70 kW 312-V three-phase induction motor drive which is specially designed and built for EVs. It incorporates a compact induction motor with sealed design, a microprocessor based controller/inverter using vector control, and a lightweight gearbox with a built-in differential.

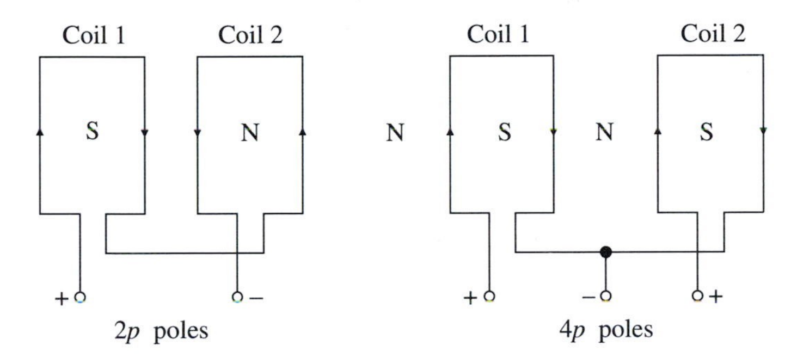

Fig. 5.37. Principle of pole-changing control.

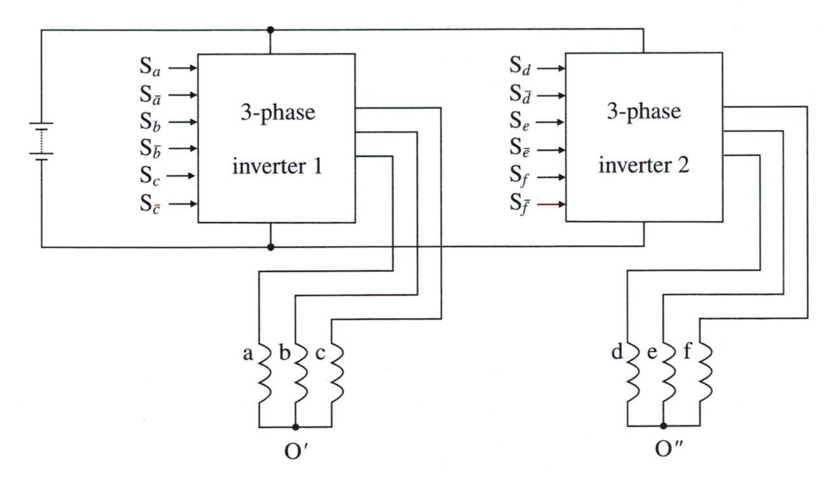

Fig. 5.38. Dual-inverter pole-changing control of an EV induction motor drive.

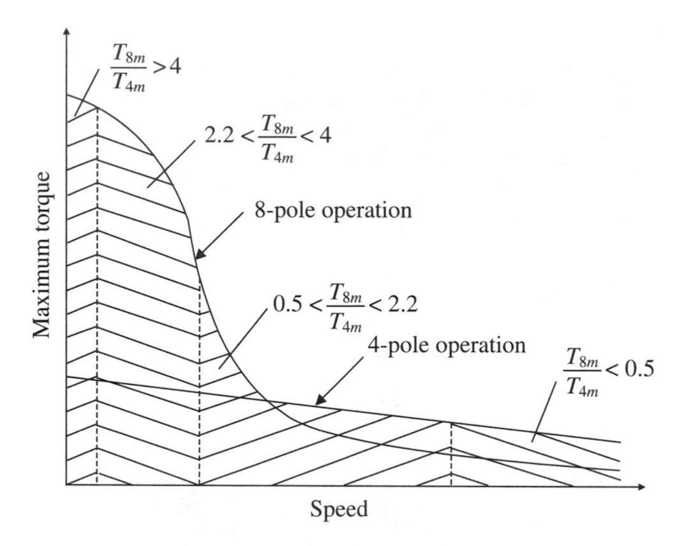

Fig. 5.39. Maximum torque characteristics of a pole-changing EV induction motor drive.

5.4 Permanent-magnet motor drives

The classification of PM motor drives is diverse. By considering the waveform feeding into the motor terminals, they can be classified as:

- PM dc motor drives and
- PM ac motor drives.

This classification is simple and obvious. Because of no brushes, commutators or slip-rings, PM ac motor drives are loosely named as PM brushless motor drives. Further considering the ac waveform feeding into the motor terminals, PM brushless motor drives includes PM synchronous motor drives and PM brushless dc motor drives. PM synchronous motor drives are fed by sinusoidal or near-sinusoidal ac waves and uses continuous rotor position feedback signals to control the commutation, whereas PM brushless dc motors are fed by rectangular ac waves and uses discrete rotor position feedback signals to control the commutation. Since the interaction between rectangular field and rectangular current in the motor can produce higher torque product than that produced by sinusoidal field and sinusoidal current, PM brushless dc motor drives possess higher power density than PM synchronous motor drives. On the other hand, this type of motor drives has a significant torque pulsation due to the commutating current during the phase commutation of the power devices. While its sinusoidal-fed counterpart produces an essentially constant instantaneous torque or so-called smooth torque like a wound-rotor synchronous motor drives (Miller, 1989).

Recently, a new type of PM brushless motor drives, namely PM hybrid motor drives has been developed for EV propulsions. The uniqueness of these motor

Controller and motor

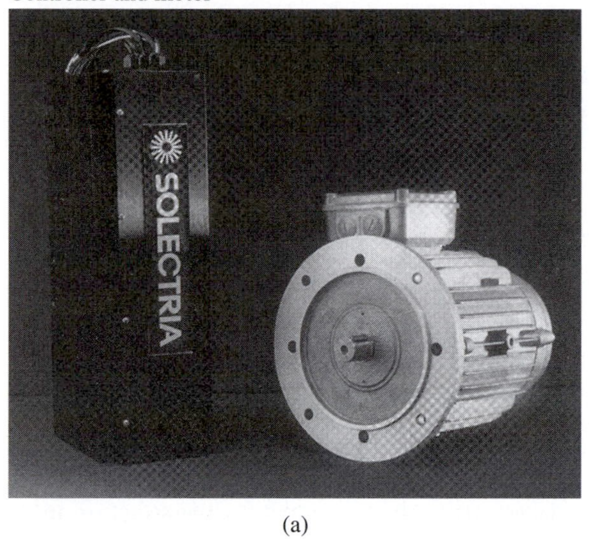

(a)

Motor and gearbox

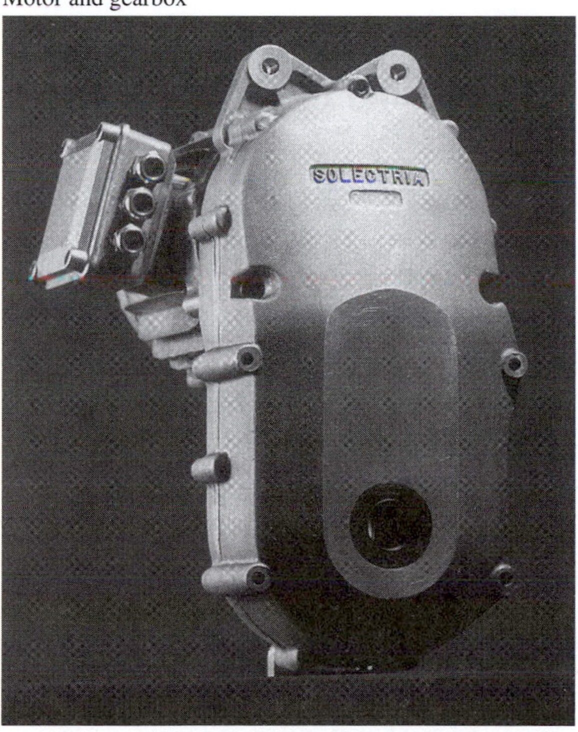

(b)

Fig. 5.40. Practical EV induction motor drive. (Photo courtesy of Solectria.)

drives is the existence of both PMs and the field winding. PMs are generally integrated into the rotor, while the field winding is usually fixed at the stator. Without employing any special control strategies, these motor drives inherently have a wider speed range than that of other PM motor drives, because the motor air-gap flux can be flexibly controlled by adjusting the dc field current. It should be noted that this flexible field control, particularly flux-weakening at high speeds, is highly desirable for constant-power operation of EVs. Because of their distinctive motor topology and flux path, these motor drives are considered as another class of PM brushless motor drives. Therefore, PM motor drives can be also classified as:

- PM dc motor drives
- PM synchronous motor drives
- PM brushless dc motor drives and
- PM hybrid motor drives.

The latter three classes are collectively called PM brushless motor drives. By eliminating those brushes and commutators in conventional dc motor drives, PM brushless motor drives can directly compete with induction motor drives (Chan, Chau *et al.*, 1996).

5.4.1 PM MATERIALS

PMs provide electric motors with life-long excitation. Figure 5.41 shows typical characteristics of viable PM materials for motors. Notice that Gauss (G) and Oersted (Oe) are non-SI units of magnetic flux density and coercivity, respectively, which are widely adopted by the field of magnetics. They are related to the SI units by $1\,\mathrm{G} = 10^{-4}\,\mathrm{T}$ and $1\,\mathrm{Oe} = 10^{3}/(4\pi)\mathrm{A/m}$. Ferrite magnets are the lowest in cost and have a straight demagnetization characteristic. They used to be widely applicable to motors, but suffer from bulky in size because of their low remanence. Alnico magnets have very high remanence but very low coercivity, thus their applications to motors are limited by the demagnetizing field that they can withstand. Although samarium–cobalt (Sm–Co) magnets have both high remanence and high coercivity, their high initial cost restrict their widespread applications to motors. Neodymium–iron–boron (Nd–Fe–B) magnets, which were introduced in 1983, have the highest remanence and coercivity. Because of their reasonable cost, Nd–Fe–B magnets have promising applications to motors (Rahman and Slemon, 1985). Another important parameter of PM materials is the maximum energy product, which is a measure of the maximum stored energy. A brief summary of typical PM properties is given in Table 5.7.

It should be noted that the PM properties are usually temperature-dependent. In general, PMs lose remanence as temperature increases. Exposure to a temperature called the Curie temperature, the magnetization of PMs is reduced to zero. Also, the demagnetization characteristic changes shape with the temperature. With a limited temperature range, the changes are reversible and approximately

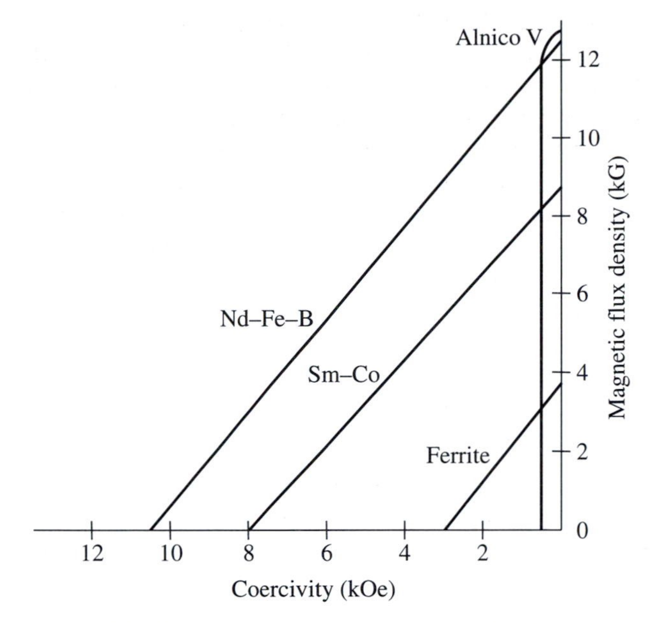

Fig. 5.41. Demagnetization curves of PMs.

Table 5.7 Properties of PMs

	Nd–Fe–B	Sm–Co	Alnico V	Ferrite
B_r(kG)	12.5	8.7	12.8	3.8
H_c(kO_e)	10.5	8.0	0.6	3.0
BH_{max}(MGOe)	36.0	18.3	5.5	3.5
μ_r	1.8	1.0	4.0	1.0
Curie temp. (°C)	310	720	800	310
Temp. coeff. (%/°C)	−0.13	−0.04	−0.03	−0.19

linear. Therefore, the operating temperature range must be taken into account during the design of PM motors.

5.4.2 PM DC MOTOR DRIVES

By replacing the field winding and pole structure with PMs, PM dc motors can be generated from conventional wound-field dc motors. Because of the space saving by PMs and the absence of field losses, PM dc motors have relatively higher power density and higher efficiency. Also, owing to the low permeability of PMs, the armature reaction of PM dc motors is usually reduced so that the commutation can be improved. These advantages enhance them to be applied for electric propulsion.

Similar to wound-field dc motor drives, PM dc motor drives are usually fed by dc choppers which have variable dc voltage control. The difference between them is due to the fact that the field excitation in PM dc motor drives is uncontrollable while the field current in wound-field dc motor drives can be independently controlled by another dc chopper. Thus, they cannot attain the operating characteristics similar to those of wound-field dc motor drives.

Same as wound-field dc motor drives, PM dc motor drives suffer from the same problem due to the use of commutators and brushes. Commutators cause torque ripples, while brushes cause friction and radio-frequency interference (RFI). Also, they need regular maintenance of commutators and brushes, making them less attractive for electric propulsion. Nevertheless, due to their simplicity, they have been accepted for low-power EV applications such as electric bikes and electric tricycles.

5.4.3 PM BRUSHLESS MOTOR DRIVES

Among those modern motor drives, PM brushless motor drives are most capable of competing with induction motor drives for electric propulsion. Their advantages are summarized below:

- Since the magnetic field is excited by high-energy PMs, the overall weight and volume can be significantly reduced for a given output power, leading to higher power density.
- Because of the absence of rotor copper losses, their efficiency is inherently higher than that of induction motors.
- Since the heat mainly arises in the stator, it can be more efficiently dissipated to surroundings.
- Since PM excitation suffers from no risk of manufacturing defects, overheating or mechanical damage, their reliability is inherently higher.
- Because of lower electromechanical time constant of the rotor, the rotor acceleration at a given input power can be increased.

The basic consideration of PM brushless motor design includes stator core outer- and inner-diameters, core length, air-gap length, PM material characteristics, number of poles, number of stator slots, stator tooth width and slot depth, number of turns of stator winding per phase, slot fill factor, topology and dimension of PMs, flux density at each part of the magnetic circuit, magnetizing current, thermal resistance at each part of the thermal circuit, speed, torque and efficiency, torque per unit weight, and weight of copper, magnetic iron core and PMs used. Some key questions are particularly useful for PM brushless motor designers: What are the limits on the motor performance? What could the designer do to get more from the motor such as by using optimizations of magnetic and thermal circuits?

System configurations

The system configuration of PM brushless motor drives for electric propulsion is similar to that of induction motor drives. Major alternatives such as single- and multiple-motor configurations as well as single- and multiple-speed transmissions can be found. Basically, the single-motor system configuration consists of a PM brushless motor, a voltage-fed inverter, an electronic controller as well as reduction and differential gears. Compared with the induction motor drive, this basic configuration has two major differences:

- The PM brushless motor is not restricted to be three-phase. In fact, a higher number of phases has the advantage of reducing phase current, hence lowering the current rating of power devices.
- Apart from producing a PWM waveform for the sinusoidal-fed PM brushless motor, the inverter may be required to generate a rectangular waveform for the rectangular-fed counterpart.

PM synchronous motors

By replacing the field winding with PM poles, PM synchronous motors can readily be generated from conventional synchronous motors. They require sinusoidal back emfs and sinusoidal stator currents for torque production. Hence, the well-known *d–q* coordinate transformation widely used in conventional synchronous motors is suitable for PM synchronous motors. These motors can be operated in either the open-loop mode or the closed-loop mode where the rotor position sensor may be used only when some sophisticated control strategies are implemented. Similar to the wound-rotor synchronous motors, the rotor of the PM synchronous motors is always in synchronism with the rotating field which depends on the applied frequency. Similar to the classical polyphase ac motors, namely the synchronous motor and the induction motor, the PM synchronous motors produce an essentially constant instantaneous torque or so-called smooth torque.

On the basis of the placement of PMs, PM synchronous motors can be categorized into surface-mounted and interior types. The surface-mounted type has the advantage of simplicity. Because of the permeability of PMs is similar to that of air, it possesses a large effective air-gap so that the armature reaction is greatly reduced. Compared with the surface-mounted type, the interior type has a higher saliency so that it has an additional reluctance torque component which is useful for constant-power operation. Moreover, by physically burying PMs inside the rotor, mechanical integrity can be maintained during high-speed operation.

When the stator of PM synchronous motors is kept similar to that of conventional synchronous motors, a special rotor configuration is shown in Fig. 5.42. This configuration adopts the flux-focusing technique to increase the air-gap flux density while maintaining the maximum magnet energy product, buries PMs

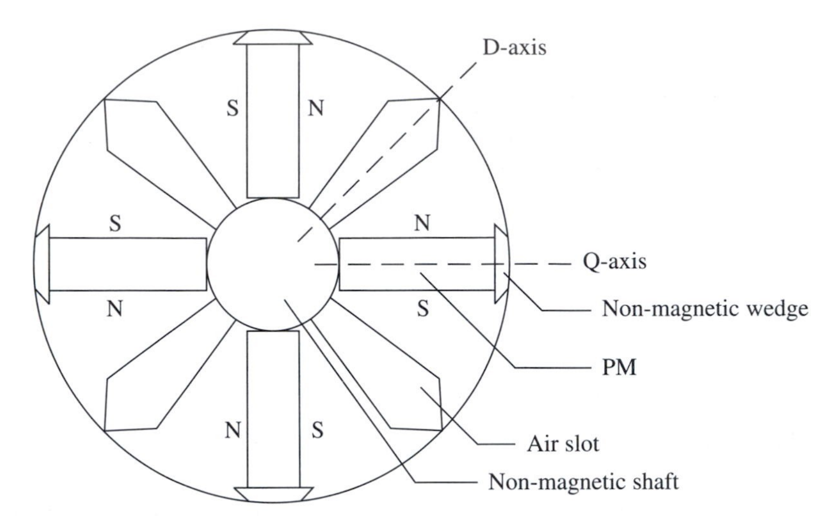

Fig. 5.42. Special rotor of a PM synchronous motor.

inside the rotor to maintain mechanical integrity during high-speed operation, and incorporates air slots inside the rotor to minimize the armature reaction. To avoid any serious harmonic distortion in the air-gap flux, a closed air slot is used. Based on this configuration, a three-phase four-pole 3-kW 32-V 4800-rpm sinusoidal-fed Nd–Fe–B PM synchronous motor has been specially designed for a 48-V battery-powered mini EV, namely the HKU Mark3 shown in Fig. 5.43 (Chau, 1991).

Fig. 5.43. HKU Mark3 mini EV.

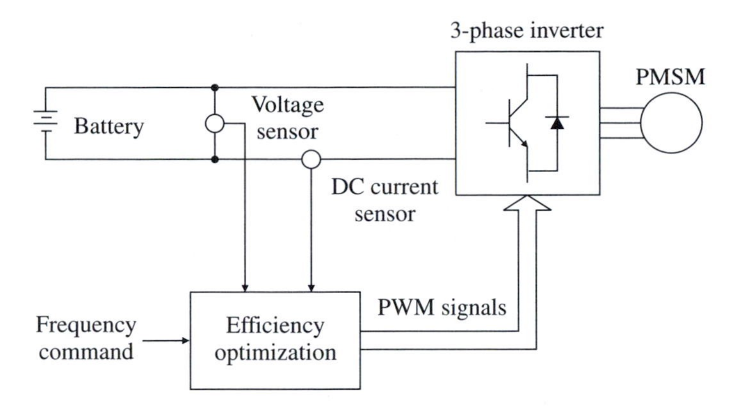

Fig. 5.44. Self-tuning efficiency-optimizing control of a PM synchronous motor drive.

Control of PM synchronous motors can be similar to that of induction motors and the control strategies for induction motors, such as variable-voltage variable-frequency (VVVF) and field-oriented control (FOC), are still applicable. Moreover, by incorporating the well-known *d–q* coordinate transformation, the well-developed flux-weakening control technique can readily be applied to these motors for constant-power operation (Sato *et al.*, 1997).

Since energy stored in the EV battery pack is so precious that a high-efficiency motor drive is of utmost importance, efficiency-optimizing control should be adopted even though PM brushless motors inherently possess higher efficiency than that of induction motors. As shown in Fig. 5.44, self-tuning efficiency-optimizing control of the above-mentioned PM synchronous motor drive has been adopted for the HKU Mark3 mini EV (Chan and Chau, 1996). In essence, the control strategy operates by measuring the voltage and current on the dc link, and adjusting the PWM voltage output of the inverter in such a way that the input power is minimum for a given output power. Thus, it optimizes the combined inverter-motor efficiency, not just the motor efficiency. This control strategy has the advantage that it does not depend on a loss model of the motor and therefore is insensitive to variations in the motor parameters, such as the temperature change in resistance, and does not require accurate modelling of complicated phenomena such as the iron loss. Figure 5.45 shows the software implementation flow chart of the control algorithm. The initial PWM output voltage is determined by the product of the frequency command and the predefined volts/hertz ratio. For a very low frequency operation, boosting voltage is required. A self-searching process is then performed by changing the voltage magnitude with an appropriate step size $|\Delta V|$. A test is made to check whether the input power has increased or decreased. An increase causes the search direction to reverse by using the block $\Delta V = -\Delta V$ while the search direction is unchanged if the input power decreases. The voltage step size should be fine enough to permit closer convergence to the optimum. If four consecutive steps are taken in the same direction, the step size is

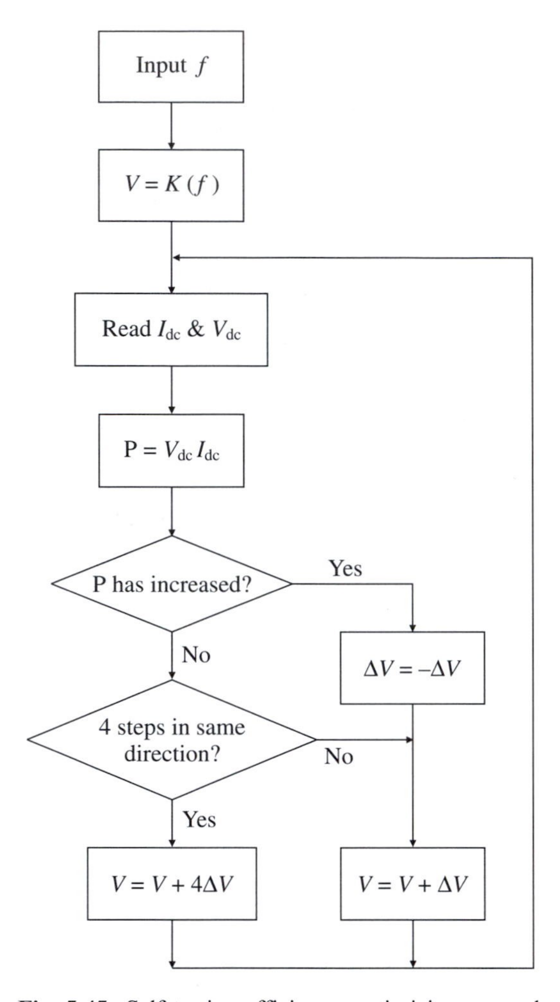

Fig. 5.45. Self-tuning efficiency-optimizing control.

enlarged by four times via the block $V = V + 4\Delta V$ so as to permit faster convergence to the optimum.

PM brushless dc motors

By inverting the stator and rotor of PM dc motors, PM brushless dc motors are generated. The most obvious advantage of these motors is the removal of brushes, leading to eliminate many problems associated with brushes. Another advantage is the ability to produce a larger torque because of the rectangular interaction between current and flux. Moreover, by specially arranging the stator winding and flux path, a phase-decoupling type of PM brushless dc motors has more superior dynamic performance and flexible controllability than that of PM synchronous motors.

Although PM synchronous motors and PM brushless dc motors are very similar in construction, their mathematical models, hence steady-state and dynamic behaviours, are different. The mathematical model of PM synchronous motors is similar to that of wound-rotor synchronous motors. As the transformation of the machine equations from the *a–b–c* phase variables to the *d–q* variables forces all sinusoidally varying inductances in the *a–b–c* frame to become constant in the *d–q* frame, the *d–q* model of PM synchronous motors can be derived. On the contrary, since the back emf in PM brushless dc motors is rectangular, the inductances do not vary sinusoidally in the *a–b–c* frame. It does not seem advantageous to transform the equations to the *d–q* frame since the inductances will not be constant after transformation. Hence, the *a–b–c* model is used for PM brushless dc motors.

Abandoning the standard stator configuration, a special class of polyphase multi-pole phase-decoupling PM brushless dc motor drive has been developed for EV propulsion (Chan *et al.*, 1994; Chan, 1998). The schematic diagram of a five-phase 22-pole phase-decoupling PM brushless dc motor drive is shown in Fig. 5.46. The motor has 20 slots in the stator, where two sides of a coil are located in

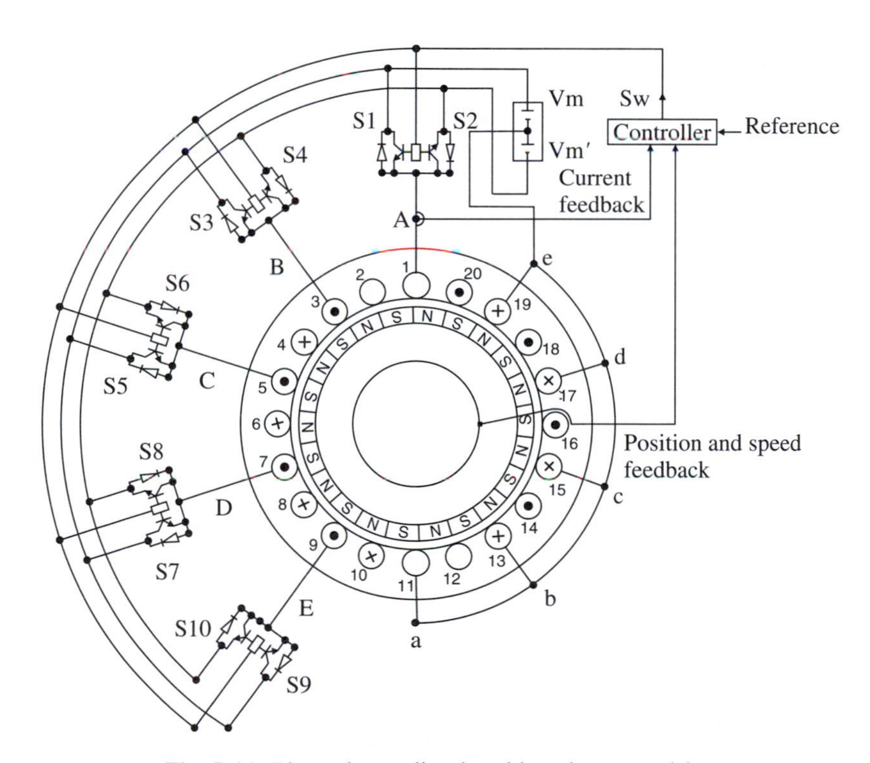

Fig. 5.46. Phase-decoupling brushless dc motor drive.

two adjacent slots, say slot 1 and slot 2, while another coil of the same phase is located in slot 11 and slot 12. These two coils are connected in series. The other phase windings can be arranged in the same way. The rotor consists of 22 pieces of PMs to form 22 poles. Two adjacent poles make up a pole pair. This indicates that the slot pitch of two adjacent slots is 11/10 pole pitch. By using this arrangement of a fractional number of slots per pole per phase, the magnetic force between the stator and any rotor position is uniform, hence eliminating the cogging torque that usually occurs in PM motors. When all currents in slots under *S*-poles flow toward the reader and all currents in slots under *N*-poles flow away from the reader, as illustrated in Fig. 5.46, the direction of torque produced on the rotor is counter-clockwise, and vice versa. According to this coordination between the stator slots and rotor poles, at any moment, there are four phase windings in the conducting state and one phase winding in the non-conducting state. For example, at the moment when the rotor position is shown in Fig. 5.46, phase *A* is in the non-conducting state. As shown in Fig. 5.47, each phase winding conducts 144° over a half-cycle, and the phase shift between adjacent phase is 36°. The conducting state depending on the feedback signal from a rotor position sensor, which consists of optodevices mounted on the motor frame (end bracket) and a toothed disk mounted on the rotor shaft. The feedback signal possesses two functions: it senses the rotor position to determine the conducting state of each winding, and provides a speed feedback signal to the controller. The direction of rotation is determined by the commutating logic, and the commutating frequency adapts to the motor

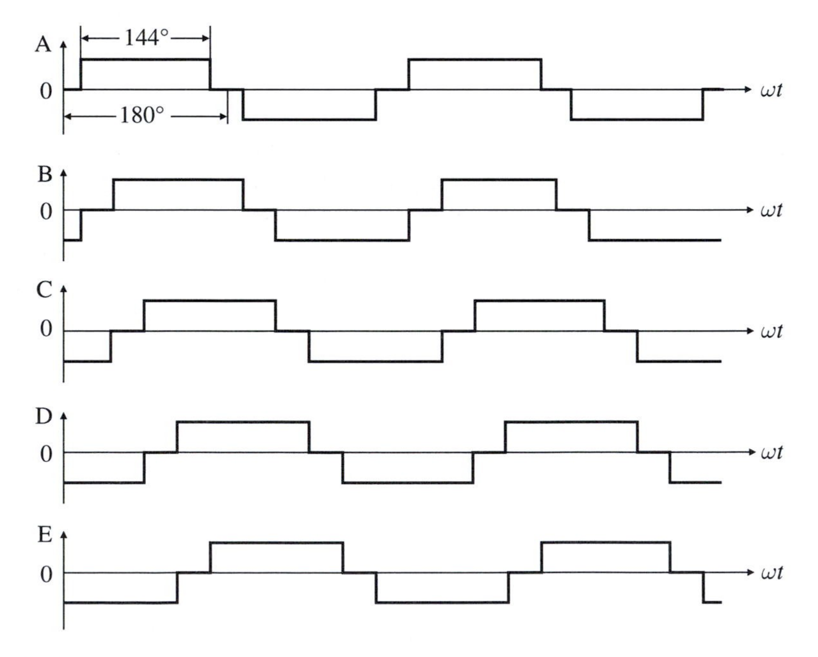

Fig. 5.47. Operating waveforms of a phase-decoupling PM brushless dc motor drive.

speed. The speed control of this motor can be easily implemented by regulating the applied voltage.

Due to the high resistivity of PMs, the induced current in the rotor can be neglected. Thus, the system equation of this phase-decoupling PM brushless dc motor is represented by:

$$\begin{bmatrix} v_1 \\ v_2 \\ \vdots \\ v_5 \end{bmatrix} = \begin{bmatrix} R & & & \\ & R & & \\ & & \ddots & \\ & & & R \end{bmatrix} \begin{bmatrix} i_1 \\ i_2 \\ \vdots \\ i_5 \end{bmatrix} + \frac{d}{dt} \begin{bmatrix} L_{11} & L_{12} & \cdots & L_{15} \\ L_{21} & L_{22} & \cdots & L_{25} \\ \vdots & \vdots & \cdots & \vdots \\ L_{51} & L_{52} & \cdots & L_{55} \end{bmatrix} \begin{bmatrix} i_1 \\ i_2 \\ \vdots \\ i_5 \end{bmatrix} + \begin{bmatrix} e_1 \\ e_2 \\ \vdots \\ e_5 \end{bmatrix}.$$

In short form, it is written as:

$$[v] = [R][i] + \frac{d}{dt}[L][i] + [e],$$

where $[v]$ is the applied voltage matrix, $[R][i]$ is the voltage drop matrix, $d[L][i]/dt$ is the transformer emf matrix, and $[e]$ is the rotational back emf matrix due to PMs. Since the inductances associated with different rotor angular positions are almost identical, it results $L_{11} = L_{22} = \cdots = L_{55} = L$, $L_{12} = L_{21} = M_{12}, \cdots,$ $L_{15} = L_{51} = M_{15}$. Because of the coil span equal to the slot pitch, the phase flux paths are independent. Hence, the mutual inductances of phase windings are negligible, $M_{12} = M_{15} = 0$. So, the system equation can be rewritten as:

$$\begin{bmatrix} v_1 \\ v_2 \\ \vdots \\ v_5 \end{bmatrix} = \begin{bmatrix} R & & & \\ & R & & \\ & & \ddots & \\ & & & R \end{bmatrix} \begin{bmatrix} i_1 \\ i_2 \\ \vdots \\ i_5 \end{bmatrix} + \frac{d}{dt} \begin{bmatrix} L & & & \\ & L & & \\ & & \ddots & \vdots \\ & & & L \end{bmatrix} \begin{bmatrix} i_1 \\ i_2 \\ \vdots \\ i_5 \end{bmatrix} + \begin{bmatrix} e_1 \\ e_2 \\ \vdots \\ e_5 \end{bmatrix}.$$

It can be seen that this five-phase PM brushless dc motor can be treated as five separate dc motors. Considering the phase shift between phases, the system equation for the j-th phase of the motor can be expressed as:

$$v_j\left(\omega t + \theta_0 - \frac{(j-1)\pi}{5}\right) = Ri_j\left(\omega t + \theta_0 - \frac{(j-1)\pi}{5}\right) + L\frac{d}{dt}i_j\left(\omega t + \theta_0 - \frac{(j-1)\pi}{5}\right)$$
$$+ e_j\left(\omega t - \frac{(j-1)\pi}{5}\right) \quad j = 1, \cdots, 5,$$

where ω is the angular frequency, and θ_0 is the advanced conduction angle (the phase shift angle of the applied voltage leading ahead the rotational emf). Hence, the electromagnetic torque is given by:

$$T(\omega t) = \frac{p}{\omega}\sum_{j=1}^{5} e_j\left(\omega t - \frac{(j-1)\pi}{5}\right) i_j\left(\omega t + \theta_0 - \frac{(j-1)\pi}{5}\right).$$

Because of phase-decoupling, each phase can be independently controlled like a conventional dc motor. Hence the control technique used for dc motors can be easily implemented for this motor, such as the dual closed-loop of speed feedback and current feedback control system. As compared with the vector control of induction motors, the control of this motor is much simpler because there is no vector transformation. The constant-torque operation of the motor drive can be readily achieved by adjusting the applied voltage. Due to the non-sinusoidal waveform in the PM brushless dc motor drive, the conventional flux-weakening control can no longer be directly applied to this motor drive type for constant-power operation. Recently a new control strategy has been developed to overcome this problem (Chan, Jiang, Xia and Chau, 1995). As shown in Fig. 5.48, the

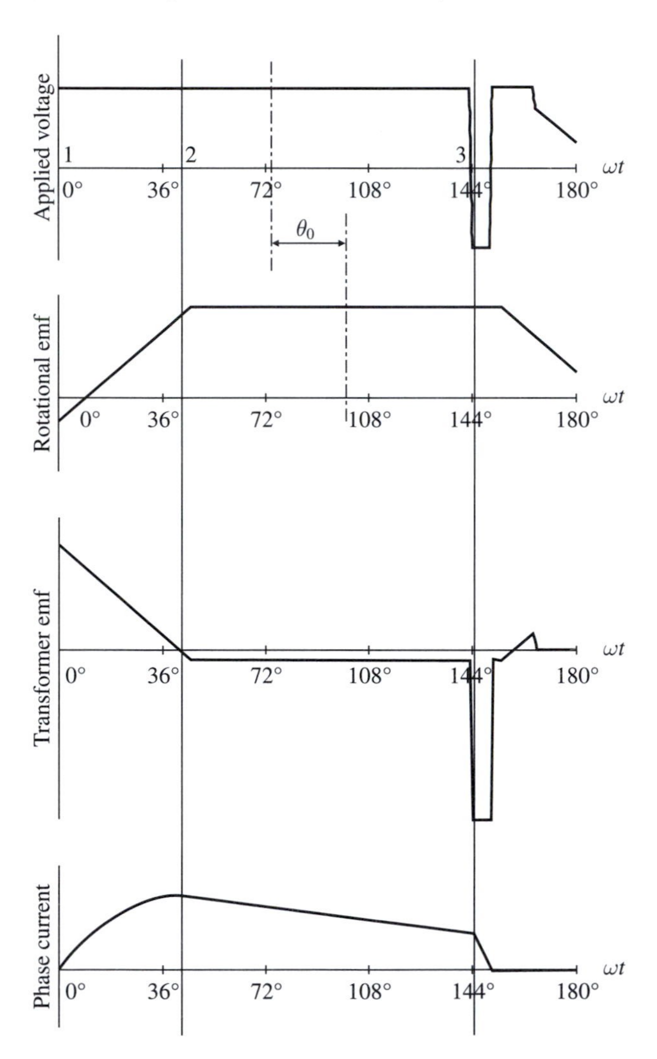

Fig. 5.48. Advanced conduction angle control of a phase-decoupling PM brushless dc motor.

Table 5.8 Technical data of a phase-decoupling PM brushless dc motor

Rated power	3.33 kW
Rated voltage	2 × 90V
Base speed	1000 rpm
Number of phases	5
Stator	
Number of slots	20
Outer diameter	175 mm
Inner diameter	86 mm
Length of core	100 mm
Winding type	Single layer
Number of coils	10
Rotor	
Number of poles	22
Outer diameter	84 mm
Inner diameter	48 mm
Magnet material	Nd–Fe–B

conduction angle is purposely advanced in such a way that the transformer emf opposes or weakens the rotational emf when the motor operates above the base speed, leading to extend the constant-power operating range. Increasingly, by selecting an appropriate advanced conduction angle, optimal efficiency can be obtained throughout the whole operating range. It should be noted that this control is so general that it can be applied to other PM brushless dc motor drives.

Based on the technical data of the motor listed in Table 5.8, the resulting torque-speed characteristics with different θ_0 are shown in Fig. 5.49. It indicates that θ_0 increases from 0° to 43° so that the motor can achieve constant-power operation for 1000 rpm to 4000 rpm. Moreover, for a particular speed, θ_0 can be further increased to a value in such a way to optimize the motor efficiency during constant-power operation.

The advantages and special features of this phase-decoupling PM brushless dc motor drive are summarized below (Chan, 1998):

- Similar to other PM brushless dc motors, this motor has the advantage that the interaction of rectangular current and rectangular flux can produce a larger torque product for the same r.m.s. values.
- Since two adjacent poles make up a pair of poles, the flux paths of different pole-pairs are independent. This multi-pole magnetic circuit arrangement enables to reduce the magnetic iron yoke, resulting in the reduction of volume and weight.
- Since the coil span of stator windings is designed to be equal to the slot pitch, the overhanging part of the coil can be significantly reduced, thus resulting in the saving of copper as well as the further reduction of volume and weight.

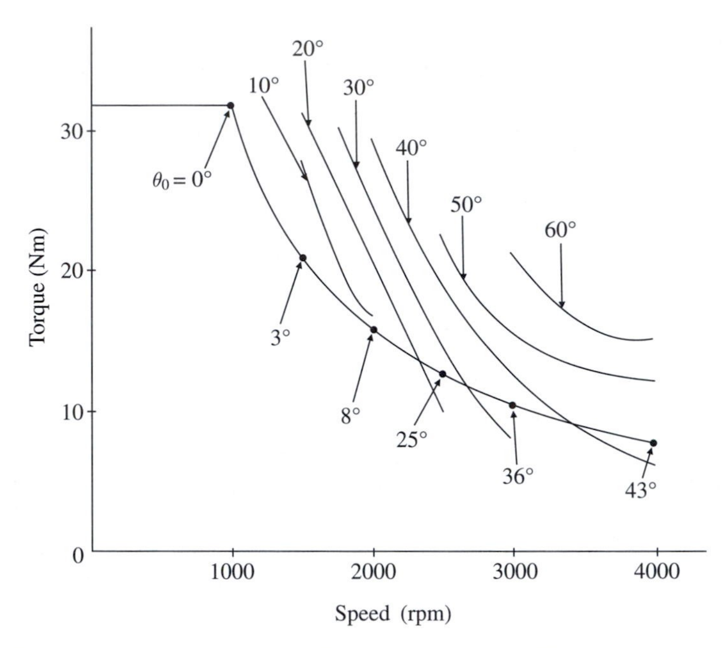

Fig. 5.49. Torque-speed characteristics of a phase-decoupling PM brushless dc motor using advanced conduction angle control.

- By using the fractional number of slots per pole per phase, the magnetic force between the stator and rotor at any rotating position is uniform, thus eliminating the cogging torque which usually occurs in the conventional PM brushless dc motors.
- Due to the phase-decoupling nature, the dynamic performance of this motor drive is inherently excellent. Increasingly, it offers the possibility of fault-tolerance operation.
- Since conventional flux-weakening control for constant-power operation is inapplicable, a new advanced conduction angle control strategy is developed which employs the transformer emf to counteract the rotational emf for constant-power operation at high speeds.
- PMs can either be mounted on the rotor surface or buried inside the rotor circuit. The surface-magnet type has the advantage of simplicity. The buried-magnet type has the advantage of mechanical integrity because the PMs are physically protected. Moreover, flux-focusing arrangement can be employed to strengthen the air-gap flux density.

PM hybrid motors

Recently, the use of additional field winding to extend the speed range of PM brushless motors has been developed. The key is to control the field current in such a way that the air-gap field provided by PMs can be weakened during high-speed constant-power operation. Due to the presence of both PMs and the field winding, these motors are so-called PM hybrid motors. In general, these PM hybrid motors can adopt either the series or shunt structure. Because of the low permeability of PMs, the series structure usually requires a relatively large mmf, and is less attractive than the shunt structure. However, these motors have the drawbacks that the structure is relatively complex.

As shown in Fig. 5.50, a PM hybrid motor drive has been newly developed for EV propulsion (Chan, Zhang *et al.*, 1996). It has a unique structure which comprises of the claw-type rotor, stationary field winding, and stator. PMs are integrated into the rotor, while the field winding and its holder are located at a stationary annular region formed by the inner and outer parts of the rotor (Chan, 1999). Thus, the components of air-gap flux, namely φ_{pm} produced by the PMs and φ_{f} produced by the field winding, are magnetically shunted in nature. The resultant air-gap flux φ_{δ} can be varied by adjusting the dc current of the field winding. Figure 5.51 shows a typical variation of the resultant air-gap flux due to the change of field current, which confirms the desired flux-weakening effect.

Since the stator winding of this motor is of the sinusoidal distribution and the waveform of air-gap flux density is essentially sinusoidal, the *d–q* model can be adopted. Hence, the system equation is given by:

$$\begin{bmatrix} u_{\mathrm{d}} \\ u_{\mathrm{q}} \\ u_{\mathrm{f}} \\ e_{\mathrm{pm}} \end{bmatrix} = \begin{bmatrix} r_{\mathrm{s}} & 0 & 0 & 0 \\ 0 & r_{\mathrm{s}} & 0 & 0 \\ 0 & 0 & r_{\mathrm{s}} & 0 \\ 0 & 0 & 0 & r_{\mathrm{s}} \end{bmatrix} \begin{bmatrix} i_{\mathrm{d}} \\ i_{\mathrm{q}} \\ i_{\mathrm{f}} \\ i_{\mathrm{pm}} \end{bmatrix} + \frac{d}{\mathrm{dt}} \begin{bmatrix} L_{\mathrm{d}} & 0 & L_{\mathrm{sf}} & 0 \\ 0 & L_{\mathrm{q}} & 0 & 0 \\ L_{\mathrm{fs}} & 0 & L_{\mathrm{f}} & 0 \\ 0 & 0 & 0 & 0 \end{bmatrix} \begin{bmatrix} i_{\mathrm{d}} \\ i_{\mathrm{q}} \\ i_{\mathrm{f}} \\ i_{\mathrm{pm}} \end{bmatrix}$$

$$+\,\omega \begin{bmatrix} 0 & -L_{\mathrm{q}} & 0 & 0 \\ L_{\mathrm{d}} & 0 & L_{\mathrm{sf}} & L_{\mathrm{spm}} \\ 0 & 0 & 0 & 0 \\ 0 & 0 & 0 & 0 \end{bmatrix} \begin{bmatrix} i_{\mathrm{d}} \\ i_{\mathrm{q}} \\ i_{\mathrm{f}} \\ i_{\mathrm{pm}} \end{bmatrix}.$$

In short form, it is written as:

$$[v] = [R][i] + \frac{d}{dt}[L][i] + \omega[G][i],$$

where $[v]$ is the voltage matrix in which u_{d}, u_{q}, u_{f} and e_{pm} are respectively the *d*-axis, *q*-axis, field winding and PM excitation voltages, and $[i]$ is the current matrix in which i_{d}, i_{q}, i_{f} and i_{pm} are respectively the *d*-axis, *q*-axis, field winding and PM excitation currents. Because of the uniqueness of the motor configuration, the inductance parameters in $[L]$ and $[G]$ are obtained using the 3-D finite element method. Hence, the torque is given by:

$$T = p[i]^T[G][i].$$

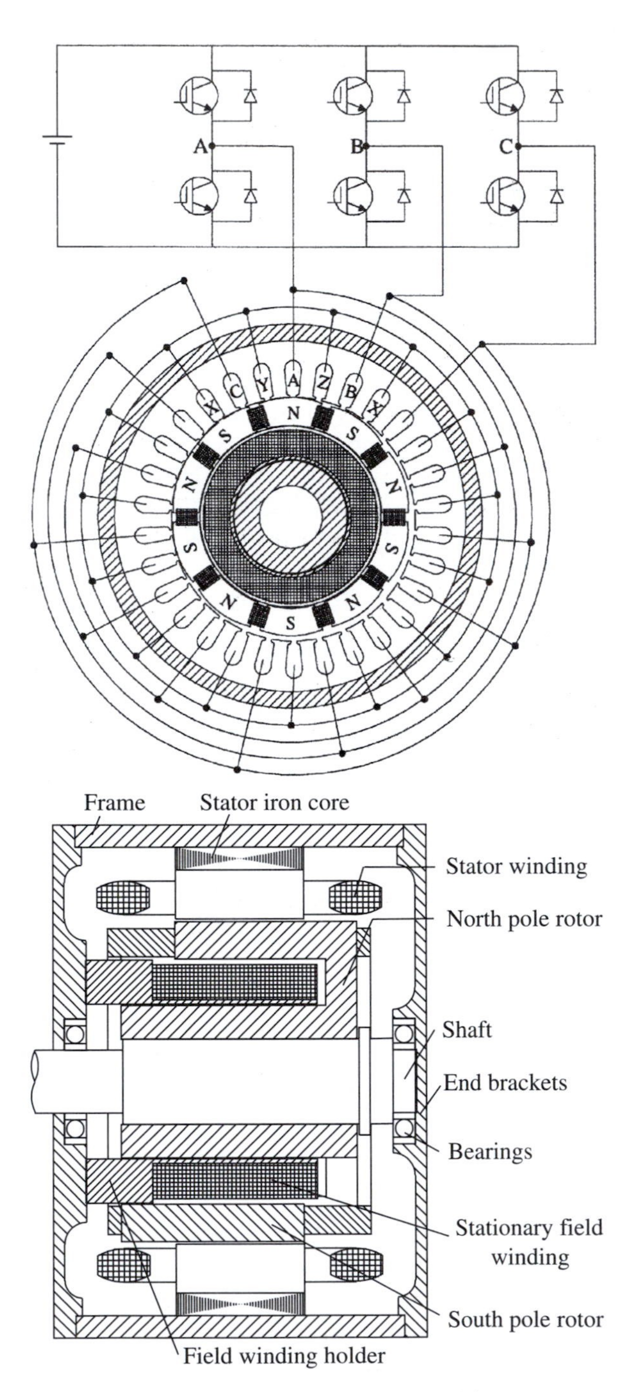

Fig. 5.50. PM hybrid motor drive.

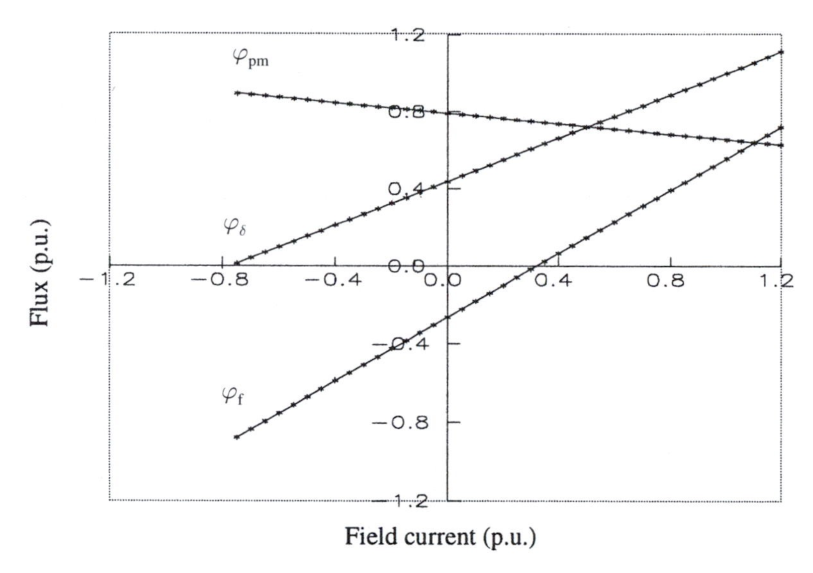

Fig. 5.51. Flux-weakening effect of a PM hybrid motor drive.

The unique configuration of the PM hybrid motor offers the control system an additional control variable, the field current i_f. In the case of low load torque, the air-gap flux can be purposely weakened to decrease the total loss of the motor so as to improve the motor efficiency. In the case of high motor speed, the flux can be also weakened to widen the constant-power operation range. For a given load power and motor speed, there is an optimal combination of the terminal voltage and field current. The block diagram of the speed control system with efficiency optimization is illustrated in Fig. 5.52. A PI regulator is used to deduce the torque reference T^* from the speed error $(\omega^* - \omega)$. Based on the motor speed and torque reference, the fuzzy logic controller functions to adjust the field current reference i_f^* and to estimate the air-gap flux reference φ_δ^*, hence controlling the motor operating at the optimal efficiency point. The vector controller functions to adjust the magnetizing component and the torque component independently, hence achieving good dynamic performance. Taking the technical data of a PM hybrid motor listed in Table 5.9, the optimized efficiency profile throughout the operating range is shown in Fig. 5.53 (Chan, Zhang *et al.*, 1997).

The advantages and special features of this motor drive are summarized as follows:

- By adopting the unique claw-type rotor structure, the leakage flux can be minimized and the construction becomes compact.
- By fixing the field winding as an inner stator, the motor axial length can be shortened, and the material consumption can be reduced.
- Due to the existence of both PMs and the field winding, the motor can be designed to achieve higher air-gap flux density and hence higher power density.

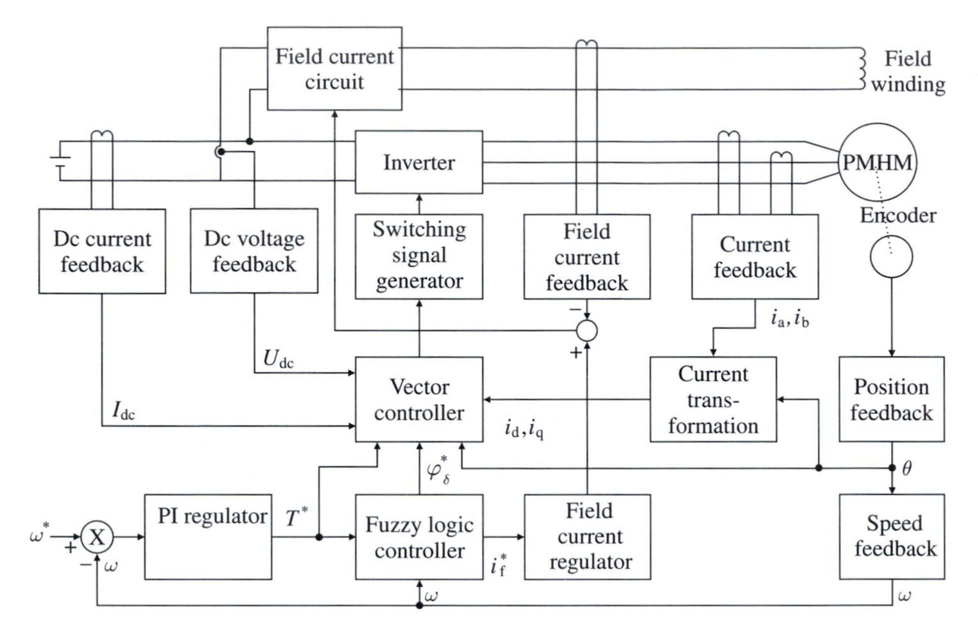

Fig. 5.52. Speed control system of a PM hybrid brushless motor drive.

Table 5.9 Technical data of a PM hybrid motor

Rated power	5 kW
Rated voltage	270 V
Rated speed	1000 rpm
Number of phases	3
Stator	
Number of slots	30
Outer diameter	210 mm
Inner diameter	148 mm
Length of core	58 mm
Winding type	Double layer
Number of coils	64
Rotor	
Number of poles	10
Outside outer diameter	146 mm
Outside inner diameter	118 mm
Inside outer diameter	65 mm
Inside inner diameter	38 mm
Magnet material	Nd–Fe–B

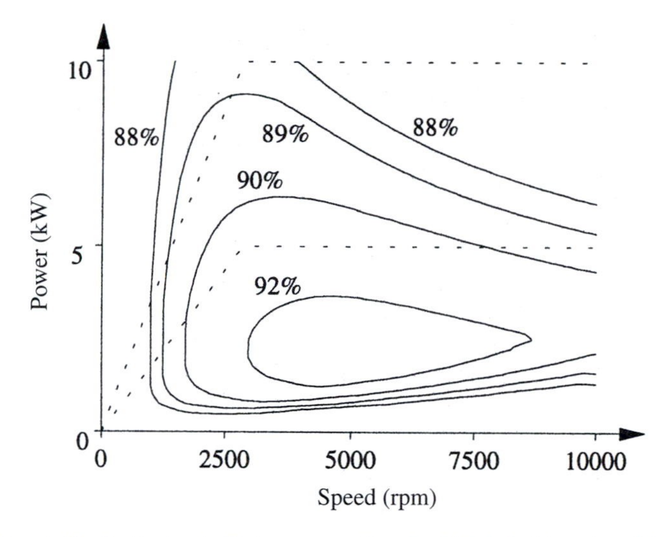

Fig. 5.53. Optimized efficiency profile of a PM hybrid motor drive.

Increasingly, the mounting of PMs adopts the flux-focusing arrangement, which allows the air-gap flux density being higher than the operating flux density of individual PMs.

- By controlling the direction and magnitude of the dc field current, the air-gap flux can be flexibly adjusted. Hence, the torque-speed characteristics can be easily shaped to meet the special requirements for EV propulsion.
- By adjusting the field current to weaken the air-gap flux produced by PMs, the speed range for constant-power operation can be significantly extended.
- By proper controlling the applied voltage and dc field current, the efficiency map of the motor drive can be optimized throughout the whole operating range. Thus, the efficiency at those operating regions for EV propulsion, such as high-torque low-speed hill climbing and low-torque high-speed cruising, can be improved.

5.5 Switched reluctance motor drives

Although the earliest recorded switched reluctance (SR) motor was built to propel a locomotive in 1838, SR motor drives could not realize their full potential until the advent of modern power electronics and powerful computing facilities. In general, a SR motor drive consists of four parts, namely the SR motor, power converter, sensor and controller as shown in Fig. 5.54. Among them, the SR motor plays a key role which converts electrical energy into mechanical motion. Figure 5.55 shows the three basic structures of SR motors which have different numbers of stator and rotor poles (Miller, 1993).

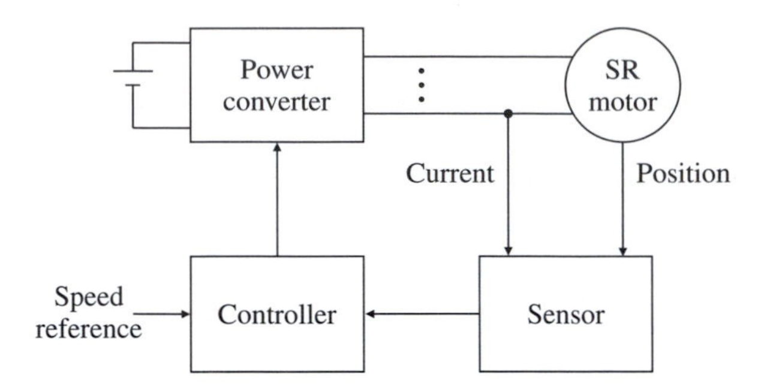

Fig. 5.54. Basic structure of SR motor drives.

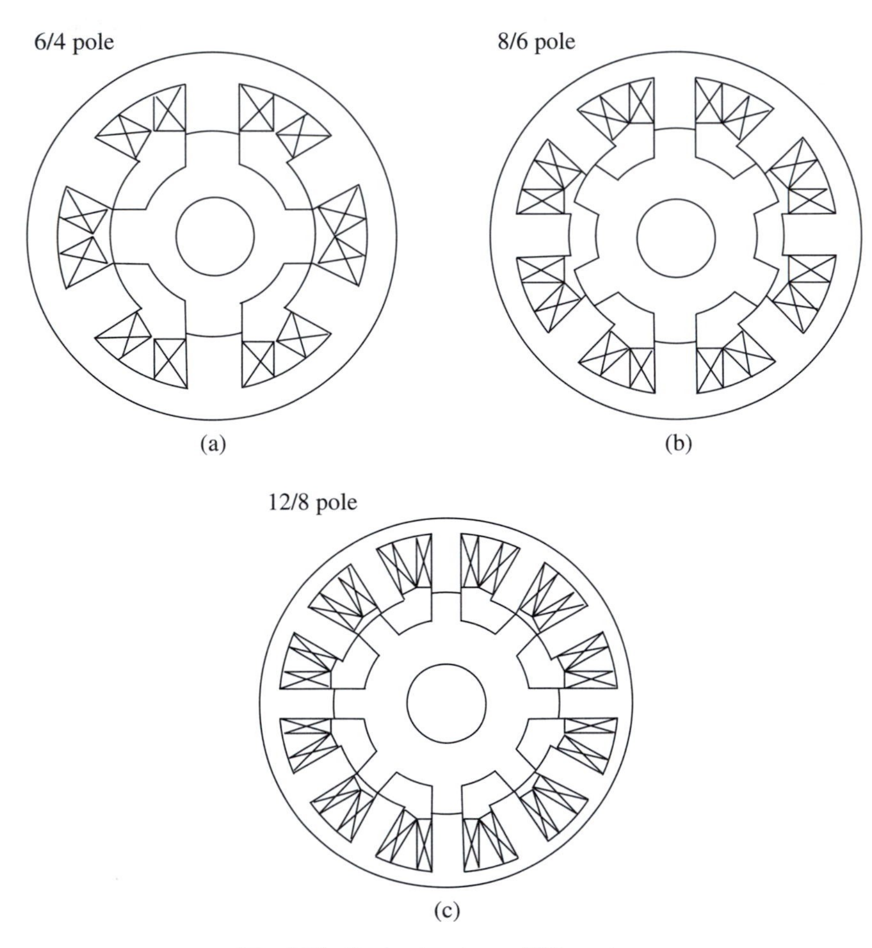

Fig. 5.55. Basic structure of SR motors.

5.5.1 PRINCIPLE OF OPERATION

Figure 5.56 shows a four-phase 8/6-pole SR motor drive, in which only one particular phase winding is sketched. Because of the salient nature of both the stator and rotor poles, the inductance L of each phase varies with the rotor position as shown in Fig. 5.57. The operating principle of the SR motor is based on the 'minimum reluctance' rule. For instance, as shown in Fig. 5.56, when the phase B winding is excited, the rotor tends to rotate clockwise so as to decrease the reluctance of the flux path until the rotor pole 2 aligns with the stator pole B where the reluctance of the flux path has a minimum value (the inductance has a maximum value). Then, the phase B is switched off and the phase A is switched on so that the reluctance torque tends to make the rotor pole 1 align with the stator pole A. The torque direction is always towards the nearest aligned position. Hence, by conducting the phase windings in the sequence of B–A–D–C according to the rotor position feedback from the position sensor, the rotor can continuously rotate clockwise.

According to the co-energy principle (Miller, 1993), the reluctance torque produced by one phase at any rotor position is given by:

$$T(\theta, i) = \frac{\partial W'(\theta, i)}{\partial \theta},$$

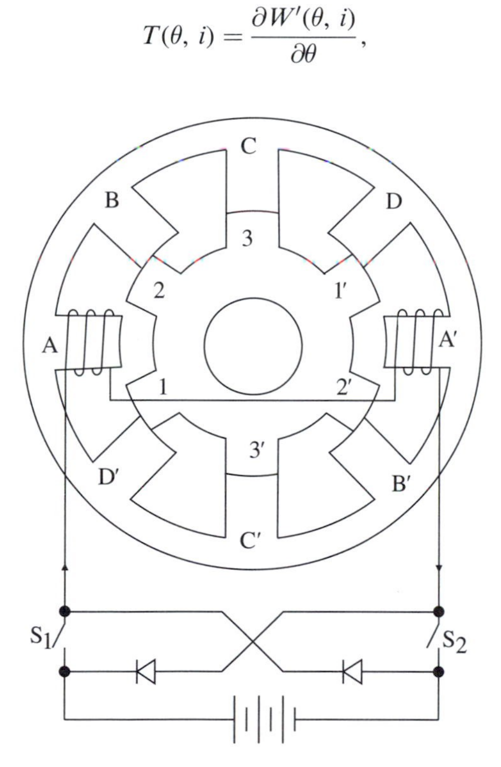

Fig. 5.56. Four-phase 8/6-pole SR motor drive.

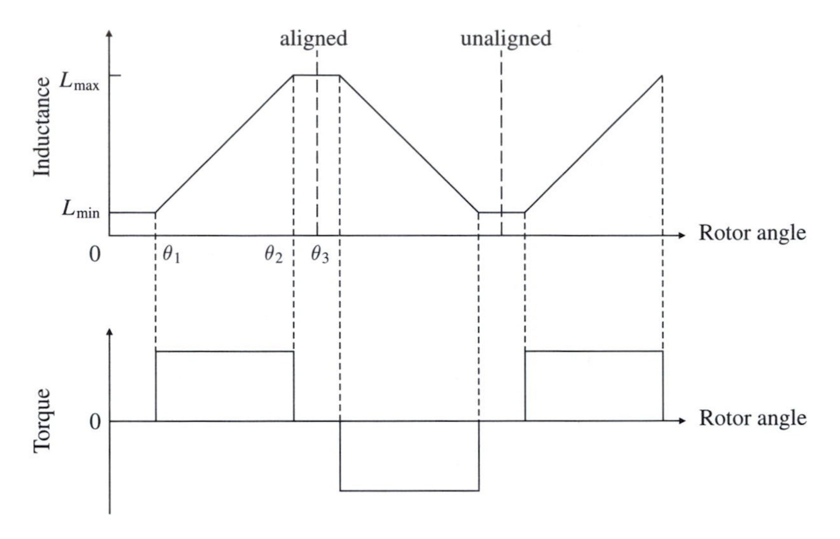

Fig. 5.57. Variations in inductance and constant-current torque vs. rotor position.

where θ is the rotor position angle, i is the phase current and $W'(\theta, i)$ is the so-called co-energy defined as the area below the magnetization curve shown in Fig. 5.58. It can be expressed as:

$$W'(\theta, i) = \int_0^i \psi(\theta, i)di.$$

Since the flux linkage $\psi(\theta, i)$ can be written as $\psi(\theta, i) = L(\theta, i)i$, it yields:

$$T(\theta, i) = \frac{1}{2}\frac{\partial}{\partial\theta}\int_0^{i^2} L(\theta, i)di^2.$$

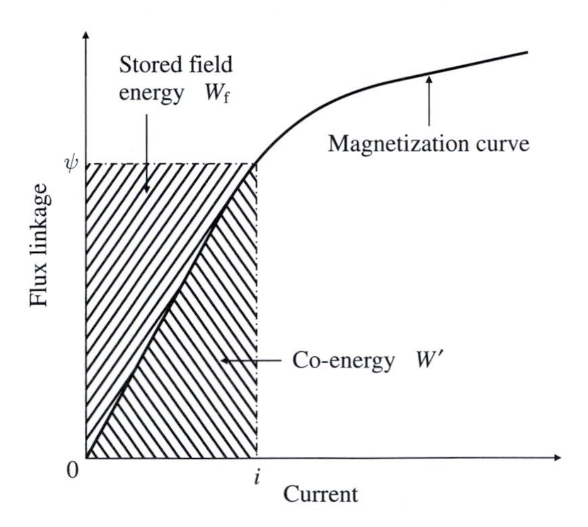

Fig. 5.58. Definition of co-energy.

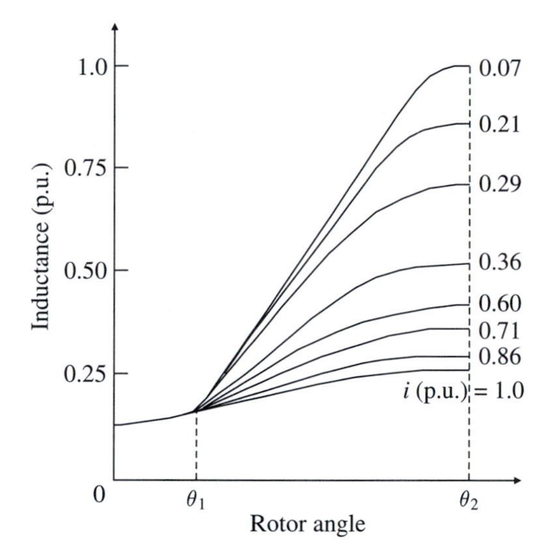

Fig. 5.59. Nonlinear inductance characteristics.

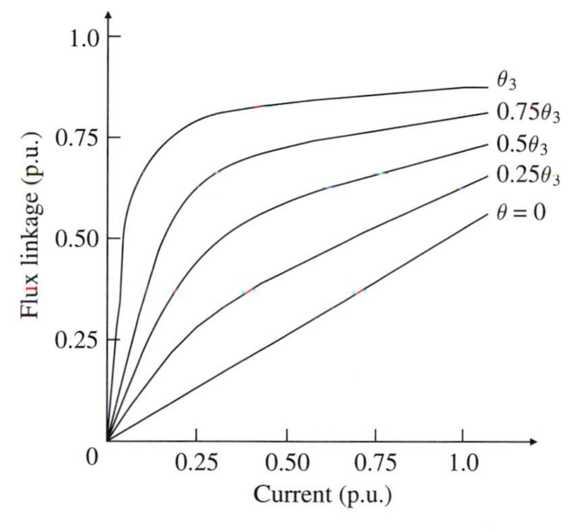

Fig. 5.60. Nonlinear flux linkage characteristics.

For the special case that the SR motor does not suffer from magnetic saturation, the inductance is independent of the phase current and the reluctance torque can be deduced as:

$$T(\theta, i) = \frac{1}{2} i^2 \frac{dL}{d\theta}.$$

The corresponding waveform under a constant phase current i is also shown in Fig. 5.57. When magnetic saturation is taken into account, the inductance is

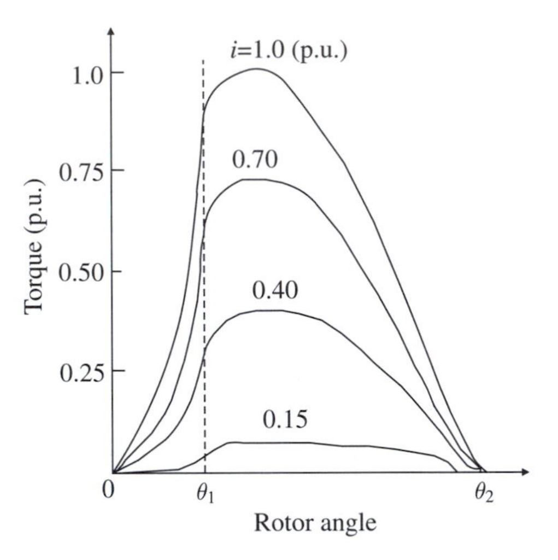

Fig. 5.61. Nonlinear torque characteristics.

a function of both the rotor position angle and the phase current. The nonlinear characteristics of inductance, flux linkage and torque are shown in Fig. 5.59 to Fig. 5.61, respectively.

From the above analysis, it can be seen that the SR motor has two significant features. One is that the direction of torque is independent of the polarity of the phase current. Another is that the motoring torque can be produced only in the direction of rising inductance ($dL/d\theta > 0$); otherwise, a negative torque (or braking torque) is produced. So, each phase can produce a positive torque only in half a rotor pole-pitch, hence creating the torque ripple. Nevertheless, this torque ripple can be alleviated by increasing the number of phases.

The voltage equation of the SR motor can be expressed as:

$$u = Ri + L\frac{di}{dt} + i\frac{dL}{d\theta}\omega,$$

where u is the phase voltage, R is the winding resistance and ω is the rotor speed.

The corresponding phase current can be represented by:

$$i = \frac{u}{\omega} f(\theta),$$

where $f(\theta)$ is a function of the structure parameters, rotor position as well turn-on and turn-off angles. Hence, the average torque of the SR motor with m phases can be obtained as:

$$T = \frac{m}{\theta_{cy}} \int_0^{\theta_{cy}} \frac{1}{2} i^2 \frac{\partial L}{\partial \theta} d\theta = \frac{mu^2}{2\theta_{cy}\omega^2} \int_0^{\theta_{cy}} f^2(\theta) \frac{\partial L}{\partial \theta} d\theta,$$

where θ_{cy} is the rotor pole pitch angle. For a given SR motor with fixed phase voltage as well as turn-on and turn-off angles, the average torque and power can respectively be rewritten as:

$$T_{av} = \frac{K}{\omega^2}$$
$$P = \frac{K}{\omega}.$$

It indicates that the SR motor torque is inversely proportional to the square of the speed while its power is inversely proportional to the speed, similar to the characteristics of a series dc motor.

The SR motor has basically two operating modes as shown in Fig. 5.62. When the speed is below the base speed ω_b, the current can be limited by chopping, so-called current chopping control (CCC). In the CCC mode, the turn-on angle θ_{on} and the turn-off angle θ_{off} are fixed and the firing angle depends only on the speed feedback. The torque can be controlled by changing the current limits, and thus the constant-torque characteristic can be achieved by CCC. During high speed operation, however, the peak current is limited by the emf of the phase winding. The corresponding characteristic is essentially controlled by phasing of switching instants relative to the rotor position, so-called angular position control (APC). In the APC mode, the constant-power characteristic can be achieved. At the critical speed ω_{sc}, both θ_{on} and θ_{off} reach their limit values. Thereafter, the SR motor can no longer keep constant-power operation and offer the natural series characteristic corresponding to $\theta_c = \theta_{cy}/2$ (θ_c is the conducting angle). Typical current and inductance waveforms in both the CCC and APC modes are shown in Fig. 5.63.

Similar to dc motor drives, the chopping frequency of SR motor drives should be above 10 kHz to minimize acoustic noise. Many converter circuits have been developed in attempts to reduce the number of power devices and take full

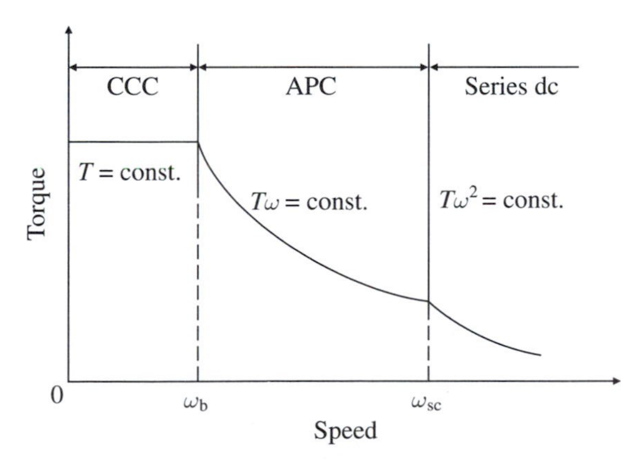

Fig. 5.62. Typical torque-speed characteristics.

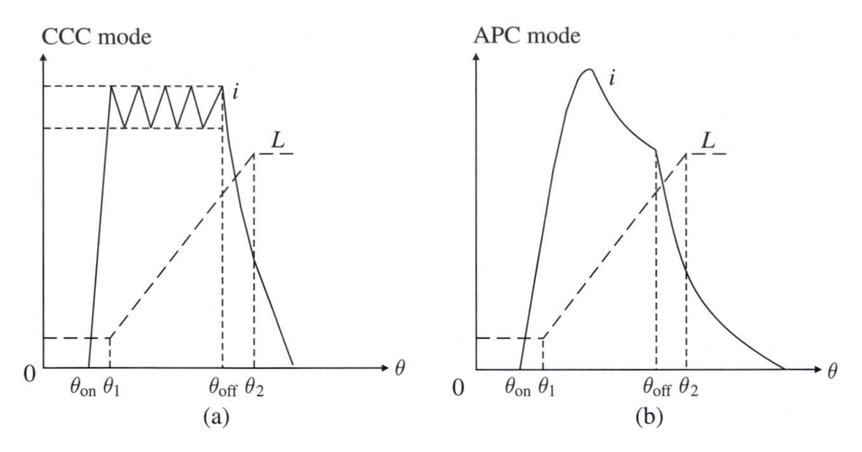

Fig. 5.63. Current and inductance waveforms in CCC and APC modes.

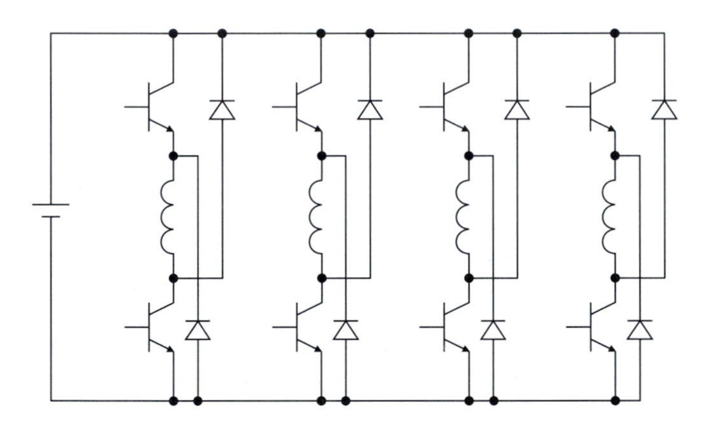

Fig. 5.64. Converter circuit of SR motor drives.

advantage of unipolar operation. However, when the device count is reduced, there is a penalty in the form of lower controllability, lower reliability, lower operating performance or extra passive components. The converter circuit shown in Fig. 5.64 is well suited for SR motor drives in EVs. It utilizes two power devices to independently control the current of each phase and two freewheeling diodes to return any stored magnetic energy to EV batteries. Since this circuit topology needs two power devices per phase, the converter cost is relatively higher than that with less power devices. However, this bridge arrangement allows control of each phase winding independent of the state of other phase windings. Thus, it is possible to allow for phase overlapping so as to increase the torque production and to extend the constant-power range for EV propulsion.

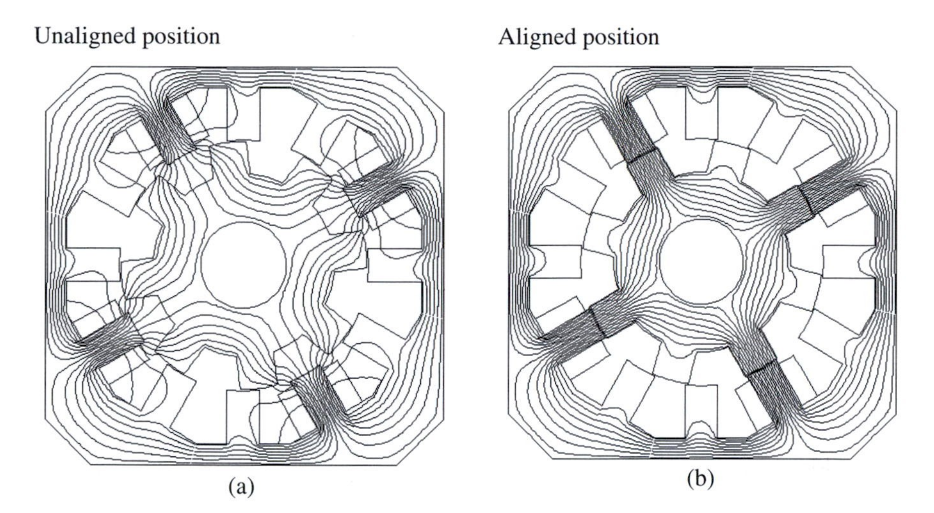

Fig. 5.65. Electromagnetic field distribution of a SR motor.

5.5.2 DESIGN OF SR MOTORS

Although SR motors possess the simplicity in construction, it does not imply any simplicity in analysis and design. Because of the heavy saturation of pole tips and the fringing effect of poles and slots, the design of SR motors suffers a great difficulty in using the magnetic circuit approach. In most cases, the electromagnetic finite element analysis is employed to determine the motor parameters and performances. Typical electromagnetic field distributions of a SR motor are shown in Fig. 5.65. Then, the optimization is based on the minimization of total losses while taking into account the pole arc constraint, height constraint and maximum flux density constraint. Nevertheless, there are some basic criteria to initialize the design process of SR motors for EV propulsion (Chan, Jiang, Zhan and Chau, 1996).

Number of phases and poles

An appropriate selection of the number of phases and poles is important to satisfy the desired motor performance. In order to allow for starting and running bidirectionally, SR motors should have at least three phases with six stator poles and four rotor poles. Among those structures shown in Fig. 5.55, the three-phase 6/4-pole SR motor offers the lowest cost and highest efficiency. However, the corresponding large torque ripple reduces the climbability of an EV. On the other hand, the four-phase 8/6-pole SR motor has a relatively higher cost and lower efficiency while possessing the smallest torque ripple to favour the climbability. The three-phase 12/8-pole SR motor is a compromised design between the three-phase 6/4-pole and four-phase 8/6-pole types. The selection of them should be based on the EV requirement and cost justification. It should be noted that the higher the number of phases and poles, the more the power devices and the higher the switching frequency are needed, leading to increase the cost and the switching loss, respectively. The numbers of both the stator and rotor poles, p_s and p_r, are governed by:

$$p_s = 2km$$
$$p_r = 2k(m \pm 1),$$

where m is the number of phases and k is a positive integer. When the rotor speed is N rpm, the commutating frequency f_{ph} of a particular phase is given by:

$$f_{ph} = p_r \frac{N}{60}.$$

To minimize the switching frequency and to decrease the iron losses in poles and yokes, the number of rotor poles should be selected as small as possible. Thus, the number of rotor poles is usually smaller than that of stator ones.

Pole arcs

To minimize the permeance associated with the minimum inductance L_{min} at the unaligned rotor position and to increase the torque overlapping, the pole arc of SR motors can be selected according to:

$$\min(\beta_s, \beta_r) > \frac{2\pi}{mp_r},$$

where β_s and β_r are the stator and rotor pole arcs, respectively. The selection of the optimum combination of pole arcs should not only consider the highest inductance ratio but also the torque ripple, starting torque and magnetic saturation.

Stator diameters and core length

When the frame of SR motors is designed, there are two important parameters to be optimized, namely the stator inner diameter D_{si} and the core length L_c. Both of them have a great influence on the volume and weight of motor materials. They can be primarily selected as:

$$D_{si} = (0.5 \sim 0.65)D_{so}$$
$$L_c = (0.5 \sim 1.0)D_{so},$$

where D_{so} is the stator outer diameter that can be selected similar to that of induction motors.

Air-gap length and rotor outer diameter

There is no doubt that a small air-gap of SR motors creates a high inductance ratio, leading to achieve high torque density, high efficiency and low converter volt-ampere requirement. However, the smaller air-gap, the higher the difficulty in machining the stator bore and rotor surface is resulted, hence increasing the motor cost. As a rule of thumb, the air-gap of SR motors should be similar to but not greater than that of induction motors with a comparable diameter. When the air-gap length is selected, the rotor outer diameter is given by:

$$D_{ro} = D_{si} - 2g_o,$$

where g_o is the air-gap length.

Height of rotor pole

When the rotor pole is unaligned with the stator pole, the phase inductance is minimum and the permeance of fringing effects is predominant. To minimize the fringing effects at the minimum inductance, the rotor pole height h_r can be governed by:

$$h_r \geq 1.1\left(\frac{\theta_{cy} - \beta_r}{2}\right)\left(\frac{D_{ro}}{2}\right)$$

$$h_r \leq 0.5(D_{si} - D_{ri}) - g_o - \left(\frac{D_{si}}{2} - g_o\right)\sin\left(\frac{\beta_r}{2}\right),$$

where D_{ri} is the rotor inner diameter.

Height of stator pole

To ensure enough space for stator windings, the stator pole height h_s is usually greater than that of the rotor, namely $h_s \geq h_r$, and is governed by:

$$h_s \leq 0.5(D_{so} - D_{si}) - \left(\frac{D_{si}}{2}\right)\sin\left(\frac{\beta_s}{2}\right).$$

Other consideration of SR motor design includes number of turns of stator winding per phase, flux density at each part of the magnetic circuit, unsaturated aligned reluctance per pole, saturated flux per phase, unaligned reluctance per pole, saturated incremental reluctance, mmf-flux characteristics, speed, torque and efficiency, thermal resistance at each part of the thermal circuit, torque per unit weight, and weight of copper and magnetic iron core used.

5.5.3 CONTROL OF SR MOTOR DRIVES

The control requirements of SR motor drives are so unique that the approaches adopted by induction motors and synchronous motors can hardly be applicable. It is also well known that the conventional control methods, such as PID control, can hardly be suitable for application to EVs. So, a fuzzy-sliding mode control (FSMC) approach, combining both fuzzy logic control (FLC) and sliding mode control (SMC), has been newly developed for a SR motor drive for EV propulsion (Chan, Zhan and Chau, 1996).

By converting the linguistic control strategy into automatic control strategy without using the mathematical model of control systems, the FLC approach can be used to deal with complex ill-defined systems. However, the work of FLC design is time-consuming and the response trajectory of the controlled system is unpredictable because fuzzy control rules are experience-oriented and the suitable membership function is generally selected by the trial and error procedure.

With the use of the SMC approach, the control system has perfect insensitivity with respect to the external disturbances as well as the system parameter variations. Thus, one can predetermine an ideal sliding surface of the state trajectory to dominate the dynamic behaviour of the controlled system. However, due to various system nonidealities such as hysteresis of switching operation, time delay of the control system and sampling of digital implementation, the state trajectory generally chatters along a nonideal sliding surface. This undesirable chattering phenomenon causes high-frequency unmodelled dynamics in the controlled system.

The FSMC approach aims to incorporate both FLC and SMC in such a way that they complement one another. Namely, the SMC functions to handle nonlinearities of the SR motor, while the FLC is used to reduce the control chattering. The speed control system of a SR motor drive for EVs is illustrated in Fig. 5.66. It consists of two loops that are the inner current loop and the outer speed loop. The FSMC input is the difference between the speed reference ω^* and speed feedback ω, while its output is the torque reference T^*. The corresponding current reference i^* is formulated according to the nonlinear torque-angle characteristics of the SR motor.

To design the FSMC system, the speed error and the speed derivative are taken as the state variables:

$$\begin{cases} x_1 = \omega - \omega^* \\ x_2 = \dot{\omega} = -\dfrac{B}{J}\omega + \dfrac{1}{J}T - \dfrac{1}{J}T_l \end{cases},$$

where B is the viscous friction coefficient, J is the moment of inertia, while T and T_l are the motor torque and load torque, respectively. By taking $u = \dot{T}^*$ as the control variable, the state equation can be expressed as:

$$\begin{pmatrix} \dot{x}_1 \\ \dot{x}_2 \end{pmatrix} = \begin{pmatrix} 0 & 1 \\ 0 & -B/J \end{pmatrix} \begin{pmatrix} x_1 \\ x_2 \end{pmatrix} + \begin{pmatrix} 0 \\ 1/J \end{pmatrix} u + \begin{pmatrix} -\dot{\omega}^* \\ -\dot{T}_l/J \end{pmatrix},$$

where $\dot{\omega}^*$ and $\dot{T}_l$ are the system disturbances. Then, the two switching surfaces, one dealing with the current control and the other with the speed control, for sliding-mode operation are defined as:

$$\begin{cases} s_i = i - i^* \\ s_\omega = x_1 + cx_2 \end{cases},$$

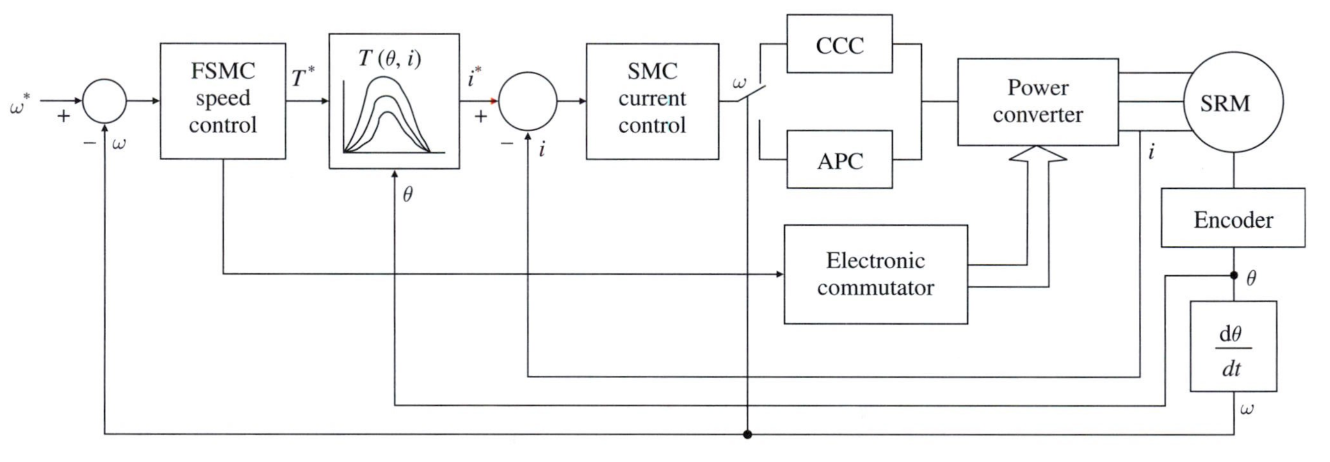

Fig. 5.66. FSMC system of a SR motor drive.

where the system dynamic behaviour is governed by the parameter c. Accordingly, the current sliding mode condition, namely $s_i\dot{s}_i < 0$, can be achieved by manipulating the phase voltage as:

$$u_{\mathrm{i}} = \begin{cases} V_{\mathrm{dc}} & s_i < 0 \\ -V_{\mathrm{dc}} & s_i > 0, \\ Ki\omega & s_i = 0 \end{cases}$$

where V_{dc} is the dc link voltage. It should be noted that this condition can be satisfied when the SR motor drive is operating in the current chopping mode. On the other hand, the speed sliding mode condition, namely $s_\omega\dot{s}_\omega 0$, is satisfied when the sliding mode control law is given by:

$$u_\omega = \begin{cases} \alpha & S_\omega < 0 \\ u_{\mathrm{eq}} & S_\omega = 0, \\ -\alpha & S_\omega > 0 \end{cases}$$

where $\alpha = \eta J/c$ and $u_{\mathrm{eq}} = -Jx_2/c$. The parameter η must be chosen large enough to guarantee the sliding mode condition.

In order to reduce the control chattering and torque ripple, the switching criteria of the speed sliding mode condition is fuzzified as:

$$\begin{aligned} R1&: \text{ if } S_\omega > 0 \text{ then } u_1 = -\alpha \\ R2&: \text{ if } S_\omega < 0 \text{ then } u_2 = \alpha \\ R3&: \text{ if } S_\omega = 0 \text{ and } x_1 \neq 0 \text{ then } u_3 = u_{\mathrm{eq1}} \\ R4&: \text{ if } S_\omega = 0 \text{ and } x_1 = 0 \text{ then } u_4 = u_{\mathrm{eq2}} \end{aligned}$$

Table 5.10 Technical data of a SR motor

Rated power	4 kW
Rated voltage	240 V
Rated speed	3000 rpm
Number of phases	3
Stator	
Outer diameter	175 mm
Inner diameter	110 mm
Air-gap	0.5 mm
Core length	168 mm
Pole height	23 mm
Pole arc	15.5°
Turns per coil	11
Rotor	
Outer diameter	109 mm
Inner diameter	38 mm
Pole arc	17.5°
Pole height	14 mm

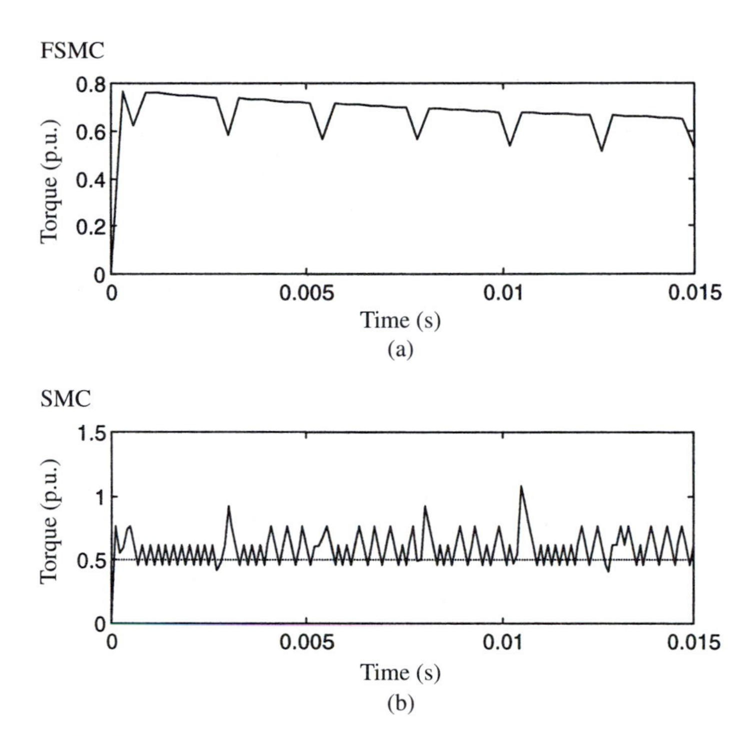

Fig. 5.67. Comparison of waveforms using FSMC and SMC.

where u_{eq1} and u_{eq2} are two different control actions, which are with two different slopes of switching lines $s_\omega = x_1 + c_i x_2$, $i = 1,2$.

A three-phase 4 kW SR motor drive, adopting FSMC, has been specially designed and built for mini EVs. The optimized design data is listed in Table 5.10. To assess the proposed FSMC approach, a comparison of torque waveforms at 1000 rpm of the SR motor drive using FSMC and SMC is shown in Fig. 5.67. The result demonstrates that the use of FSMC can provide a significant reduction in the torque ripple.

References

Bose, B.K. (1992). *Modern Power Electronics: Evolution, Technology, and Applications*. IEEE Press, New York.

Chan, C.C. (1993). An overview of electric vehicle technology. *Proceedings of IEEE*, **81**, 1201–13.

Chan, C.C. and Chau, K.T. (1991). A new PWM algorithm for battery-source 3-phase inverters. *Electric Machines and Power Systems*, **19**, 43–54.

Chan, C.C. and Chau, K.T. (1996). An advanced permanent magnet motor drive system for battery-powered electric vehicles. *IEEE Transactions on Vehicular Technology*, **45**, 180–8.

Chan, C.C. and Chau, K.T. (1997). An overview of power electronics in electric vehicles. *IEEE Transactions on Industrial Electronics*, **44**, 3–13.

Chan, C.C., Chau, K.T., Chan, D.T.W. and Yao, J.M. (1997). Soft switching inverters in electric vehicle. *Proceedings of the 14th International Electric Vehicle Symposium*, CD-ROM.

Chan, C.C., Chau, K.T., Jiang, J.Z., Xia, W., Zhu, M. and Zhang, R. (1996). Novel permanent magnet motor drives for electric vehicles. *IEEE Transactions on Industrial Electronics*, **43**, 331–9.

Chan, C.C. (1998). Adjustable flux permanent magnet brushless dc motor. UK Patent No. GB 2284104.

Chan, C.C. (1998). A permanent magnet brushless dc motor. UK Patent No. GB 2289991.

Chan, C.C. (1999). Improvement in a permanent magnet hybrid brushless dc motor. UK Patent No. GB 2291274.

Chan, C.C., Chau, K.T. and Yao, J.M. (1997). Soft-switching vector control for resonant snubber based inverters. *Proceedings of IEEE International Conference on Industrial Electronics*, pp. 605–10.

Chan, C.C., Jiang, J.Z., Chen, G.H. and Chau, K.T. (1993). Computer simulation and analysis of a new polyphase multipole motor drive. *IEEE Transactions on Industrial Electronics*, **40**, 570–6.

Chan, C.C., Jiang, J.Z., Chen, G.H., Wang, X.Y. and Chau, K.T. (1994). A novel polyphase multipole square-wave permanent magnet motor drive for electric vehicles. *IEEE Transactions on Industry Applications*, **30**, 1258–66.

Chan, C.C., Jiang, J.Z., Xia, W. and Chau, K.T. (1995). Novel wide range speed control of permanent magnet brushless motor drives. *IEEE Transactions on Power Electronics*, **10**, 539–46.

Chan, C.C., Jiang, Q., Zhan, Y.J. and Chau, K.T. (1996). A high-performance switched reluctance drive for P-star EV project. *Proceedings of 13th International Electric Vehicle Symposium*, **II**, 78–83.

Chan, C.C., Jiang, Q. and Zhou, E. (1995). A new method of dimension optimization of switched reluctance motors. *Proceedings of Chinese International Conference on Electrical Machines*, pp. 1004–9.

Chan, C.C., Leung, W.S. and Ng, C.W. (1990). Adaptive decoupling control of induction-motor drives. *IEEE Transactions on Industrial Electronics*, **37**, 41–7.

Chan, C.C. and Wang, H.Q. (1996). New scheme of sliding-mode control for high performance induction motor drives. *IEE Proceedings—B*, **143**, 177–85.

Chan, C.C., Zhan, Y.J. and Chau, K.T. (1996). Fuzzy variable structure control of switched reluctance motor drive for EVs. *Proceedings of 13th International Electric Vehicle Symposium*, **II**, 573–8.

Chan, C.C., Zhang, R., Chau, K.T. and Jiang, J.Z. (1996). A novel brushless PM hybrid motor with a claw-type rotor topology for electric vehicles. *Proceedings of 13th International Electric Vehicle Symposium*, **II**, 579–84.

Chan, C.C., Zhang, R., Chau, K.T. and Jiang, J.Z. (1997). Optimal efficiency control of PM hybrid motor drives for electric vehicles. *Proceedings of IEEE Power Electronics Specialists Conference*, pp. 363–8.

Chau, K.T. (1991). Computer-aided design of a permanent magnet motor. *Electric Machines and Power Systems*, **19**, 501–11.

Chau, K.T., Chan, C.C. and Wong, Y.S. (1998). Advanced power electronic drives for electric vehicles. *Electromotion*, **5**, 42–53.

Chau, K.T., Cheng, M. and Chan, C.C. (1999). Performance analysis of 8/6-pole doubly salient permanent magnet motor. *Electric Machines and Power Systems*, **27**, 1055–67.

Chau, K.T., Ching, T.W., Chan, C.C. and Chan, D.T.W. (1997). A novel two-quadrant zero-voltage transition converter for DC motor drives. *Proceedings of IEEE International Conference on Industrial Electronics*, pp. 517–22.

Chau, K.T., Jiang, S.Z. and Chan, C.C. (2000). Reduction of current ripple and acoustic noise in dual-inverter pole-changing induction motor drive. *Proceedings of IEEE Power Electronics Specialists Conference*, pp. 67–72.

Chen, S. and Lipo, T.A. (1996). A novel soft switched PWM inverter for AC motor drives. *IEEE Transactions on Power Electronics*, **11**, 653–9.

Cheng, M., Chau, K.T., Chan, C.C. and Zhou, E. (2000). Performance analysis of split-winding doubly salient permanent magnet motor for wide speed operation. *Electric Machines and Power Systems*, **28**, 277–88.

Cheng, M., Chau, K.T., Chan, C.C., Zhou, E. and Huang, X. (2000). Nonlinear varying-network magnetic circuit analysis for doubly salient permanent magnet motors. *IEEE Transactions on Magnetics*, **26**, 339–48.

Ching, T.W., Chau, K.T. and Chan, C.C. (1998). A new zero-voltage-transition converter for switched reluctance motor drives. *Proceedings of IEEE Power Electronics Specialists Conference*, pp. 1295–301.

Cho, J.G., Kim, W.H., Rim, G.H. and Cho, K.Y. (1997). Novel zero transition PWM converter for switched reluctance motor drives. *Proceedings of IEEE Power Electronics Specialists Conference*, pp. 887–91.

Divan, D.M. (1986). The resonant DC link converter—a new concept in static power conversion. *Proceedings of IEEE Industry Application Society Annual Meeting*, pp. 648–56.

Dubey, G.K. (1989). *Power Semiconductor Controlled Drives*. Prentice-Hall, Englewood Cliffs, New Jersey.

Lai, J.S. (1997). Resonant snubber-based soft-switching inverters for electric propulsion drives. *IEEE Transactions on Industrial Electronics*, **44**, 71–80.

Li, Q., Zhou, X. and Lee, F.C. (1996). A novel ZVT three-phase rectifier/inverter with reduced auxiliary switch stresses and loss. *Proceedings of IEEE Power Electronics Specialists Conference*, **1**, 153–8.

Liao, Y., Liang, F. and Lipo, T.A. (1995). A novel permanent magnet motor with doubly salient structure. *IEEE Transactions on Industry Applications*, **31**, 1069–78.

Miller, T.J.E. (1989). *Brushless Permanent-Magnet and Reluctance Motor Drives*. Oxford University Press, Oxford, USA.

Miller, T.J.E. (1993). *Switched Reluctance Motors and Their Control*. Hillsboro, Oxford University Press, Magna Physics and Oxford, Ohio.

Mori, M., Kido, Y., Mizuno, T., Ashikaga, T., Matsuda, I. and Kobayashi, T. (1996). Development of an inverter-fed six-phase pole change induction motor for electric vehicles. *Proceedings of the 13th International Electric Vehicle Symposium*, **II**, 511–17.

Murai, Y., Cheng, J. and Yoshida, M. (1997). A soft-switched reluctance motor drives circuit with improved performances. *Proceedings of IEEE Power Electronics Specialists Conference*, pp. 881–6.

Rahman, M.A. and Qin, R. (1997). A permanent magnet hysteresis hybrid synchronous motor for electric vehicles. *IEEE Transactions on Industrial Electronics*, **44**, 46–53.

Rahman, M.A. and Slemon, G.R. (1985). Promising applications of neodymium boron iron magnets in electrical machines. *IEEE Transactions on Magnetics*, **21**, 1712–6.

Sato, Y., Fujita, K., Yanase, T. and Kinoshita, S. (1997). New control methods for high performance PM motor drive systems. *Proceedings of the 14th International Electric Vehicle Symposium*, CD-ROM.

Solectria AC Drive System. Solectria, 2000.

Unnewehr, L.E. and Nasar, S.A. (1982). *Electric Vehicle Technology*. John Wiley & Sons, New York.

West, J.G.W. and Jack, A. (2000). Designing motors and generators for automotive applications, *Intertech Seminar Notes*.

Yamada, K., Watanabe, K., Kodama, T., Matsuda, I. and Kobayashi, T. (1996). An efficiency maximizing induction motor drive system for transmissionless electric vehicle. *Proceedings of the 13th International Electric Vehicle Symposium*, **II**, 529–36.

Zhan, Y.J., Chan, C.C. and Chau, K.T. (1999). A novel sliding-mode observer for indirect position sensing of switched reluctance motor drives. *IEEE Transactions on Industrial Electronics*, **46**, 390–7.

6 Energy sources

The mission of all EV energy sources is to supply electrical energy for propulsion. Thus, the energy capacity EC of these sources is usually represented by using the unit in Wh, which is defined by:

$$\mathrm{EC} = \int_0^t v(t)i(t)dt,$$

where $v(t)$ is the instantaneous source voltage in V, $i(t)$ is the instantaneous discharging current in A, and t is the discharging period in h. If $v(t)$ is not considered, it will be termed as the coulometric capacity CC in Ah as defined by:

$$\mathrm{CC} = \int_0^t i(t)dt.$$

The theoretical energy capacity in Wh can well represent the energy content of most EV energy sources except the electrochemical batteries. The reason is simply because batteries cannot be discharged down to zero voltage; otherwise, they may be permanently damaged. So, as shown in Fig. 6.1, a cut-off voltage needs to be defined at the 'knee' of the discharging curve of a particular battery at which a battery is considered to be fully discharged, so-called 100% depth-of-discharge (DOD). Thus, the energy capacity and coulometric capacity of batteries available before reaching the cut-off voltage are termed as the usable energy capacity and usable coulometric capacity, respectively. It should be noted that the energy capacity is more important and useful than the coulometric capacity for EV

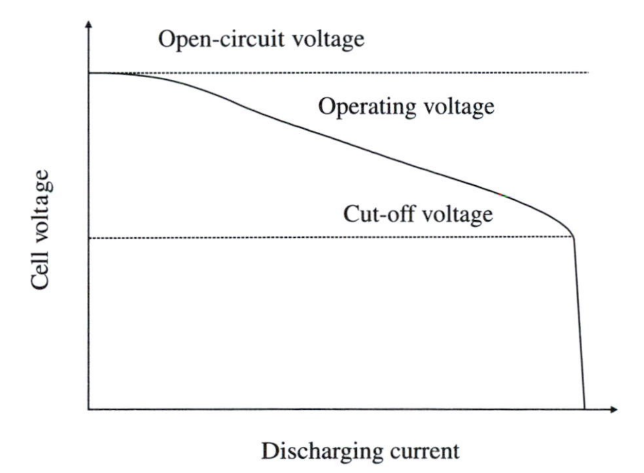

Fig. 6.1. Cut-off voltage of batteries.

energy sources. Nevertheless, the coulometric capacity has been widely employed to describe the capacity of batteries. Sometimes, the coulometric capacity is even loosely named as the 'capacity' for general battery applications.

For batteries, the usable coulometric capacity and hence usable energy capacity generally vary with their discharging current, operating temperature and ageing. The discharging/charging current is usually expressed by the discharging/charging rate:

$$I = kC_n,$$

where I is the discharging/charging current, n represents the C rate at which the battery coulometric capacity was rated, C is the rated coulometric capacity, and k is a multiple or fraction of C. For examples, the $C/5$ rate for a battery rated at 5 Ah denotes that the discharging current is of $kC_n = (1/5) \times 5 = 1$ A; a 10 Ah battery is discharged at 2 A can be denoted by the rate of $I/C_n = (2/10)C = 0.2\ C$ or $C/5$. Thus, the usable energy capacity or coulometric capacity of batteries should be quoted together with the value of C rate, and they generally decrease with increasing the C rate.

In contrast to the DOD, the parameter used to describe the residual coulometric capacity of batteries is termed as the state-of-charge (SOC). It is defined as the percentage ratio of residual coulometric capacity to usable coulometric capacity. Similar to the coulometric capacity, this SOC is affected by the discharging rate, operating temperature and ageing.

Energy densities of an energy source refer to the usable energy capacity per unit mass or volume. The gravimetric energy density is usually named as the specific energy in Wh/kg. The volumetric one is loosely named as the energy density in Wh/l. The former is more important than the latter for EV energy sources because the corresponding weight will greatly affect the precious driving range of EVs while the volume will only affect the usable space. So, the specific energy is a key parameter to assess the suitability of an EV energy source for the desired driving range. Since the usable energy capacity greatly varies with the C rate, the corresponding specific energy and energy density are usually quoted together with the C rate.

Power densities of an energy source denote the deliverable rate of energy per unit mass or volume, so-called the specific power (gravimetric power density) in W/kg and power density (volumetric power density) in W/l. Again, we normally pay more attention on specific power rather than power density for EV applications. Increasingly, the specific power is a key parameter to assess the suitability of an EV energy source for the desired acceleration and hill-climbing capability. For batteries, the specific power generally varies with the level of DOD. So, the corresponding specific energy is usually quoted together with the percentage of DOD.

Cycle life is a key parameter to describe the life of EV energy sources based on the principle of energy storage. It is usually defined as the number of deep-discharge cycles before failure. Since this cycle life is greatly affected by the DOD, it is usually quoted together with the percentage of DOD. For example, a battery can be claimed to offer 400 cycles at 100% DOD or 1000 cycles at 50% DOD. In case the EV energy source is based on energy generation rather than energy storage, the corresponding life is usually counted by the service lifetime in h or kh.

Energy efficiency of EV energy sources is usually defined as the output energy over the input energy. When it is an energy storage source, this energy efficiency is simply a ratio of the output electrical energy during discharging to the input electrical energy during charging. Typically, the energy efficiency of a battery is in the range of 55–75%. It should be noted that this energy efficiency is different from the charge efficiency which is actually defined as the ratio of discharged Ah to charged Ah. The typical charge efficiency of a battery is 65–90%. For EV applications, the energy efficiency is definitely more important than the charge efficiency.

Cost is a sensitive parameter for the EV energy sources because it usually make them unattractive as compared to those energy sources for ICEVs. It consists of two main components, namely the initial (manufacturing) cost and running (maintenance) costs. The former is generally dominant the latter. The corresponding unit is usually in $/kWh. At present, the initial cost fluctuates with different batteries, probably in the range of US$ 120–1200/kWh.

There is no doubt that energy sources are the most limiting element in the development of EVs. The general requirements for EV energy sources are high specific energy and high specific power to pursue better driving range and performance, long cycle life to enable comparable with vehicle life, high efficiency and cost effectiveness to achieve economical vehicle operation, and maintenance free to relieve vehicle services.

There are numerous types of energy sources that have been proposed for EVs (Hoolboom and Szabados, 1994; Robinson, 1995). Those being actively considered to be vital for EVs are categorised as rechargeable electrochemical batteries (loosely called batteries), fuel cells, ultrahigh-capacitance capacitors (loosely called ultracapacitors), and ultrahigh-speed flywheels. Among them, the batteries, capacitors and flywheels are energy storage systems in which electrical energy is stored during charging, whereas the fuel cells are energy generation systems in which electricity is generated by chemical reaction. At present and in the near future, the batteries have been identified to be the major EV energy source because of their technological maturity and reasonable cost. Recently, both the fuel cells and capacitors have received substantial attention, and have shown promising application to EVs in mid term. In long term, the use of flywheels as the energy source for EVs may be an alternative solution (Chan *et al.*, 1999; Chau *et al.*, 1999).

6.1 Batteries

The basic element of each battery is the electrochemical cell. A connection of a number of cells in series forms a battery. Figure 6.2 shows the basic principle of the electrochemical cell or batteries in which both the positive electrode (P) and

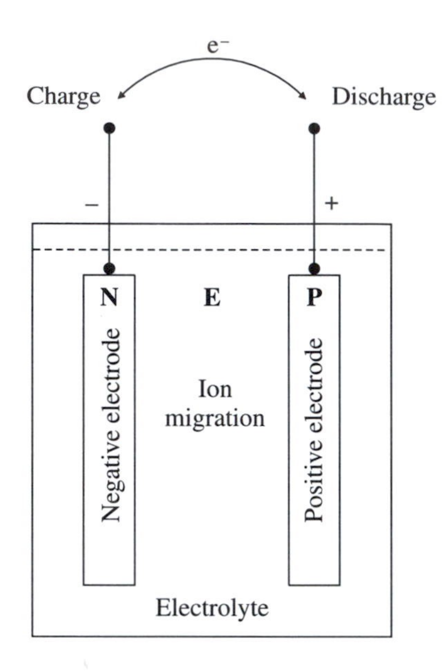

Fig. 6.2. Basic principle of batteries.

negative electrode (N) are immersed in the electrolyte (E). During discharge, the negative electrode performs oxidation reaction which drives electrons to the external circuit, while the positive electrode carries out reduction reaction which accepts electrons from the external circuit. During charge, the process is reversed so that electrons are injected into the negative electrode to perform reduction while the positive electrode releases electrons to carry out oxidation.

At the present time and in the foreseeable future, batteries have been agreed to be the major energy source for EVs. The US Advanced Battery Consortium (USABC), which is the organization formed in the United States by the Department of Energy, the Electric Power Research Institute, Ford, General Motors, Chrysler and the battery manufacturers to fund research on advanced battery technology, has set the mid-term and long-term performance goals of EV batteries. As summarized in Table 6.1, these performance goals for EV batteries are very demanding ones and it is obvious that no existing battery technology is capable of meeting all these criteria because the USABC aims to make an EV as close in performance to an ICEV as possible.

Those viable EV batteries consist of the valve-regulated lead–acid (VRLA), nickel–cadmium (Ni–Cd), nickel–zinc (Ni–Zn), nickel–metal hydride (Ni–MH), zinc/air (Zn/Air), aluminium/air (Al/Air), sodium/sulphur (Na/S), sodium/nickel chloride ($Na/NiCl_2$), lithium–polymer (Li–Polymer) and lithium–ion (Li–Ion) types. As shown in Fig. 6.3, these batteries are classified into lead–acid, nickel-based, zinc/halogen, metal/air, sodium-β, and ambient-temperature lithium categories (Linden, 1984; 1995).

Table 6.1 Performance goals of USABC

Performance goals	Mid-term	Long-term
Primary		
Specific energy ($C/3$ discharge rate) (Wh/kg)	80 (100 desired)	200
Energy density ($C/3$ discharge rate) (Wh/l)	135	300
Specific power (80% DOD/30s) (W/kg)	150 (200 desired)	400
Power density (W/l)	250	600
Life (years)	5	10
Cycle life (80% DOD) (cycles)	600	1000
Ultimate price (US$/kWh)	<150	<100
Operating temperature (°C)	−30–65	−40–85
Recharge time (h)	<6	3 to 6
Fast recharge time (40–80% SOC) (h)	0.25	
Secondary		
Efficiency ($C/3$ discharge, 6 h charge) (%)	75	80
Self-discharge (%)	<15 (48 h)	<15 (month)
Maintenance	no maintenance	
Abuse resistance	tolerance	
Thermal loss (for high-temperature batteries)	3.2 W/kWh	

DOD: Depth-of-discharge
SOC: State-of-charge

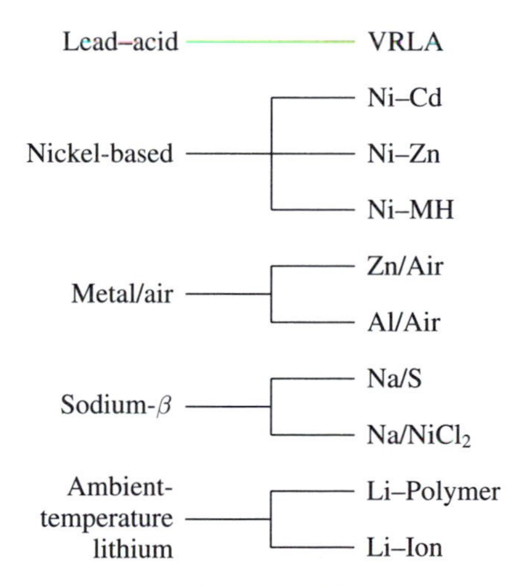

Fig. 6.3. Classification of EV batteries.

6.1.1 LEAD–ACID BATTERY

The lead–acid (Pb–Acid) battery was invented in 1860 (Berndt, 1997). It has been a successful commercial product for over a century. The wide use of the Pb–Acid battery is mainly due to its mature technology and low price. Nevertheless, new designs and fabrication processes are still being introduced at significant rates to meet the USABC's performance criteria of EV batteries.

The Pb–Acid battery has the nominal cell voltage of 2 V, specific energy of 35 Wh/kg, energy density of 70 Wh/l, and specific density of 200 W/kg. It uses metallic lead as the negative electrode and lead dioxide as the positive electrode. The electrolyte is a sulphuric acid solution. The overall electrochemical reactions are:

$$Pb + PbO_2 + 2H_2SO_4 \leftrightarrow 2PbSO_4 + 2H_2O.$$

On discharge, both lead and lead dioxide are converted into lead sulphate. On charge, the reactions are reversed. Notice that the electrolyte, sulphuric acid, participates in the electrochemical reactions and its concentration changes with the state-of-charge (SOC). In fact, the open-circuit voltage or equilibrium voltage of the Pb–Acid battery cell depends only on the acid concentration and is independent of the amount of lead, lead dioxide or lead sulphate present in the cell so long as all three substances are available. This cell voltage E_0 is generally governed by the Nernst equation $E_0 = (0.84 + \rho)$ V, where ρ is the acid density in kg/l. It should be noted that the dependence of the open-circuit voltage on the acid concentration causes the disadvantage that the discharge voltage does not remain constant even at low discharge rates. Also, the open-circuit voltage is affected by the temperature.

The nominal voltage of the Pb–Acid cell is 2 V. On discharge, the cutoff voltage at moderate rates is 1.75 V and may be as low as 1.0 V at extremely high rates at low temperatures. On charge, the charging current should be controlled to maintain the cell voltage lower than the gassing voltage (about 2.4 V). Otherwise, the overcharge reactions begin, resulting in the production of hydrogen and oxygen gases with the loss of water:

$$2H_2O \rightarrow 2H_2 + O_2.$$

In the sealed type Pb–Acid battery, a special porous separator is so employed in the cell that the evolved oxygen is transferred from the negative electrode to the positive electrode and then re-combines with hydrogen to form water. Thus, it provides a definite advantage of maintenance-free operation. Moreover, the immobilization of the gelled electrolyte or absorbed electrolyte with absorptive glass mat separators can allow the battery to operate in different orientations without spillage. A well-accepted sealed type Pb–Acid battery is so-called the VRLA* battery.

As the Pb–Acid battery has maintained its prime position for more than a century, there are a number of advantages contributing to this outstanding position: proven technology and mature manufacturing, low cost (lowest of all secondary batteries), high cell voltage (highest of all aqueous electrolyte batteries), good high-rate performance that suitable for EV applications, good low- and high-temperature performances, high energy efficiency (75–80%), and available in a variety of sizes and designs. On the other hand, the Pb–Acid battery still

* VALVE REGULATED LEAD ACID

suffers from a number of disadvantages and needs continual development. For examples, its specific energy and energy density are relatively low (typically 35 Wh/kg and 70 Wh/l), its self-discharge rate is relatively high (about 1% per day at 25 °C), its cycle life is relatively short (about 500 cycles), and it is unsuitable for long-term storage because of electrode corrosion by sulphation.

Different Pb–Acid batteries with improved performance are being developed for EVs. Improvements of the VRLA battery in specific energy over 40 Wh/kg and energy density over 80 Wh/l with the possibility of rapid recharge have been attained. One of these advanced VRLA batteries is the Electrosource's Horizon battery. It adopts the lead wire woven horizontal plate and hence offers competitive advantages of high specific energy and energy density (43 Wh/kg and 84 Wh/l), high specific power (285 W/kg), long cycle life (over 600 cycles for on-road EV applications), rapid recharge capability (50% capacity in 8 min and 100% in less than 30 min), low cost (US$2000–3000 a vehicle), mechanical ruggedness (robust structure of horizontal plate), maintenance-free (sealed battery technology), and environmental friendliness (green manufacturing processes and 98% recyclable). Other advanced Pb–Acid battery technology includes bipolar designs and micro tubular grid designs.

At present, the Pb–Acid battery is still the most attractive energy source for EVs. Apart from the Electrosource, there are many other suppliers of the advanced VRLA battery such as the GS, Hawker, Johnson Controls, Panasonic, Sonnenschein, Trojan and YUASA. Actually, since the last two decades, the Pb–Acid battery has been widely adopted by many EVs including the Bedford CF, Chrysler Voyager, Citroën C15, Daihatsu Hijet, Ford Ranger, GM EV1, Mazda Bongo Friendee, Mercedes-Benz MB308, Peugeot J5, Suzuki Alto, US Electricar and Volkswagen Citystromer.

6.1.2 NICKEL-BASED BATTERIES

There are many kinds of electrochemical batteries using nickel oxyhydroxide as the active material for the positive electrode, including the Ni–Cd, Ni–Zn, and Ni–MH (Berndt, 1997). Among them, the Ni–Cd battery has been well accepted for EV applications because of its proven technology and good performance. The Ni–MH battery is a fast-growing star which has shown high potentiality to supersede the Ni–Cd battery in the near-term EV market. On the other hand, the Ni–Zn battery is still under continual development which exhibits potentiality to compete with the others in mid-term EV applications.

Ni–Cd battery

For more than 80 years, the Ni–Cd battery has been successfully utilized in heavy-duty industrial applications. Due to the resurgence of interest in EVs in the late 1970s and early 1980s, it led to further development of the Ni–Cd battery for EV applications (Morrow, 1995).

The Ni–Cd battery possesses the nominal parameters of 1.2 V, 56 Wh/kg, 110 Wh/l and 225 W/kg. Its active materials are metallic cadmium for the negative electrode and nickel oxyhydroxide for the positive electrode. The alkaline electrolyte is an aqueous potassium hydroxide solution. The electrochemical reactions of discharge and charge are:

$$Cd + 2NiOOH + 2H_2O \leftrightarrow Cd(OH)_2 + 2Ni(OH)_2.$$

On discharge, metallic cadmium is oxidized to form cadmium hydroxide and nickel oxyhydroxide is reduced to nickel hydroxide under consumption of water. On charge, the reverse reactions occur. In contrast to the sulphuric acid electrolyte used in the Pb–Acid battery, the potassium hydroxide electrolyte in the Ni–Cd battery is not significantly changed in density or composition during discharge and charge.

The Ni–Cd battery technology has gained enormous technical importance because of the advantages of high specific power (over 220 W/kg), long cycle life (up to 2000 cycles), highly tolerant of electrical and mechanical abuse, flat voltage profile over a wide range of discharge currents, rapid recharge capability (about 40–80% in 18 min), wide operating temperature range (−40–85 °C), low self-discharge rate (<0.5% per day), excellent long-term storage due to negligible corrosion, and available in a variety of sizes and designs. Of course, the Ni–Cd have some disadvantages which offset its wide acceptance for EV applications. They are high of initial cost (2–4 times Pb–Acid), relatively low cell voltage (nominally 1.2 V compared to 2 V for Pb–Acid), and the carcinogenicity and environmental hazard of cadmium.

The Ni–Cd battery can generally be divided into two major categories, namely the vented and sealed types. The vented type consists of many alternatives. The vented sintered-plate type is a more recent development of the Ni–Cd battery, having a higher specific energy but being more expensive. It is characterized by its flat discharge profile and superior high-rate and low-temperature performance. Similar to the sealed Pb–Acid battery, the sealed Ni–Cd battery incorporates a specific cell design feature to prevent a build-up of pressure in the cell caused by gassing during overcharge. As a result, the battery can be sealed and requires no maintenance other than recharging, but may have a memory effect.

The major manufacturers of the Ni–Cd battery for EV applications are SAFT and VARTA. Recent EVs powered by the Ni–Cd battery have included the Chrysler TE Van, Citroën AX, Mazda Roadster, Mitsubishi EV, Peugeot 106, Renault Clio, and HKU U2001. Presently, the Ni–MH battery is a strong competitor of the Ni–Cd battery.

Ni–Zn battery

Starting from the 1930s, the Ni–Zn battery technology has been investigated. Despite this history, the Ni–Zn battery has not achieved commercial importance because of the limited life of the zinc electrode. On the basis of advances being made in the improvement of its cycle life, it promises to have commercial significance because of its high specific energy and low material cost (Anan and Adachi, 1994).

The Ni–Zn battery nominally operates at 1.6 V and delivers 60 Wh/kg, 120 Wh/l and 300 W/kg. It uses zinc as the negative electrode and nickel oxyhydroxide as the positive electrode. The electrolyte is an alkaline potassium hydroxide solution. The discharge and charge reactions are:

$$Zn + 2NiOOH + 2H_2O \leftrightarrow Zn(OH)_2 + 2Ni(OH)_2.$$

On discharge, metallic zinc in the negative electrode is oxidized to form zinc hydroxide and nickel oxyhydroxide in the positive electrode is reduced to nickel hydroxide. On charge, the reactions are reversed.

Compared with conventional and other nickel-based batteries, the Ni–Zn battery has the advantages of higher specific energy and specific power than the Ni–Cd (60 Wh/kg and 300 W/kg), high cell voltage (highest of the nickel-based family), lower projected cost than the Ni–Cd (100–300 US$/kWh), non-toxicity (more environmental friendliness than the Ni–Cd), tolerance of overcharge and overdischarge, capable of high discharge and recharge rates, and wide operating temperature range (−20–60 °C). However, the major and serious drawback of the Ni–Zn battery is its short cycle life (about 300 cycles). It is mainly due to the partial solubility of zinc species in the electrolyte that the dissolution and redeposition of the zinc do not occur at the same location and with the same morphology. This 'shape change' of the zinc electrode results in capacity decay. Developments of improved electrodes, separators and electrolytes have renewed interest in this technology, and effort is in progress to increase the cycle life.

At present, significant interest in the Ni–Zn battery is directed at EV applications. YUASA and Kyushu Electric Power have jointly developed a Ni–Zn battery module for EVs. Hence, a battery system consisting of three modules in series has been installed in a 94-kg electric scooter which can deliver specific energy of 65 Wh/kg and specific power of 150 W/kg for an urban driving range of 20 km.

Ni–MH battery

The Ni–MH battery has been on the market since 1992. Its characteristics are similar to those of the Ni–Cd battery. The principal difference between them is the use of hydrogen, absorbed in a metal hydride, for the active negative electrode material in place of the cadmium. Because of superior specific energy to the Ni–Cd and free from toxicity or carcinogenicity such as cadmium, the Ni–MH battery is superseding the Ni–Cd battery.

At present, the Ni–MH battery technology has the nominal voltage of 1.2 V and attains the specific energy of 65 Wh/kg, energy density of 150 Wh/l and specific power of 200 W/kg. Its active materials are hydrogen in the form of a metal hydride for the negative electrode and nickel oxyhydroxide for the positive electrode. The metal hydride is capable of undergoing a reversible hydrogen desorbing–absorbing reaction as the battery is discharged and recharged. An aqueous solution of potassium hydroxide is the major component of the electrolyte. The overall electrochemical reactions are:

$$MH + NiOOH \leftrightarrow M + Ni(OH)_2.$$

When the battery is discharged, metal hydride in the negative electrode is oxidized to form metal alloy and nickel oxyhydroxide in the positive electrode is reduced to nickel hydroxide. During charge, the reverse reactions occur.

A key component of the Ni–MH battery is the hydrogen storage metal alloy which should be formulated to obtain a material that is stable over a large number of cycles. There are two major types of these metal alloys being used for the Ni–MH battery. These are the rare-earth alloys based around lanthanum nickel, known as the AB_5, and alloys consisting of titanium and zirconium, known as the AB_2. The AB_2 alloys typically have a higher capacity than the AB_5 alloys. However, the trend is to use the AB_5 alloys because of better charge retention and stability characteristics.

Since the Ni–MH battery is still under continual development, its advantages based on present technology are summarized as: highest specific energy and energy density of nickel-based batteries (65 Wh/kg and 150 Wh/l), environmental friendliness (cadmium free), flat discharge profile (similar to Ni–Cd), and rapid recharge capability (similar to Ni–Cd). However, it still suffers from the problem of high initial cost. Also, it may have a memory effect and be exothermic on charge.

The Ni–MH battery has been considered as an important near-term choice for EV applications. A number of battery manufacturers, such as GM Ovonic, GP, GS, Panasonic, SAFT, VARTA and YUASA, have actively engaged in the development of this battery technology specially for powering EVs. For example, a GP's Ni–MH battery can offer 70 Wh/kg, 170 Wh/l and 180 W/kg. Since 1993, Ovonic Battery has installed its Ni–MH battery in the Solectria GT Force EV for testing and demonstration. A 19-kWh battery has delivered over 65 Wh/kg, 175 Wh/l and 150 W/kg to enable the EV achieving the maximum speed of 134 km/h, acceleration from zero to 80 km/h in 14 s and a city driving range of 206 km.

6.1.3 METAL/AIR BATTERIES

The electrochemical coupling of a reactive negative electrode to an air electrode provides the metal/air battery with an inexhaustible positive electrode reactant. As the battery utilizes ambient air and the most common metal, it is noted to have potentially very high specific energy and energy density. Due to this performance potential, a significant amount of effort has gone into the metal/air battery development.

Both electrically rechargeable and mechanically rechargeable metal/air battery configurations have been developed. Conventional electrical recharge requires either a third electrode to sustain oxygen evolution or a bifunctional electrode to allow for both oxygen reduction and evolution. In the mechanically rechargeable design, the discharged metal electrode is physically removed and replaced with a fresh one and can use a relatively simple unifunctional air electrode which needs to operate only in the discharge mode.

The general advantages of rechargeable metal/air batteries, including the electrically or mechanically rechargeable zinc/air (Zn/Air), and mechanically rechargeable aluminium/air (Al/Air), are very high specific energy and energy density (as high as 600 Wh/kg and 400 Wh/l for Al/Air), low cost (only common metal and ambient air), environmental friendliness, flat discharge voltage, and capacity independent of load and temperatures. In addition, those mechanically rechargeable batteries have two distinct advantages which are very essential for EV applications: fast and convenient refuelling (comparable to petrol refuelling with a few minutes), and centralized recharging/recycling (most efficient and environmentally sound use of electricity). On the other hand, there are some general disadvantages associated with rechargeable metal/air batteries. They are with low specific power (at most 105 W/kg for Zn/Air), relatively limited temperature range, carbonation of alkali electrolyte due to carbon dioxide in air, and evolution of hydrogen gas from corrosion in electrolyte.

Zn/Air battery

The Zn/Air battery has been developed as an electrically rechargeable battery and as a mechanically rechargeable battery. Although both of them have been applied to EV applications, the trend is to use the mechanically rechargeable one.

The electrically rechargeable Zn/Air battery nominally operates at 1.2 V and delivers the specific energy of 180 Wh/kg, energy density of 160 Wh/l and specific power of 95 W/kg. The negative electrode consists of zinc particles and the positive electrode is a bifunctional air electrode. The electrolyte is potassium hydroxide. The simplified electrochemical reactions can be described as:

$$2Zn + O_2 \leftrightarrow 2ZnO.$$

On discharge, zinc is firstly oxidized to potassium zincate dissolved in the electrolyte and then to a precipitation of zinc oxide. On charge, the reactions are reversed. However, on charge–discharge cycling, the zinc electrode generally suffers from the shape change problems.

The mechanically rechargeable Zn/Air battery can avoid the need for a bidirectional air electrode and the shape change problems of the zinc electrode resulting from charge–discharge cycling. Hence, it can offer higher specific energy and specific power, namely 230 Wh/kg and 105 W/kg. The mechanically refuelling system being considered for EV applications is to remove and replace the depleted zinc negative electrode cassettes robotically at a fleet servicing location or at a public service station. The discharged fuel is then electrochemically recharged at central facilities that serve regional distribution networks. The recharging process includes four steps. Firstly, the discharged cassettes are mechanically taken apart and the zinc oxide discharge product is removed. Secondly, zinc oxide is dissolved in a potassium hydroxide solution to form a zincate solution. Thirdly, the zincate solution is electrolysed in an electrowinning bath. Finally, the electro-won zinc is compacted onto the negative electrode cassettes.

A mechanically rechargeable Zn/Air battery has been developed by Electric Fuel for various electric passenger cars and commercial vehicles. A 160-kWh Zn/Air battery was installed and tested in a Mercedes-Benz 180E van in 1994. The corresponding range at a constant speed of 64 km/h was found to be 689 km. Moreover, Deutsche Post AG, the German Postal Service, has sponsored an US$18 million extensive field test of 20 Mercedes-Benz MB410 vans and 44 Opel Corsa Combo light pick-up trucks powered by the Electric Fuel's mechanically rechargeable Zn/Air battery. This MB410 van, equipped with a 150-kWh battery, has exhibited the maximum speed of 120 km/h and the range of 480 km while carrying a full cargo of 1700 kg.

Al/Air battery

Recent development on the metal/air battery involving the more active metal has focused on aluminium because of its geological abundance (the third most abundant element in the earth's crust), potentially low cost and relative ease of handling. Since the Al/Air battery has a very high charging potential to be electrically recharged in an aqueous system in which water is preferentially electrolysed, the development has been directed to the mechanically rechargeable system.

The Al/Air battery has the nominal voltage of 1.4 V. The negative electrode is of aluminium metal and the positive electrode is only a simple unifunctional air electrode for discharge mode of operation. The electrolyte can be either a saline solution or an alkaline potassium hydroxide solution. The discharge reaction of this mechanically rechargeable battery is:

$$4Al + 3O_2 + 6H_2O \rightarrow 4Al(OH)_3.$$

The Al/Air battery with a saline electrolyte is attractive only for low power applications. On the other hand, the alkaline Al/Air battery can offer high specific energy and energy density of 250 Wh/kg and 200 Wh/l and is suitable for high power applications. Nevertheless, the corresponding specific power is as low as 7 W/kg.

Because of its exceptionally low specific power, the Al/Air battery is seldom used as the sole energy source for EVs. To take the advantage of very high specific energy and energy density, it is commonly used in conjunction with another electrically rechargeable battery as an EV range extender.

6.1.4 SODIUM-β BATTERIES

The sodium-β battery technology include designs based on the Na/S and $Na/NiCl_2$ chemistries because of their two common features, namely liquid sodium as one reactant and β-alumina ceramic as the electrolyte. The Na/S battery has been under active development for over 25 years. Some 10 years after the first presentation of the Na/S, the $Na/NiCl_2$ battery was introduced which offers potentially easier solutions to some of the problems that have confronted the development of the Na/S battery.

Na/S battery

The Na/S battery operates at 300–350 °C with the nominal cell voltage of 2 V, specific energy of 170 Wh/kg, energy density of 250 Wh/l and specific power of 390 W/kg. Based on the battery configuration, the performance characteristics are reduced to 100 Wh/kg, 150 Wh/l and 200 W/kg. The active materials are molten sodium for the negative electrode and molten sulphur/sodium polysulphides for the positive electrode. The β-alumina ceramic electrolyte functions as a sodium ion-conducting solid medium and an excellent separator for the molten electrodes to prevent any direct self-discharge. The electrochemical reactions of the Na/S battery are:

$$2\mathrm{Na} + x\mathrm{S} \leftrightarrow \mathrm{Na_2S}_x \quad (x = 5\text{–}2.7).$$

On discharge, sodium is oxidized to form sodium ions which migrate through the electrolyte and combine with the sulphur that is being reduced in the positive electrode to form sodium pentasulphide. Then the sodium pentasulphide is progressively converted into polysulphides with higher sulphur compositions (Na_2S_x) where x is from 5 to 2.7. On charge, these reactions are reversed.

Continual interests in the development of the Na/S battery for over 25 years are motivated by its attractive features of high specific energy and energy density (100 Wh/kg and 150 Wh/l), high specific power (200 W/kg), high energy efficiency (over 80%), flexible operation (functional over wide range of operating conditions), and insensitivity to ambient conditions (sealed high-temperature operation). However, there are some limitations and several important improvements are still required before commercializing the Na/S battery. They are the safety problem (high reactivity and corrosiveness of molten active materials), inadequate freeze–thaw durability (weak ceramic electrolyte subjected to mechanical stress), and need of thermal management (additional energy and thermal insulation).

$Na/NiCl_2$ battery

In the $Na/NiCl_2$ battery, the active materials are molten sodium for the negative electrode and solid nickel chloride for the positive electrode. In addition to the β-alumina ceramic electrolyte as used in the Na/S, there is a secondary electrolyte (sodium-aluminium chloride) in the positive electrode chamber. The secondary electrolyte functions to conduct sodium ions from the primary β-alumina electrolyte to the solid nickel chloride positive electrode. The corresponding electrochemical reactions are:

$$2Na + NiCl_2 \leftrightarrow Ni + 2NaCl.$$

On discharge, the solid nickel chloride is converted to the nickel metal and sodium chloride crystal. On charge, these reactions are reversed. The Na/$NiCl_2$ battery operates at 250–350 °C with the nominal cell voltage of 2.5 V. Based on the battery configuration, the performance characteristics are the specific energy of 86 Wh/kg, energy density of 149 Wh/l and specific power of 150 W/kg. Comparing to the Na/S, it takes the advantages of higher open-circuit cell voltage (2.5 V for the Na/$NiCl_2$ vs. 2 V for the Na/S), wider operating temperature (250–350 °C for the Na/$NiCl_2$ whereas 300–350 °C for the Na/S), safer products of reaction (less corrosive than molten Na_2S_x), more reliable failure mode (sodium reacts with the secondary electrolyte to short-circuit the cell on electrolyte failure), and better freeze–thaw durability (smaller temperature difference). However, the Na/$NiCl_2$ battery still has the disadvantages of potentially higher cost (relatively high cost of nickel), and lower specific power (150 W/kg for the Na/$NiCl_2$ and 200 W/kg for the Na/S).

AEG ZEBRA has been the major developer of the Na/$NiCl_2$ battery. A ZEBRA battery (model Z11) can offer the specific energy of 86 Wh/kg, energy density of 149 Wh/l and specific power of 150 W/kg. The ZEBRA battery has been installed in the Mercedes-Benz 190E (a total weight of 1575 kg) for evaluation, which has achieved the maximum speed of 115 km/h, acceleration from 0 to 50 km/h in 13 s and the range in urban traffic of up to 100 km.

6.1.5 AMBIENT-TEMPERATURE LITHIUM BATTERIES

There are a number of different approaches being taken in the design of rechargeable ambient-temperature lithium batteries. The major approach is to use metallic lithium for the negative electrode and a solid inorganic intercalation material for the positive electrode. The electrolyte can be a solid polymer, leading to name as the lithium–polymer (Li–Polymer) battery. Another approach that is becoming more popular is the use of a lithiated carbon material in place of metallic lithium. As lithium ions move back and forth between the positive and negative electrodes during discharge and charge, it leads to name as the lithium–ion (Li–Ion) battery (Broussely *et al.*, 1995).

Li–Polymer battery

The Li–Polymer battery uses lithium metal and a transition metal intercalation oxide (M_yO_z) for the negative and positive electrodes, respectively. This M_yO_z possesses a layered structure into which lithium ions can be inserted or from where they can be removed on discharge and charge, respectively. A thin solid polymer electrolyte (SPE) is used, which offers the merits of improved safety and flexibility in design. The general electrochemical reactions are:

$$x\text{Li} + M_yO_z \leftrightarrow \text{Li}_xM_yO_z.$$

On discharge, lithium ions formed at the negative electrode migrate through the SPE and are inserted into the crystal structure at the positive electrode. On charge, the process is reversed. By using a lithium foil negative electrode and vanadium oxide (V_6O_{13}) positive electrode, the $Li/SPE/V_6O_{13}$ cell is the most attractive one within the family of Li–Polymer. It operates at the nominal voltage of 3 V and has the specific energy of 155 Wh/kg, energy density of 220 Wh/l and specific power of 315 W/kg. The corresponding advantages are high cell voltage (3 V), very high specific energy and energy density (155 Wh/kg and 220 Wh/l), very low self-discharge rate (about 0.5% per month), capability of fabrication in a variety of shapes and sizes, and safer design (reduced activity of lithium with solid electrolyte). However, it has a drawback of relatively weak low-temperature performance due to its temperature dependence of ionic conductivity.

Starting from 1993, 3M and Hydro-Québec have jointly developed this Li–Polymer battery technology for EVs. They have produced a 20-V module, offering 155 Wh/kg, 220 Wh/l and 315 W/kg. The corresponding cycle life is 600 cycles, targeted at 1000 cycles.

Li–Ion battery

Since the first announcement of the Li–Ion battery in 1991, the Li–Ion battery technology has seen an unprecedented rise to what is now considered to be the most promising rechargeable battery of the future. Although still in the stage of development, the Li–Ion battery has already gained acceptance for EV applications.

The Li–Ion battery uses a lithiated carbon intercalation material (Li_xC) for the negative electrode instead of metallic lithium, a lithiated transition metal intercalation oxide ($Li_{1-x}M_yO_z$) for the positive electrode and a liquid organic solution or a solid polymer for the electrolyte. Lithium ions are swinging through the electrolyte between the positive and negative electrodes during discharge and charge. The general electrochemical reactions are described as:

$$Li_xC + Li_{1-x}M_yO_z \leftrightarrow C + LiM_yO_z.$$

On discharge, lithium ions are released from the negative electrode, migrate via the electrolyte and are taken up by the positive electrode. On charge, the process is reversed. Possible positive electrode materials include $Li_{1-x}CoO_2$, $Li_{1-x}NiO_2$ and $Li_{1-x}Mn_2O_4$, which have the advantages of stability in air, high voltage and reversibility for the lithium intercalation reaction.

The $Li_xC/Li_{1-x}NiO_2$ type, loosely written as $C/LiNiO_2$ or simply called as the nickel-based Li–Ion battery, has the nominal cell voltage of 4 V, specific energy of 120 Wh/kg, energy density of 200 Wh/l and specific power of 260 W/kg. The cobalt-based type has higher specific energy and energy density, but with a higher cost and a significant increase of the self-discharge rate. The manganese-based type has the lowest cost and its specific energy and energy density lie between those of the cobalt-based and nickel-based types. It is anticipated that the development of the Li–Ion battery will ultimately move to the manganese-based type because of the low cost, abundance and environmental friendliness of the manganese-based material. The general advantages of the Li–Ion battery are highest cell voltage (as high as 4 V), high specific energy and energy density (90–130 Wh/kg and 140–200 Wh/l), safest design of lithium batteries (absence of metallic lithium), and long cycle life (about 1000 cycles). However, it still suffers from a drawback of relatively high self-discharge rate (as high as 10% per month).

Many battery manufacturers, such as GS, Hitachi, Panasonic, SAFT, SONY and VARTA, have actively engaged in the development of the Li–Ion battery. Starting from 1993, SAFT has focused on the nickel-based Li–Ion battery. A 10.5-V $C/LiNiO_2$ module can offer the specific energy of 126 Wh/kg and specific power of 262 W/kg. On the other hand, GS has focused on the manganese-based type and its 14.4-V $C/LiMn_2O_4$ module can offer 90 Wh/kg and 450 W/kg. The first application of the Li–Ion battery to EVs should be the Nissan FEV in 1995. Then, it has also been applied to the Nissan Prairie Joy and Altra. Based on SONY Li–Ion modules having 90 Wh/kg and 300 W/kg, the Altra EV has achieved a top speed of 120 km/h and a range of 192 km upon city driving.

6.1.6 EVALUATION OF BATTERIES

Detailed chemistries of the aforementioned batteries can be found in relevant battery handbooks (Linden, 1995). Some of their important parameters, including specific energy, energy density, specific power, cycle life and projected cost with respect to the USABC's goals are shown in Table 6.2. It should be noted that these

Table 6.2 Key parameters of EV batteries

	Specific energy[a] (Wh/kg)	Energy density[a] (Wh/l)	Specific power[b] (W/kg)	Cycle life[b] (Cycles)	Projected cost[d] (US$/kWh)
VRLA	30–45	60–90	200–300	400–600	150
Ni–Cd	40–60	80–110	150–350	600–1200	300
Ni–Zn	60–65	120–130	150–300	300	100–300
Ni–MH	60–70	130–170	150–300	600–1200	200–350
Zn/Air	230	269	105	NA[c]	90–120
Al/Air	190–250	190–200	7–16	NA[c]	NA
Na/S	100	150	200	800	250–450
$Na/NiCl_2$	86	149	150	1000	230–350
Li–Polymer	155	220	315	600	NA
Li–Ion	90–130	140–200	250–450	800–1200	>200
USABC	200	300	400	1000	<100

NA: Not available
[a]At *C*/3 rate
[b]At 80% DOD
[c]Mechanical recharging
[d]For reference only

parameters are only for indicative purposes since the data may have wide variations among different battery manufacturers. Even for the same manufacturer, different models of the same battery may also have significant variations because of different trade-offs among the specific energy, specific power and cycle life. Moreover, these data always change with the advancement of battery technology. From the table, it can be found that none of them can fully satisfy the USABC's long-term goals which aim to enable EVs directly competing with ICEVs.

Nevertheless, in order to meet the California mandate of 10% zero-emission vehicles by 2003, the development of EV batteries has to be continued and accelerated. Table 6.3 summarizes the key features, including advantages, disadvantages and potentiality, of the aforementioned batteries. It can be found that those batteries with near-term high potentiality are the VRLA, Ni–Cd and Ni–MH. Since the features of the Ni–MH are superior to those of the Ni–Cd except maturity, the Ni–Cd is being superseded by the Ni–MH. Actually, some manufacturers used to produce the Ni–Cd for EV applications have redirected their efforts to the Ni–MH. Thus, in near term, the VRLA is still popular due to its maturity and cost-effectiveness, whereas the Ni–MH is attractive because of its good performances. On the other hand, those batteries with mid-term high potentiality include the Ni–Zn, Zn/Air, Na/$NiCl_2$, Li–Polymer and Li–Ion. The Li–Ion

Table 6.3 Key features of EV batteries

	Key advantages/disadvantages for EV applications	Potentiality
VRLA	mature, low cost, fast rechargeable, high specific power/low specific energy	near-term very high
Ni–Cd	mature, fast rechargeable, high specific power/high cost, low specific energy	near-term high
Ni–Zn	high specific energy, high specific power, low cost/short cycle life	mid-term high
Ni–MH	high specific energy, high specific power, fast rechargeable/high cost	near-term very high
Zn/Air	mechanically rechargeable, low cost, very high specific energy/low specific power, cannot accept regenerative energy	mid-term very high
Al/Air	mechanically rechargeable, low cost, very high specific energy/very low specific power, cannot accept regenerative energy	near-term low
Na/S	high specific energy, high specific power/high cost, safety concerns, need of thermal management	mid-term moderate
Na/$NiCl_2$	high specific energy/high cost, need of thermal management	mid-term high
Li–Polymer	very high specific energy, high specific power/weak low-temperature performance	mid-term high
Li–Ion	very high specific energy, very high specific power/high cost	mid-term very high

has been identified by many battery manufacturers to be the most promising mid-term EV battery. Its key obstacle is high initial cost which should be greatly reduced upon mass production. The Zn/Air is also very promising because of its excellent specific energy and fast mechanical refuelling. However, this mechanically rechargeable battery cannot accept energy resulting from regenerative braking. Since the major drawback of the Ni–Zn, namely short cycle life, is being alleviated in recent development, it may have the potential to compete with the Ni–MH in mid term. The $Na/NiCl_2$ is relatively the acceptable high-temperature battery for EV applications. It is promising in mid term provided that the battery performances can be further improved. The Li–Polymer has demonstrated to exhibit good performances for EV applications. It is promising in mid term providd that more battery manufacturers are involved to accelerate its research and development.

6.2 Fuel cells

Since the use of the battery as a sole energy source for EVs generally suffers from intrinsically limited ranges and relatively long recharging times, it offers an opportunity to fuel cells for EV applications. The key advantage of fuel cells over batteries is that a fuel cell powered EV can give a driving range comparable to an ICEV because its range is determined only by the amount of fuel available in the fuel tank, and is independent of the size of fuel cells. Actually, the size of fuel cells is only governed by the required power level of EVs. Other major advantages of fuel cells are that their reactant feeding time is generally much shorter than the recharging time of batteries (except for those mechanically rechargeable ones), their lifetime is generally much longer than that of batteries, and they generally require less maintenance than batteries.

The development of fuel cells has covered over 160 years since the invention in 1839. The first practical application was in the 1960s as a spacecraft power source. The fuel cell is an electrochemical device which converts the free-energy change of an electrochemical reaction into electrical energy. In contrast to a battery, the fuel cell generates electrical energy rather than stores it, and continues to do so as long as a fuel supply is maintained. Its advantageous features are efficient conversion of fuel to electrical energy, quiet operation, zero or low emissions, waste heat recoverable, rapid refuelling, fuel flexibility, durable and reliable.

As shown in Fig. 6.4, a fuel cell basically consists of three major components, namely the anode (A), cathode (C) and electrolyte (E). The anode (fuel electrode) provides a common interface for the fuel and electrolyte, catalyses the fuel oxidation reaction, and drives electrons to the external circuit. On the other hand, the cathode (oxygen electrode) provides a common interface for the oxygen and electrolyte, catalyses the oxygen reduction reaction, and receives electrons from the external circuit. Between the anode and cathode, the electrolyte functions to transport one of the ionic species involved in the fuel and oxygen electrode reactions, and also prevents the conduction of electrons.

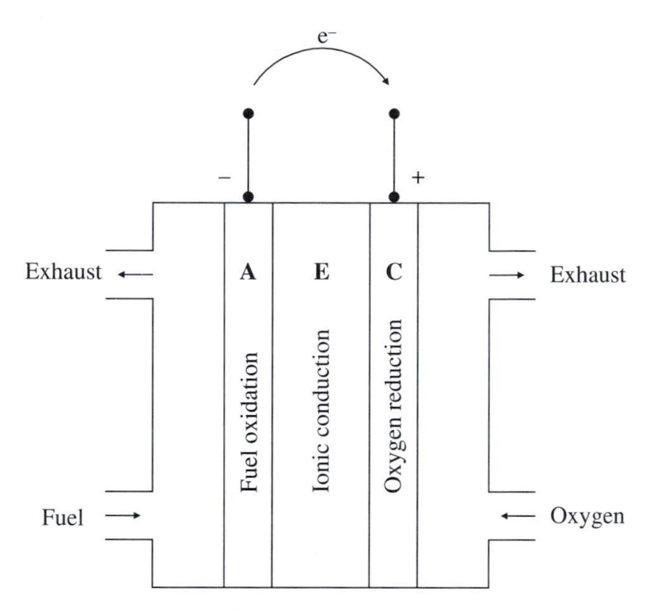

Fig. 6.4. Basic principle of fuel cells.

Hydrogen seems to be an ideal nonpolluting fuel for the fuel cell because it has the highest energy content per unit of weight of any fuel, and the by-product as a result of the fuel cell reaction is just plain water:

$$2H_2 + O_2 \rightarrow 2H_2O.$$

Since hydrogen is not a primary fuel, it is generally derived from various primary fuels such as hydrocarbons (CH_4 to $C_{10}H_{22}$), methanol and coal by means of a fuel processor. There are three major ways of storing hydrogen. Firstly, it can be stored as a compressed gas, so-called compressed hydrogen gas (CHG). Similar to the compressed natural gas, the CHG can be stored at 20–34.5 MPa in fibre-glass-reinforced aluminium containers. Secondly, it can be chilled below its boiling point (−253 °C) to form liquid hydrogen, which is then stored in cryogenic containers. Thirdly, it can be brought to react with some metals such as magnesium and vanadium to form metal hydrides. The reaction is reversible, depending on the temperature of dissolution (up to about 300 °C). Table 6.4 shows the theoretical energy contents of some prominent fuels, including hydrogen stored in various forms, liquid methanol and liquid petrol. The CHG storage offers the advantages of lightweight, low cost, mature technology and fast refuelling capability, but suffers from bulky size and safety concerns. The liquid hydrogen offers both high specific energy and fast refuelling capability, but has the drawbacks of expensive production and distribution costs as well as high volatility. Although the metal hydrides can provide the merits of compact size and inherent safety, they

Table 6.4 Theoretical energy contents of prominent fuels

	Specific energy (Wh/kg)	Energy density (Wh/l)
Compressed hydrogen gas[a]	33600	600
Liquid hydrogen[b]	33600	2400
Magnesium hydride	2400	2100
Vanadium hydride	700	4500
Methanol	5700	4500
Petrol	12400	9100

[a]At ambient temperature and 20 MPa
[b]At cryogenic temperature and 0.1 MPa

suffer from either too high temperature of dissociation such as magnesium hydride (287 °C) or relatively low specific energy such as vanadium hydride (700 Wh/kg).

Because of a vast number of variables among the fuel cell systems, such as the type of fuel, type of electrolyte, type of fuelling and operating temperatures, many classifications have appeared in the literature. Having done some of streamlining over the years, they are generally classified by the type of electrolyte, namely acid, alkaline, molten carbonate, solid oxide and solid polymer. Instead of using hydrogen as the fuel, carbon monoxide and methanol have also been adopted by some fuel cells. However, the by-product of these fuel cells becomes carbon dioxide, rather than plain water (Blomen and Mugerwa, 1993).

6.2.1 ACID FUEL CELLS

The acid fuel cell is generally characterized by the ionic conduction of hydrogen ions, and platinum or platinum alloys as electrocatalysts for both the anode and cathode. As shown in Fig. 6.5, its operating principle can be described by the following electrochemical reactions:

$$\text{Anode} \quad H_2 \rightarrow 2H^+ + 2e^-$$
$$\text{Cathode} \quad O_2 + 4H^+ + 4e^- \rightarrow 2H_2O$$

In the early development of acid fuel cells, many different acids were investigated to be the electrolyte, such as the sulphuric acid, hydrofluoric acid and phosphoric acid. Finally, the phosphoric acid won this competition because of its attractive features of stable operation at temperatures up to at least 225 °C, reasonably high conductivity at temperatures above 150 °C, as well as efficient rejection of product water and waste heat at the operating temperature. In fact, the phosphoric acid fuel cell (PAFC) is the only acid fuel cell ready for applications. The PAFC generally operates at 150–210 °C and at atmospheric or slightly higher pressures, offering the power density of 0.2–0.25 W/cm^2. Its projected life can be over 40 kh, while the projected cost should be over US$1000/kW. The major disadvantage of the PAFC is its dependence on noble metal electrocatalysts.

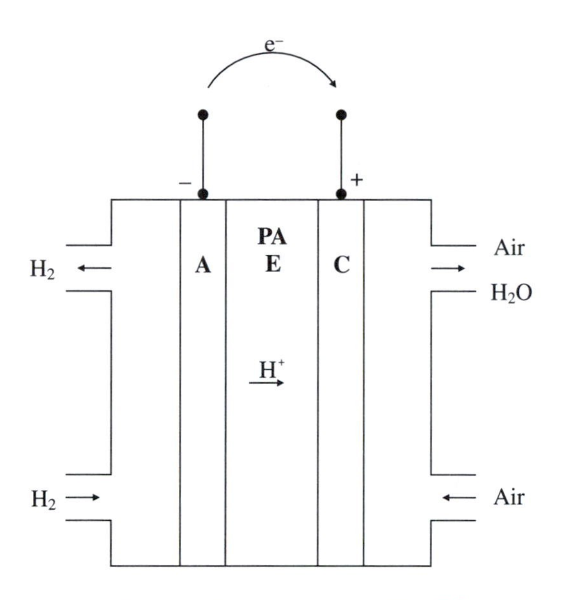

Fig. 6.5. Basic principle of PAFC.

6.2.2 ALKALINE FUEL CELLS

As shown in Fig. 6.6, the operating principle of an alkaline fuel cell (AFC) can be described by the following electrochemical reactions:

$$\begin{aligned} &\text{Anode} \quad H_2 + 2OH^- \rightarrow 2H_2O + 2e^- \\ &\text{Cathode} \quad O_2 + 2H_2O + 4e^- \rightarrow 4OH^- \end{aligned}.$$

Potassium hydroxide has been the electrolyte of choice for the AFC because of its high conductivity of hydroxide ions. The AFC generally operates at around 60–100 °C and at atmospheric pressure. Typically, it offers the power density of 0.2–0.3 W/cm^2. Its projected life and cost are over 10 kh and US$200/kW, respectively.

Compared to the PAFC, the AFC is more attractive for EV applications because its projected cost is lower by almost 5 times. The reason is due to the fact that the AFC can adopt low-cost non-noble metal or oxide electrocatalysts, such as nickel for the anode and lithiated nickel oxide for the cathode, to provide reasonable performances. Also, the lower working temperature further enhances the AFC to be more attractive than the PFAC for EV applications. However, there are two major challenges for widespread applications of the AFC. Firstly, since it generally operates at less than 100 °C, the necessary methods for product water rejection and heat removal are critical. Secondly, carbon dioxide must be completely removed from the inlet hydrogen and air before their entry into the cell. Even a small amount of carbon dioxide is sufficient to carbonate the electrolyte and form solid deposits in the porous electrode.

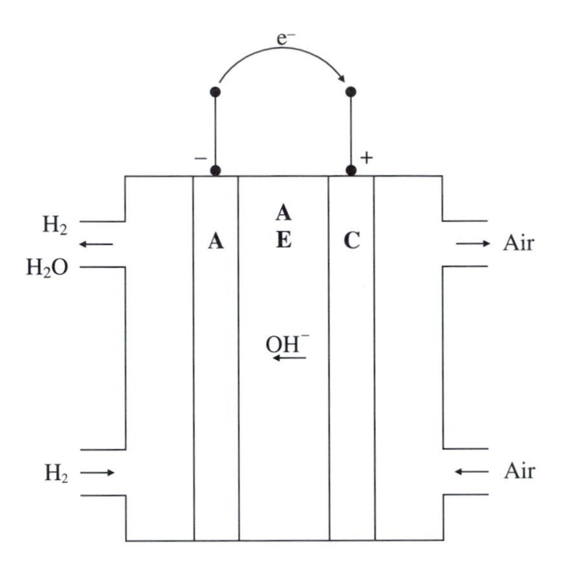

Fig. 6.6. Basic principle of AFC.

6.2.3 MOLTEN CARBONATE FUEL CELLS

In the molten carbonate fuel cell (MCFC), nickel–chromium alloy is used as the anode, lithiated nickel oxide as the cathode, and molten alkali metal (lithium, potassium or sodium) carbonate as the electrolyte. As shown in Fig. 6.7, the corresponding electrochemical reactions are given by:

$$\begin{array}{ll} \text{Anode} & H_2 + CO_3^{2-} \rightarrow H_2O + CO_2 + 2e^- \\ \text{Cathode} & O_2 + 2CO_2 + 4e^- \rightarrow 2CO_3^{2-} \end{array}$$

A consequence of these reactions is that carbon dioxide should be recycled from the anode to the cathode. The transfer can be carried out either by burning the anode exhaust with excess air and then removing the water vapour, or by using a product exchange device to separate carbon dioxide from the anode exhaust. The MCFC generally operates at 600–700 °C and at atmospheric pressure. It typically offers the power density of 0.1–0.2 W/cm^2, projected life of over 40 kh and projected cost of US$1000/kW.

Because of high-temperature operation, the MCFC has the advantages that the electrode reactions work well without using noble metal or oxide electrocatalysts, and the corresponding waste heat is available at a relatively high temperature (so-called high-grade heat). On the other hand, the MCFC suffers from a major drawback that high-temperature operation imposes severe constraints on materials suitable for EV applications.

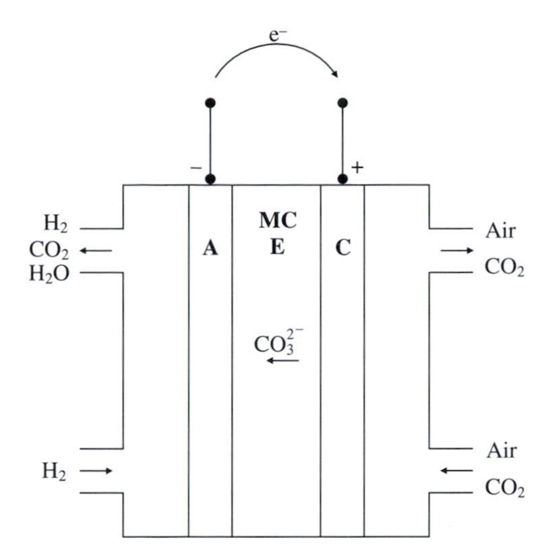

Fig. 6.7. Basic principle of MCFC.

6.2.4 SOLID OXIDE FUEL CELLS

The solid oxide fuel cell (SOFC) is an all-solid-state fuel cell which employs a solid nonporous metal oxide electrolyte to allow for ionic conduction by the migration of oxygen ions through the lattice of the crystal. Yttria-stabilised zirconia is commonly used as the electrolyte, nickel-zirconia cermet as the anode and strontium-doped lanthanum manganite as the cathode. As shown in Fig. 6.8, the corresponding electrochemical reactions are described as:

$$\text{Anode} \quad H_2 + O^{2-} \rightarrow H_2O + 2e^-$$
$$\text{Cathode} \quad O_2 + 4e^- \rightarrow 2O^{2-}$$

The SOFC generally operates at 900–1000 °C and at atmospheric pressure, offering the power density of 0.24–0.3 W/cm^2. Its projected life and cost are over 40 kh and US$1500/kW, respectively.

Because of high-temperature operation, the SOFC offers those advantages of the MCFC, namely non-noble metal or oxide electrocatalysts and high-grade heat. In addition, it does not suffer from the constraint of the MCFC to recycle carbon dioxide from the anode exhaust to the cathode inlet. However, similar to that of the MCFC, it has the disadvantage that high-temperature operation limits its application for EVs.

6.2.5 SOLID POLYMER FUEL CELLS

The solid polymer fuel cell (SPFC), also named as the proton exchange membrane fuel cell (PEMFC) by some fuel cell developers, uses a solid polymer membrane as

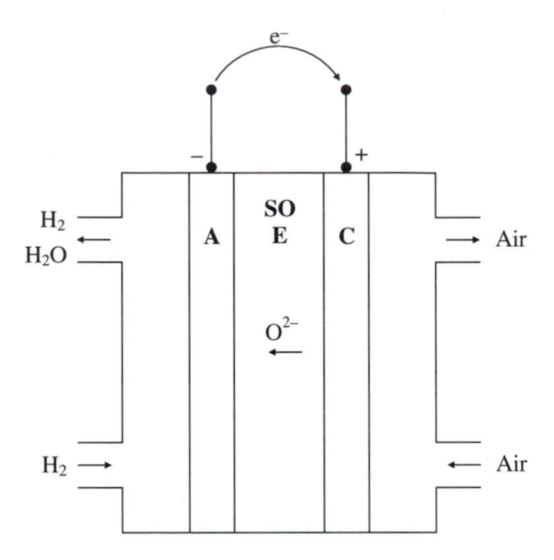

Fig. 6.8. Basic principle of SOFC.

the electrolyte. This membrane is sandwiched between two platinum-electrocatalysed porous electrodes, namely the anode and cathode. As shown in Fig. 6.9, the electrochemical reactions are as follows:

$$\begin{array}{ll} \text{Anode} & H_2 \rightarrow 2H^+ + 2e^- \\ \text{Cathode} & O_2 + 4H^+ + 4e^- \rightarrow 2H_2O \end{array}.$$

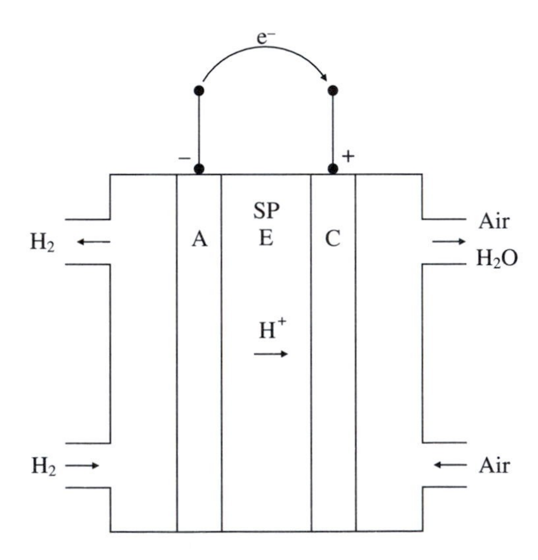

Fig. 6.9. Basic principle of SPFC.

The SPFC works at a lower temperature than the PAFC, MCFC and SOFC. Typically, it operates at 50–100 °C and at atmospheric or slightly higher pressures. With low platinum loading electrodes, it can offer the power density of 0.35–0.6 W/cm^2. The corresponding projected life and cost are over 40 kh and US$200/kW, respectively.

The SPFC has five definite advantages to favour for EV applications. Firstly, the power density of the SPFC is the highest among all available types of fuel cells. The higher the power density, the small the size of the fuel cell needs to be installed for the desired power demand of an EV. Secondly, its low-temperature operation and hence rapid start-up are desirable for an EV. Thirdly, its electrolyte, being solid, does not change, move about or vaporize from the cell. Fourthly, since the only liquid in the cell is water, the possibility of any corrosion is essentially eliminated. Finally, in contrast to the AFC which is very sensitive to carbon dioxide, the SPFC is insensitive to carbon dioxide in the inlet hydrogen and air. However, similar to the PAFC, the major drawback is its dependence on noble metal eletrocatalysts.

6.2.6 DIRECT METHANOL FUEL CELLS

Instead of using hydrogen, methanol can be directly used as the fuel for a fuel cell, so-called the direct methanol fuel cell (DMFC). There are some definite motivations of using the DMFC. Firstly, methanol is the simplest organic fuel which can most economically and efficiently be produced on a large scale from the relatively abundant fossil fuels, namely coal and natural gas. Secondly, it is the most electroactive organic fuel. Thirdly, methanol is a liquid fuel which can be easily stored, distributed and marketed for EV applications, whereas hydrogen always suffers from the difficulty of storage and distribution.

In the DMFC, both the anode and cathode adopt platinum or platinum alloys as electrocatalysts, while the electrolyte can be trifluoromethane sulphonic acid or proton exchange membrane (PEM). As shown in Fig. 6.10, the principle of operation can be described as:

$$\begin{array}{rl} \text{Anode} & CH_3OH + H_2O \rightarrow CO_2 + 6H^+ + 6e^- \\ \text{Cathode} & O_2 + 4H^+ + 4e^- \rightarrow 2H_2O \\ \text{Net} & 2CH_3OH + 3O_2 \rightarrow 2CO_2 + 4H_2O \end{array}.$$

The DMFC is relatively immature among the aforementioned fuel cells. At the present status of DMFC technology, it generally operates at 50–100 °C and at atmospheric pressure, offering the power density of 0.04–0.23 W/cm^2. Its projected life and cost are over 10 kh and US$200/kW, respectively. There are still many challenges for research and development of the DMFC such as the power level and cost that should be addressed before its practical application to EVs.

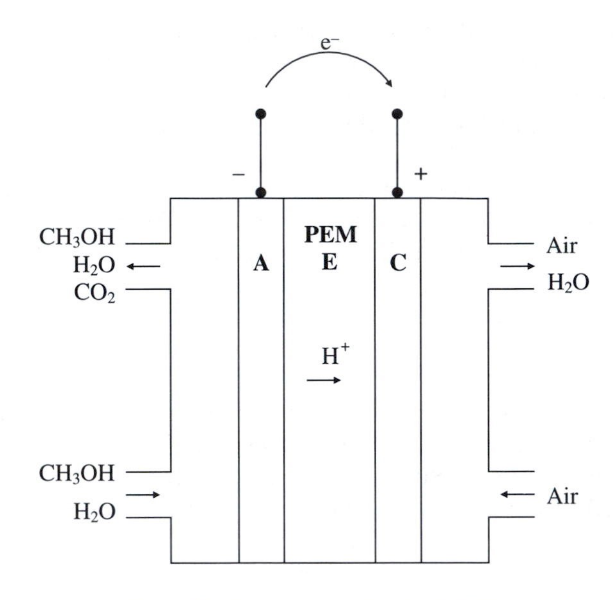

Fig. 6.10. Basic principle of DMFC.

6.2.7 EVALUATION OF FUEL CELLS

Detailed chemistries of the aforementioned fuel cells can be found in relevant reference books (Blomen and Mugerwa, 1993). Typical characteristics of the aforementioned fuel cells are summarized in Table 6.5. Accordingly, both the MCFC and SOFC suffer from very high-temperature operation, respectively over 600 °C and 900 °C, making them practically difficult to be applied to EVs. For the DMFC, the corresponding technology is still immature although it has been developed for over 30 years. Also, its available power level and power density are too low for practical application to EVs. The others, namely the PAFC, AFC and SPFC, are all technically possible for EV applications—termed EV fuel cells.

Among those EV fuel cells, the PAFC is less attractive than the AFC because of its relatively higher working temperature and higher projected cost. In fact, the AFC has ever been implemented in practical EVs. In 1966, Union Carbide in the United States developed a 32-kW 32-module AFC system which was used to

Table 6.5 Typical characteristics of fuel cells

	PAFC	AFC	MCFC	SOFC	SPFC	DMFC
Working temp. (°C)	150–210	60–100	600–700	900–1000	50–100	50–100
Power density (W/cm^2)	0.2–0.25	0.2–0.3	0.1–0.2	0.24–0.3	0.35–0.6	0.04–0.23
Projected life (kh)	40	10	40	40	40	10
Projected cost (US$/kW)	1000	200	1000	1500	200	200

power a 3227-kg experimental delivery van (namely the GM Electrovan). This first fuel cell van could achieve the top speed of 112 km/h, acceleration from zero to 96 km/h in 30 s, and the range of 160–240 km. In 1976, Elenco, a Belgian-Dutch consortium, also launched an AFC program for EVs. Incorporating a series–parallel connection of 24 cells, an 11-kW AFC system (later extended to 15 kW) was successfully tested in a delivery-type EV, namely the Volkswagen electric van. Moreover, a European project, namely the Eureka, was launched in 1994, aiming at the technological demonstration of a fuel cell & battery hybrid energy source system in an existing city bus. The adopted hybrid energy source consisted of a 78-kW Elenco AFC and a 60-kW SAFT Ni–Cd battery. The waste heat of the AFC was used for heating up the bus in the winter time. The hydrogen fuel was stored in liquid form in a cryogenic container in the bus. Based on 45-kg liquid hydrogen, the bus (about 25 t including passengers) could achieve the range of 250–300 km, corresponding to 12–15 h of city driving.

With the advancement of SPFC technology, the SPFC takes advantages over the AFC for EV applications. The major reasons are due to its higher power density and longer projected life while maintaining the low working temperature and economical projected cost. Thus, recent research and development on fuel cells for EVs have been focused on the SPFC technology (Cornu *et al.*, 1994; Vermeulen, 1994). Starting from 1993, a Canadian firm, namely Ballard Power Systems, has pioneered the development of CHG-fuelled SPFC technology for EVs (Howard, 1994). Including the container and auxiliaries, its CHG-fuelled SPFC system has achieved the specific energy of about 500 Wh/kg, which is much higher than the maximum value offered by any batteries. However, the corresponding system specific power has been only of 50–180 W/kg, limiting its sole application to those EVs desiring high acceleration rate and hill-climbing capability. Nevertheless, Ballard Power Systems and Daimler-Benz jointly produced a CHG-fuelled SPFC powered bus in 1997, namely the NEBUS. This bus has installed 21-kg CHG in seven fibreglass-reinforced aluminium tanks and employed five 25-kW SPFC stacks, hence delivering the power of 250 kW and guaranteeing the range of 250 km. At present, the major challenge is how to significantly reduce the material cost of solid polymer membrane and platinum-electrocatalysed electrodes. The available prototype cost of the SPFC was of about US$2700–5400/kW which is extremely expensive compared to the ICE cost of about US$11–27/kW. Of course, the best way to considerably reduce this cost is to go into production volumes.

By retaining the definite advantage of liquid fuel while avoiding those shortcomings of the DMFC, the concept of methanol-fuelled SPFC system is becoming more and more attractive for EVs. As shown in Fig. 6.11, methanol and water are firstly mixed, vaporized, and then converted into hydrogen and carbon dioxide gases via an on-board reformer. The resulting hydrogen gas is fed to the SPFC to generate the desired electricity and the reusable pure water. The purifier functions to prevent any undesirable reformer by-products such as carbon monoxide gas from poisoning the precious electrocatalysts of the SPFC. Although this technology seems to be contradictory to the pursuit of zero-emission vehicles, it is still

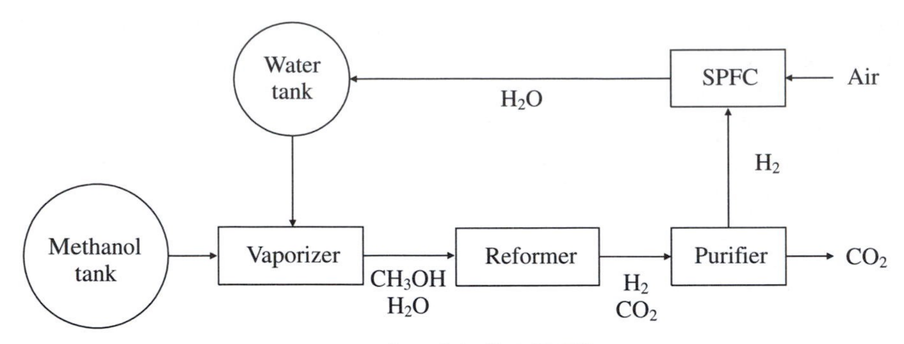

Fig. 6.11. Methanol-fuelled SPFC system.

environmentally friendly as it does not generate harmful emissions such as carbon monoxide, nitrogen oxides and hydrocarbons. Recently, Daimler-Benz and Ballard Power Systems has presented the first methanol-fuelled SPFC powered EV, namely the NECAR 3, which can travel over 400 km using 38 l of liquid methanol. Toyota has also announced that its fuel cell RAV4 EV has achieved the range of 500 km per tank of methanol. Even so, these methanol-fuelled SPFC EVs are far from economically viable within the near future.

Further extending the concept of liquid-fuelled fuel cell EVs, research on extracting hydrogen from petrol using an on-board reformer has been launched. The argument of this research is simple—hundreds of billion dollars have been invested in the way petrol is distributed, and it is impossible to change this infrastructure just because there are fuel cell EVs that run in hydrogen or methanol. No matter this argument is agreeable or not, the success of this concept can definitely move the fuel cell EVs approaching to reality. Recently, Chrysler has decided to realize this concept by demonstrating a petrol-fuelled fuel cell EV within the next few years. Definitely, there are still much to be done in research and development of petrol-fuelled fuel cell EVs before they become commercially viable.

6.3 Ultracapacitors

Because of frequent start/stop operation of EVs, the discharge profile of the battery is highly variable. The average power required from the battery is relatively low while the peak power of relatively short duration required for acceleration or hill climbing is much higher. The ratio of the peak power to the average power can be as high as 16:1 for a high-performance EV. In fact, the amount of energy involved in the acceleration and deceleration transients is roughly 2/3 of the total amount of energy over the entire vehicle mission in the urban driving. Therefore, based on present battery technology, the design of batteries has to carry out the trade-offs among the specific energy, specific power and cycle life.

The difficulty of simultaneously obtaining high values of specific energy, specific power and cycle life has lead to some suggestions that EVs may best be powered by a pair of energy sources. The main energy source, usually a battery, is optimized for the range while the auxiliary source for acceleration and hill climbing. This auxiliary source can be recharged from the main source during less demanding driving or regenerative braking. An auxiliary energy source which has received wide attention is the ultracapacitor.

6.3.1 FEATURES OF ULTRACAPACITORS

In the foreseeable development of the ultracapacitor, it cannot be used as a sole energy source for EVs because of its exceptionally low specific energy. Nevertheless, there are a number of advantages that can be resulted from using the ultracapacitor as an auxiliary energy source. The promising application is the so-called battery & ultracapacitor hybrid energy system for EVs. Hence, the specific energy and specific power requirements of the EV battery can be decoupled, thus affording an opportunity to design the battery that is optimized for the specific energy and cycle life with little attention being paid to the specific power. Due to the load levelling effect of the ultracapacitor, the high-current discharge from the battery is minimized so that the available energy, endurance and life of the battery can be significantly increased. Moreover, compared to the battery, the ultracapacitor can provide much faster and more efficient energy recovery during regenerative braking of EVs. Therefore, as a combined effect of load levelling and efficient energy recovery, the vehicle range can be greatly extended. Notice that system integration and optimization should be made to coordinate the battery, ultracapacitor, electric motor and power converter. The power converter and corresponding controller should take care both the electric motor and ultracapacitor.

6.3.2 DESIGN OF ULTRACAPACITORS

Before the use of the ultracapacitor in a hybrid configuration with the battery for EV applications, some basic concepts need to be addressed. The battery should deliver the output power approximately equal to the required average power over the vehicle mission, whereas the ultracapacitor should deliver the power exceeding the average power level and should recover the regenerative braking energy. When the power demand exceeds the output power deliverable by the ultracapacitor, the battery should provide the necessary power. The ultracapacitor should be kept in charge by the recovered energy and, for any difference, by the energy from the battery during periods of low-power demand. When the ultracapacitor is in the full-charge situation, the recovered energy should be directly used to recharge the battery. A two-quadrant dc–dc converter should be placed between the battery and the ultracapacitor to control the power split between them, and to limit the rate at which the battery recharges the ultracapacitor during periods of low-power demand. Without this converter, the requirement that both the battery and ultracapacitor are at the same voltage results in the ultracapacitor providing or accepting significant power only during times in which the battery voltage is changing rapidly, thus reducing the function of load levelling.

The double-layer capacitor technology is the major approach to achieve the ultracapacitor concept. The basic principle of a double-layer capacitor is illustrated in Fig. 6.12. When a voltage is applied across the electrodes, a double layer is formed by the dipole orientation and alignment of electrolyte molecules over the entire surface of the electrodes. This polarization is used to store energy in the capacitor according to:

$$C = \frac{\varepsilon A}{d}$$
$$E = \frac{1}{2} C V^2,$$

where ε is the effective dielectric constant, d is the separation distance, A is the electrode surface area, C is the capacitance, V is the applied voltage and E is the stored energy. By adopting high-dielectric materials, short separation distances and large electrode surface areas, the capacitance can be greatly increased. At the present status of ultracapacitor technology, the corresponding electrode materials may be carbon/metal fibre composites, doped conducting polymer films on carbon cloth or mixed metal oxide coatings on metal foil, while the electrolyte materials may be aqueous/organic solution or solid polymer (Trippe *et al.*, 1993; Burke, 1994).

The size of the required ultracapacitor can roughly be estimated by using simple energy calculations. Based on the aforementioned criteria of the battery

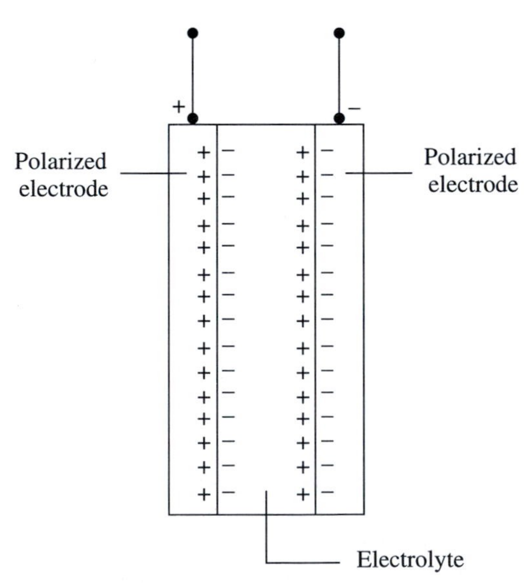

Fig. 6.12. Basic principle of ultracapacitors.

& ultracapacitor hybrid energy source for EVs that the output from the battery should be ideally limited to what is needed for constant-speed operation while the rest output is supplied from the ultracapacitor, the maximum stored energy W_{cap} in the ultracapacitor should be proportional to the maximum energy required for both acceleration and hill climbing. The corresponding energy equation is given by:

$$W_{cap} = \left(\frac{1}{2}mv^2 + mgh\right)\Big/\eta,$$

where m is the total mass of the vehicle, v is the maximum vehicle speed, g is the gravitational acceleration, h is the maximum height difference of the vehicle mission and η is the system energy efficiency of the use of the ultracapacitor. Hence, the required capacitance C can be calculated by:

$$W_{cap} = \frac{1}{2}C\left(V_{max}^2 - V_{min}^2\right),$$

where V_{max} and V_{min} are respectively the maximum and minimum voltage limits of the ultracapacitor.

For example, an EV, with the weight of 1800 kg and the maximum speed of 120 km/h, is considered to be powered by the battery & ultracapacitor hybrid energy system for urban and suburban operation which has the maximum height difference of 50 m. Assuming that the system energy efficiency of the use of the ultracapacitor is 90%, W_{cap} can be calculated as 2 MJ or 581 Wh. Selecting the working voltage of the ultracapacitor from 28 to 56 V, the required capacitance is determined as 1700 F. Given that an advanced ultracapacitor module has the rated voltage of 56 V, rated capacitance of 95 F, weight of 37 kg and volume of 18 l, we need to install 18 modules in parallel, leading to increase the weight and size by 296 kg and 144 l, respectively.

6.3.3 EVALUATION OF ULTRACAPACITORS

According to the goals set by the US Department of Energy for the inclusion of ultracapacitors in EVs, the near-term specific energy and specific power should be better than 5 Wh/kg and 500 W/kg, respectively, while the advanced performance values should be over 15 Wh/kg and 1600 W/kg. So far, none of the available ultracapacitors can fully satisfy these goals. Nevertheless, research and development of ultracapacitors for EV applications are actively engaged by some companies.

Maxwell Technologies has claimed that its Power Cache PC7223 ultracapacitor cell offers the highest capacitance (namely 2700 F at 2.3 V) in mass production. As listed in Table 6.6, its specific energy and specific power can achieve the values of 2.48 Wh/kg and 732 W/kg. By connecting 28 cells of PC7223 in series, the 56-V 95-F ultracapacitor module can store energy of 150 kJ while offering the specific energy and specific power of 1.12 Wh/kg and 330 W/kg, respectively. Panasonic

Table 6.6 Specifications of a 2700-F ultracapacitor

Dimensions (mm)	164 × 62 × 62
Weight (kg)	0.8
Volume (l)	0.623
Rated capacitance (F)	2700
Nominal voltage (V)	2.3
Rated current (A)	400
Specific energy (Wh/kg)	2.48
Specific power[a] (W/kg)	732
Operating temperature (°C)	−20–60
Series resistance (mΩ)	0.85

[a]With 6-s discharge from 2.3 V to 1.5 V

has also successfully produced a series of ultracapacitor cells (up to 1800 F at 2.3 V).

Recently, the FIAT Cinquecento Elettra, incorporating with the Sonnenschein Pb–Acid battery (9 kWh, 300 kg) and the Alsthom Alcatel ultracapacitor (250 Wh, 60 kg), has been used to evaluate the energy saving of the battery & ultracapacitor energy storage system (Anerdi *et al.*, 1995). Without changing the total weight of the whole energy storage system, a simulation study has shown that there are energy savings of up to 40% and 20% on urban and suburban driving conditions, respectively. An experimental prototype has also been testified, showing that an energy saving of 14% can be achieved over the whole ECE Cycle (Fält *et al.*, 1995). Increasingly, based on 40 × 2 Panasonic ultracapacitor cells (1600 F at 2.3 V), Chugoku Electric Power and Tokyo R&D has implemented a VRLA & ultracapacitor hybrid energy system in the Mazda Bongo Friendee for evaluation. Hence, the merit of using the ultracapacitor for load levelling the VRLA battery can be confirmed for EV applications. Nevertheless, the ultracapacitor's performances (especially the specific energy) should be significantly improved and its cost needs to be greatly reduced before widely applying to EVs.

6.4 Ultrahigh-speed flywheels

The use of flywheels for storing energy in mechanical formats is not a new concept. More than 25 years ago, the Oerlikon Engineering Company in Switzerland made the first passenger bus solely powered by a massive flywheel. This flywheel, weighed 1500 kg and operated at 3000 rpm, was recharged by electricity at each bus stop. The traditional flywheel is a massive steel rotor with hundreds of kg that spins at the order of ten hundreds of rpm. On the contrary, the advanced flywheel is a lightweight composite rotor with tens of kg and rotates at the order of ten thousands of rpm, so-called the ultrahigh-speed flywheel.

The concept of ultrahigh-speed flywheels appears to be a feasible means for fulfilling the stringent energy storage requirements for EV applications, namely high specific energy, high specific power, long cycle life, high energy efficiency, quick recharge, maintenance free and cost effective. When the flywheel is conceived as an auxiliary energy source in a hybrid configuration with the main energy source in an EV, it stores energy in mechanical form during periods of cruising or regenerative braking while generates electrical energy to meet the peak power demands during periods of starting, acceleration or hill climbing. Apart from providing load levelling for the main energy source, the ultrahigh-speed flywheel can potentially act as the sole energy source in an EV.

6.4.1 FEATURES OF ULTRAHIGH-SPEED FLYWHEELS

Basically, ultrahigh-speed flywheels are characterized by the features of high specific energy, high specific power, high efficiency for conversion between electrical and mechanical energies, and easily adaptable to long-term energy storage. The potential benefits of using the ultrahigh-speed flywheel as an auxiliary energy source in a hybrid configuration with the battery for EVs are similar to those of using the ultracapacitor. Firstly, the requirements on specific energy and specific power of the battery can be decoupled, affording to optimize the battery's specific energy density and hence cycle life. Secondly, as the high-rate power demand and high-current discharge are greatly reduced by the load levelling effect of the flywheel, the usable energy, endurance and cycle life of the battery can be increased. Thirdly, the flywheel can allow rapid interim recharges with high efficiency during periods of low power demand or regenerative braking. Due to the combined effect of load levelling of the main energy source and improved energy recovery during regenerative braking, the vehicle range can be remarkably extended.

Instead of using a battery or fuel cell, an EV can potentially be powered solely by the ultrahigh-speed flywheel. The corresponding long-term potential benefits for EV applications are possible. It can potentially provide higher specific energy and higher specific power than any batteries. Perhaps, its specific power may be even higher than the ICE. It should also solve the problem of limited cycle life suffered by other energy sources because the cycle life of the flywheel is practically unlimited or at least longer than the vehicle life.

6.4.2 DESIGN OF ULTRAHIGH-SPEED FLYWHEELS

The major technologies that continually fuel the development of ultrahigh-speed flywheels for EV applications include flywheel materials for ultrahigh-speed operation, electric machines for efficient bidirectional conversion between electrical and mechanical energies, and power converters for efficient bidirectional conversion between the desired modes of electrical energy.

In order to explain the difference between the traditional massive flywheel and the recent lightweight ultrahigh-speed flywheel, it is straightforward to consider the following equation describing the mechanical energy E stored in a flywheel:

$$E = \frac{1}{2} J\omega^2,$$

where J is the moment of inertia and ω is the rotational speed. Since the mechanical energy depends on the square of speed but only linearly with the inertia, there is a significant gain in energy storage even when the increase in speed is offset by the decrease in inertia with the same ratio. For example, a flywheel of 50 kg running at 50 000 rpm can store 10 times that of 500 kg at 5000 rpm.

Although the higher the rotational speed the more the energy can be stored, there is a limit in which the tensile strength σ of the material constituting the flywheel cannot withstand the stress resulting from the centrifugal force. As the maximum stress acting on the flywheel depends on its geometry, specific density ρ and rotational speed, the maximum benefit can be obtained by adopting the flywheel material having the maximum ratio σ/ρ. Notice that the theoretical specific energy of the flywheel is proportional to this ratio. Table 6.7 summarizes these characteristics of some composite materials for ultrahigh-speed flywheels.

A constant-stress principle can be employed for the design of ultrahigh-speed flywheels. To achieve the maximum energy storage, every element in the rotor should be equally stressed to its maximum limit. It results in a shape of gradually decreasing thickness that theoretically approaches zero as the radius approaches infinity. On the other hand, to charge and discharge the flywheel (actually the rotor of an electric machine), the permanent magnet (PM) brushless machine has been accepted to be the most appropriate type. Apart from possessing high power density and high efficiency, the PM brushless motor has a unique advantage that no heat is generated inside the PM rotor which is particularly essential for the rotor to work in a vacuum environment to minimize the windage loss. Meanwhile, the corresponding stator winding should be liquid-cooled by circulating through the hollow conductors whereas the stator core is cooled by conduction through the stator winding and through the vessel wall. Apart from encasing the rotor in a vacuum shell to minimize the windage loss, magnetic bearings can be employed to further minimize the friction loss. These magnetic bearings have the added

Table 6.7 Composite materials for ultrahigh-speed flywheels

	Tensile strength σ (MPa)	Specific density ρ (kg/m^3)	Ratio σ/ρ (Wh/kg)
E-glass	1379	1900	202
Graphite epoxy	1586	1500	294
S-glass	2069	1900	303
Kevlar epoxy	1930	1400	383

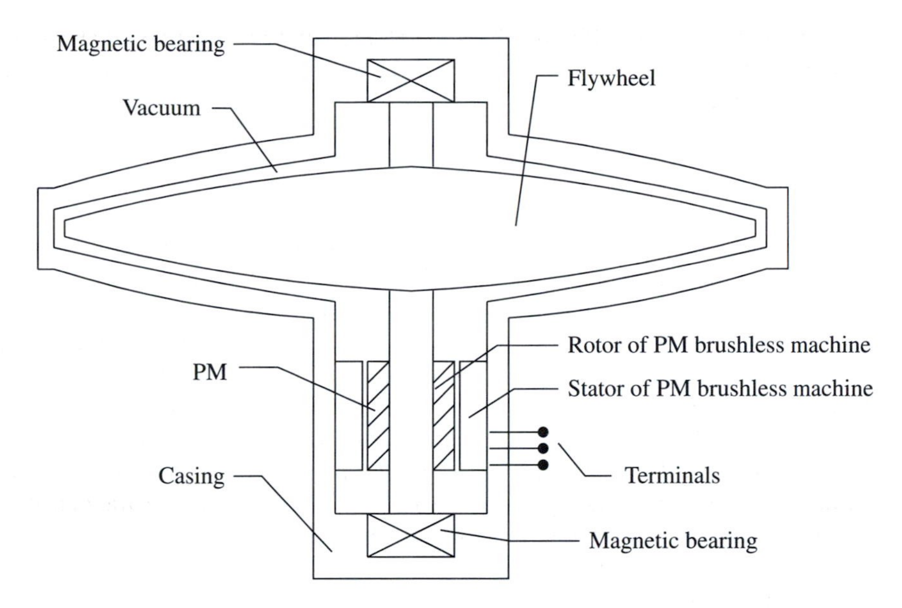

Fig. 6.13. Basic principle of ultrahigh-speed flywheels.

attraction of extending the flywheel life because there are no mechanical contacts and hence no wear-and tear problems. Figure 6.13 illustrates the basic principle of a typical ultrahigh-speed flywheel in which composite material, constant-stress principle, PM brushless machine and magnetic bearings have been employed (Hoolboom and Szabados, 1994).

Having considered the electric machine for the conversion between electrical and mechanical energies, another important consideration is the power converter for the conversion between modes of electrical energy. This power converter must be capable of bidirectional power flow to supply and extract energy from the PM brushless machine, hence the flywheel. Moreover, it should provide high power density and high efficiency. Recently, active research has been undergone which employs various soft-switching techniques to significantly increase the operating frequency with practically zero switching losses, leading to improve the power density and efficiency of the power converter.

Different from applying the ultrahigh-speed flywheel for energy storage in stationary plants, its application to EVs suffers from two special problems. Firstly, gyroscopic forces occur whenever an EV departs from its straight-line course, such as in turning and in pitching upward or downward from road grades. These forces essentially reduce the manoeuvrability of the EV. Secondly, if there is an accident on the flywheel, its stored energy in mechanical form will be released in a very short period of time. The corresponding power released will be very high which can cause severe damage to the vehicle. For example, if an 1-kWh flywheel (which acts as an auxiliary energy source for load levelling) breaks apart in 1–5 s, it will generate a huge power of 720–3600 kW. So, the containment in case of failure is presently the most significant obstacle to implement the ultrahigh-speed flywheel in the EV.

The simplest way to alleviate the gyroscopic forces is to employ multiple smaller flywheels. By operating them in pairs (one half spinning clockwise while another half anticlockwise), the net gyroscopic effect on the EV becomes theoretically zero. Practically, it still has some problems on the distribution and coordination of these flywheels. Also, the overall specific energy and specific power of all flywheels may be smaller than a single one. Similarly, the simplest way to minimize the damage due to the breakage of the ultrahigh-speed flywheel is to adopt multiple small modules, but suffering from the possible reduction of specific energy and specific power. Recently, a new failure containment has been proposed. Instead of diminishing the thickness of the rotor's rim to zero based on the maximum stress principle, the rim's thickness is purposely enlarged. Hence, the neck area just before the rim (virtually a mechanical fuse) will break first at the instant that the rotor suffers from a failure. Due to the use of this mechanical fuse, only the mechanical energy stored in the rim needs to be released or dissipated in the casing upon failure (Fijalkowski and Krosnicki, 1995).

Similar to that of the ultracapacitor, the size estimation of the ultrahigh-speed flywheel can be roughly estimated using simple energy calculations. Based on the battery & ultrahigh-speed flywheel hybrid configuration of an EV, the maximum stored energy W_{fly} in the flywheel is proportional to the maximum energy required for both acceleration and hill climbing. The corresponding energy equation is given by:

$$W_{\mathrm{fly}} = \left(\frac{1}{2}mv^2 + mgh\right) \Big/ \eta,$$

where m is the total mass of the vehicle, v is the maximum vehicle speed, g is the gravitational acceleration, h is the maximum height difference and η is the system energy efficiency of the use of the ultrahigh-speed flywheel. Hence, the flywheel inertia J can be calculated by:

$$W_{\mathrm{fly}} = \frac{1}{2}J\left(\omega_{\mathrm{max}}^2 - \omega_{\mathrm{min}}^2\right),$$

where ω_{max} and ω_{min} are respectively the maximum and minimum speed limits of the flywheel. In case the flywheel is a simple circular disc-shaped rotor, the size can be easily determined by:

$$J = \frac{\pi}{2}\rho r^4 t,$$

where ρ is the specific density, r is the radius and t is the thickness.

For example, an EV, with the weight of 1800 kg and the maximum speed of 120 km/h, is considered to be powered by the battery & ultrahigh-speed flywheel hybrid energy system for urban and suburban operation which has the maximum height difference of 50 m. Assuming that the system energy efficiency of the use of the ultrahigh-speed flywheel is 90%, W_{fly} can be calculated as 2 MJ or 581 Wh. Selecting the speed range of the flywheel as 40 000–80 000 rpm, the required inertia is determined as 0.076 kgm^2. By using a circular disc-shaped rotor, made of Kelvar epoxy (1400 kg/m^3), the corresponding radius and thickness can be obtained as 14 and 9 cm, respectively.

6.4.3 EVALUATION OF ULTRAHIGH-SPEED FLYWHEELS

The Lawrence Livermore National Laboratory (LLNL) in the United States has actively engaged on the development of ultrahigh-speed flywheels, including small modules of 1 kWh for on-board EV applications and large modules of 2–25 kWh for stationary applications. Other active flywheel developers are the Ashman Technology, AVCON, Northrop Grumman, Power R&D, Rocketdyne/Rockwell, Trinity Flywheels, and US Flywheel Systems. On the basis of present technologies, the whole ultrahigh-speed flywheel system can achieve the specific energy of 10–150 Wh/kg and specific power of 2–10 kW/kg. LLNL has built a prototype (20-cm diameter and 30-cm height) which can achieve 60 000 rpm, 1 kWh and 100 kW.

Recently, the FIAT Cinquecento Elettra has been used to evaluate the practical potentiality of the ultrahigh-speed flywheel (Anerdi *et al.*, 1994). By operating it as an auxiliary energy source for load levelling the Pb–Acid battery, the simulation results have indicated that there is a saving in energy for over 20%. For operation as a sole energy source, the driving range per charge is projected to be over 200 km. Further development of these flywheels has been focused on the optimization of weight, size and cost, as well as the practical application to EVs. Nevertheless, because of the aforementioned two special problems (gyroscopic forces and failure containment), the use of flywheels as an on-board energy storage system for EVs still has much to be done.

Instead of facing the two challenges as an on-board energy storage system, the ultrahigh-speed flywheel can be used as a stationary energy storage system (25-kWh capacity and 130-kW capability) to provide rapid recharging for EVs. The reason is so clear that the flywheel offers the ability to release high power for rapid recharging while minimizing the corresponding peak power demand on our power system network. This new way of application to EVs is receiving a great attention, and should be much easier to achieve than the use as an on-board energy storage system.

6.5 Hybridization of energy sources

Based on the aforementioned facts, none of the available energy sources, including batteries, fuel cells, ultracapacitors and ultrahigh-speed flywheels, can fulfil all demands of EVs to enable them competing with petrol-powered vehicles. In essence, these energy sources cannot provide high specific energy and high specific power simultaneously. Instead of limiting to use a sole energy source, EVs can adopt the concept of using multiple energy sources, so-called the hybridization of energy sources (Wong *et al.*, 1999; Wong *et al.*, 2000). Since both the control and packaging complexities of this concept increase with the number of energy sources involved, only the hybridization of two sources (one for high specific energy while the other for high specific power) is considered to be viable. The advantages resulting from the use of this hybrid energy system for EVs are summarized below:

- Since the EV requirements on energy and power can be decoupled, it affords an opportunity to design EV sources such as batteries and fuel cells for high specific energy, whereas to optimize other sources such as ultracapacitors and ultrahigh-speed flywheels for high specific power.
- Since there is no need to carry out trade-offs between the pursuits of specific energy and specific power, the cycle life and production cost of these sources can readily be lengthened and minimized, respectively.
- The unique advantages of various EV energy sources can be fully utilized, such as the maturity and low cost of batteries, the outstanding specific energy and high fuel efficiency of fuel cells, the enormous specific power and instantaneous charge/discharge capability of ultracapacitors, as well as the outstanding specific power and practically unlimited cycle life of ultrahigh-speed flywheels.

The principle of operation of this hybrid energy system, consisting of both high specific energy and high specific power sources, is illustrated in Fig. 6.14. Firstly, during the normal driving condition of EVs, the high specific energy source normally supplies the necessary energy to the electric motor via the power converter. To enable the system ready for sudden power demand, this source can precharge the high specific power source in the light-load period. Secondly, during the acceleration or hill-climbing condition, both the sources need to simultaneously supply the desired energy to the electric motor. Thirdly, during the braking or downhill condition, the electric motor operates as a generator so that the regenerative energy flows back to recharge the high specific power source via the power converter. If this source can not fully accept the regenerative energy, the surplus can be diverted to recharge the high specific energy source provided that it is energy receptive.

6.5.1 NEAR-TERM HYBRIDS

Based on the available technology of various energy sources, there are several viable hybridization schemes for EVs in the near term: battery & battery hybrids, battery & ultracapacitor hybrids, and fuel cell & battery hybrids.

In the battery & battery hybrids, one battery provides high specific energy while another battery offers high specific power. Taking into account the maturity and cost, the Zn/Air & VRLA hybrid seems to be a natural choice. It combines the merits of the 230 Wh/kg of Zn/Air for long driving range and the 300 W/kg of VRLA for acceleration and hill climbing. This choice also overcomes the incapability of the mechanically rechargeable Zn/Air which cannot accept the precious

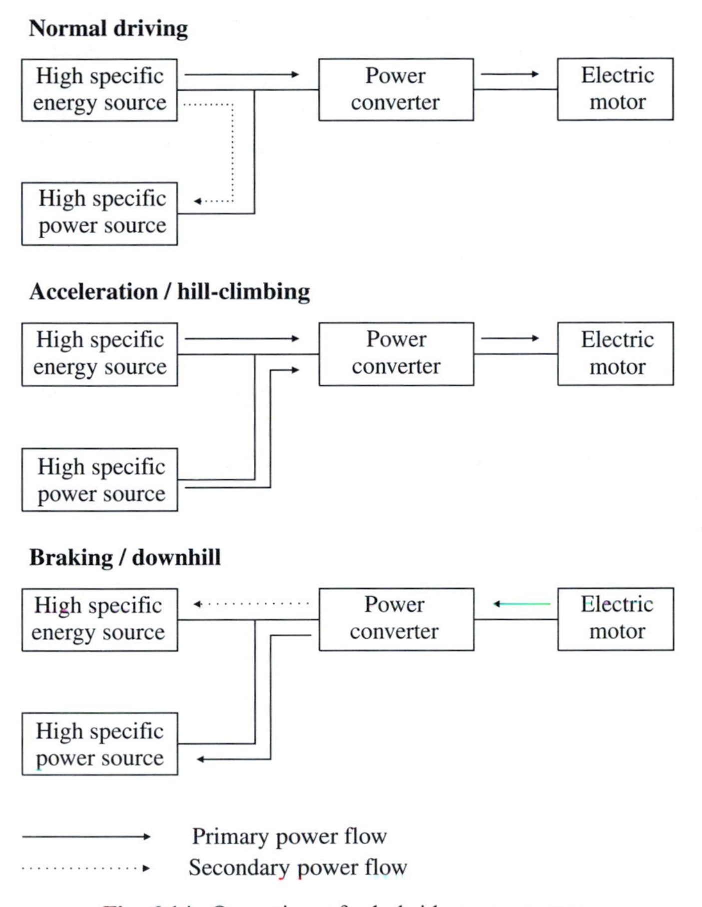

Fig. 6.14. Operation of a hybrid energy system.

regenerative energy during braking or downhill. Other possible choice for this scheme of hybridization can be the Zn/Air & Ni–MH hybrid, or the Zn/Air & Li–Ion hybrid.

Since the ultracapacitor can offer the specific energy far below the minimum requirement for EV applications, it has to work together with other EV energy sources. Naturally, it collaborates with various batteries to form the battery & ultracapacitor hybrids. During the hybridization, an additional two-quadrant dc–dc converter is usually placed between the battery source and the ultracapacitor source, because the working voltage of the ultracapacitor source is quite low (generally less than 100 V even many ultracapacitors have already been internally stacked up). Recently, the VRLA & ultracapacitor hybrid has already received attention such that the average energy consumption is supplied by the VRLA while the peak power demand is by the ultracapacitor source. The ultracapacitor source is recharged during regenerative braking or from the VRLA at periods of low-power demand. Other viable combinations of this scheme of hybridization are the Ni–MH & ultracapacitor hybrid, and the Li–Ion & ultracapacitor hybrid.

Although the fuel cell can offer outstanding specific energy, it suffers from low specific power and incapability of accepting regenerative energy. Thus, the fuel cell & battery hybrid is a good collaboration in which the battery can be purposely selected to compensate those shortcomings of the fuel cell. The SPFC & VRLA, SPFC & Ni–MH and SPFC & Li–Ion hybrids are typical choices since these three batteries are of high specific power and rapid recharge capability.

6.5.2 LONG-TERM HYBRIDS

In long term, the ultracapacitor should be improved to such a level that its specific energy is sufficiently high to provide all necessary instantaneous energy for acceleration and hill climbing, as well as to accept all regenerative energy during braking and downhill. Similarly, the ultrahigh-speed flywheel should also be able to attain such specific energy in the long term. Consequently, they can possibly replace batteries to hybridize with the fuel cell to form the fuel cell & ultracapacitor hybrid and the fuel cell & ultrahigh-speed flywheel hybrid.

References

Anan, F. and Adachi, K. (1994). Development of Ni–Zn batteries for EV's. *Proceedings of the 12th International Electric Vehicle Symposium*, pp. 537–46.

Anerdi, G., Ancarani, A., Bianchi, R., Cipriani, M., Quaglia, D., Brusaglino, G., Ravello, V. and Andrieu, X. (1995). Supercapacitor as a buffer to enhance energetical, operational and economical effectiveness of electric vehicle systems. *Proceedings of EVT Conference*, pp. 75–83.

Anerdi, G., Brusaglino, G., Ancarani, A., Bianchi, R., Quaglia, G., Barberis, U., Ravera, C., Mellor, P.H., Howe, D. and Zegers, P. (1994). Technology potential of flywheel storage and application impact on electric vehicles. *Proceedings of the 12th International Electric Vehicle Symposium*, pp. 37–47.

Berndt, D. (1997). *Maintenance-Free Batteries: Lead–Acid, Nickel/Cadmium, Nickel/Hydride: A Handbook of Battery Technology*. Somerset, Research Studies Press, England and John Wiley & Sons, New York.

Blomen, L.J.M.J. and Mugerwa, M.N. (1993). *Fuel Cell Systems*. Plenum Press, New York.

Broussely, M., Flament, P., Morin, C. and Sarre, G. (1995). Lithium-carbon liquid electrolyte battery system for electric vehicle. *Proceedings of EVT Conference*, pp. 41–50.

Burke, A.F. (1994). Electrochemical capacitors for electric vehicles—technology update and implementation considerations. *Proceedings of the 12th International Electric Vehicle Symposium*, pp. 27–36.

Chan, C.C., Lo, E.W.C. and Shen, W.X. (1999). An overview of battery technology in electric vehicles. *Proceedings of the 16th International Electric Vehicle Symposium*, CR-ROM.

Chau, K.T., Wong, Y.S. and Chan, C.C. (1999). An overview of energy sources for electric vehicles. *Energy Conversion and Management*, **40**, 1021–39.

Cornu, J.P., Peski, V., Geeter, E.D., Broeck, H.V.D., Dufour, A., Marcenaro, B., Bout, P. and Woortman, M.G. (1994). Hybrid fuel cell battery city bus technology demonstration project. *Proceedings of the 12th International Electric Vehicle Symposium*, pp. 91–9.

Fijalkowski, B.T. and Krosnicki, J.W. (1995). High-density mechanical energy-storing superhigh-speed autodrive- and/or autoabsorbable flywheel. *Proceedings of EVT Conference*, pp. 89–98.

Fält, J., Hylander, J. and Sjöberg, M. (1995). Laboratory tests on a control system for a super-capacitor to be applied in an electric vehicle. *Proceedings of EVT Conference*, pp. 183–93.

Hoolboom, G.J. and Szabados, B. (1994). Nonpolluting automobiles. *IEEE Transactions on Vehicular Technology*, **43**, 1136–44.

Howard, P.F. (1994). Ballard zero emission fuel cell bus engine. *Proceedings of the 12th International Electric Vehicle Symposium*, pp. 81–90.

Linden, D. (1984). *Handbook of Batteries and Fuel Cells*. McGraw-Hill, New York.

Linden, D. (1995). *Handbook of Batteries*, (2nd edition). McGraw-Hill, New York.

Morrow, H. (1995). Nickel–cadmium batteries for EVs. *Electric & Hybrid Vehicle Technology*, UK & International Press, 91–4.

Robinson, T. (1995). *Electric & Hybrid Vehicle Technology '95*. UK & International Press, Surrey.

Trippe, A.P., Burke, A.F. and Blank, E. (1993). Improved electric vehicle performance with pulsed power capacitors. *SAE SP-969*, pp. 89–93.

Vermeulen, I.F. (1994). Characterisation of PEM fuel cells for electric drive trains. *Proceedings of the 12th International Electric Vehicle Symposium*, pp. 439–48.

Wong, Y.S., Chau, K.T. and Chan, C.C. (1999). Hybridisation of energy sources in electric vehicles. *Proceedings of the 16th International Electric Vehicle Symposium*, CD-ROM.

Wong, Y.S., Chau, K.T. and Chan, C.C. (2000). Optimization of energy hybridization in electric vehicles. *Proceedings of the 17th International Electric Vehicle Symposium*, CD-ROM.

7 EV Auxiliaries

Similar to ICEVs, EVs need many auxiliary devices or subsystems to enhance the vehicle's manoeuvrability and driver's comfortability. Some of these auxiliaries are dedicated to EVs such as the battery charger, battery management system (BMS) and regenerative braking system, whereas the others are common to those of ICEVs such as the air-conditioner, auxiliary power supply, power steering system, navigation system, lighting, defroster, screen-wiper and radio. Even common to both EVs and ICEVs, most EV auxiliaries (including the air-conditioner, auxiliary power supply, power steering system and navigation system) have their unique features and operation (Chan and Chau, 1997).

Incorporating all EV auxiliaries relating to electrical energy control, the energy management system (EMS) plays an important role in modern EVs. As shown in Fig. 7.1, the EMS makes use of sensory inputs from various EV subsystems to select the battery charging scheme, to indicate the battery SOC, to predict the remaining driving range, to manage the battery operation, to modulate temperature control inside the EV compartment, to adjust lighting brightness, and to recover regenerative braking energy for battery charging. It should be noted that

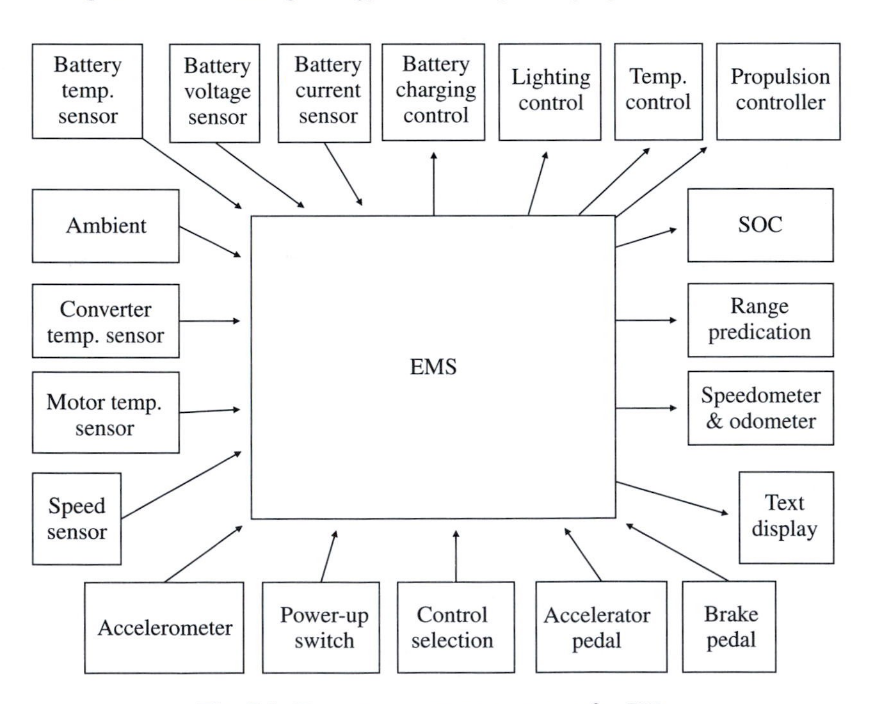

Fig. 7.1. Energy management system for EVs.

a major subset of the EMS is called the BMS which handles the battery's indication, measurement, prediction and overall management.

7.1 Battery characteristics and chargers

For a practical EV, the battery charger is one of the key auxiliaries, which functions to recharge the discharged batteries of the EV. This EV charger can be on-board or off-board. The on-board one should be designed with small size and lightweight to minimize its drawback to the driving range, hence dedicated to low-power slow-charging purposes. On the other hand, the off-board charger has no limitation on its size and weight, hence it is devoted to offer high-power fast-charging purposes. Modern EVs with the on-board charger can also be recharged through the off-board charger.

Before dealing with the battery charger and BMS, it is necessary to understand the discharging and charging characteristics of those EV batteries (Linden, 1995). In fact, the design of battery chargers is mainly based on the battery charging characteristics, whereas the design of battery indicators mainly depends on the discharging characteristics. In the followings, discussions will be focused on four major rechargeable EV batteries, namely the Pb–Acid, Ni–Cd, Ni–MH and Li–Ion.

7.1.1 BATTERY DISCHARGING CHARACTERISTICS

During the discharging process, the electrochemical redox reactions inside the battery are equivalent to convert the stored chemical energy into usable electrical energy. This process is usually characterised by the battery's discharging current and voltage level. Figure 7.2 shows some typical discharging characteristics (cell voltage vs. discharging time with varying *C* rate) of a Pb–Acid battery, which indicate that the cell voltage generally decreases with the *C* rate (a representation of the discharging current), and also droops with the discharging time (Bode, 1977). Other battery types also exhibit similar characteristics but with a lesser extent. Also, the usable capacity of different batteries (including the Pb–Acid, Ni–Cd, Ni–MH and Li–Ion) generally decreases with the *C* rate as illustrated in Fig. 7.3. For example, if a 100-Ah battery rated at the $C/5$ rate can be discharged at 20 A for 5 h, it will be discharged at $C/10$ rate (or 10 A) for more than 10 h, hence delivering more than the rated capacity. On the contrary, it will be discharged at the $1C$ rate (or 100 A) for less than 1 h, hence delivering less than 100 Ah. Moreover, Fig. 7.4 shows the discharging characteristics (cell voltage vs. DOD) at the $C/5$ rate of different batteries in which the cell voltage reaches the cut-off voltage at the 100% DOD.

Besides the discharging rate, the operating temperature also plays an important role on the discharging characteristics, especially the cell voltage and the usable capacity. As shown in Figs. 7.5 and 7.6, both of the characteristics of cell voltage and usable capacity generally decrease with the reduction in temperatures. The

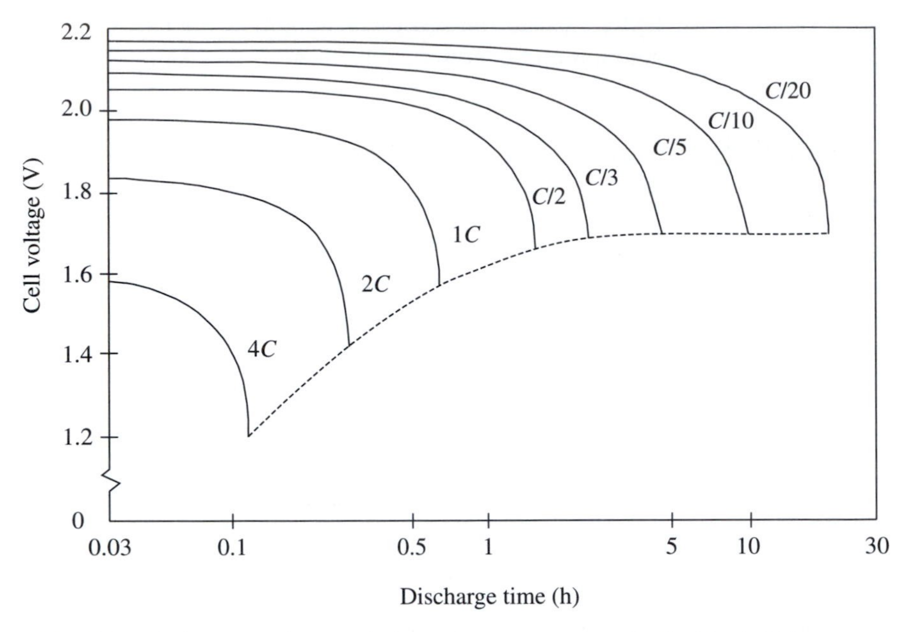

Fig. 7.2. Discharging characteristics (cell voltage vs. time) of a Pb–Acid battery.

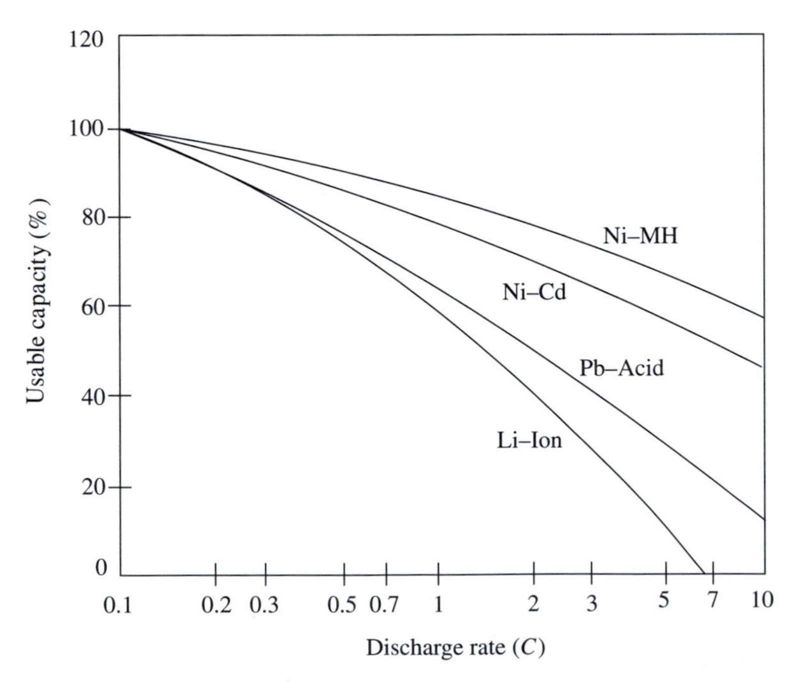

Fig. 7.3. Discharging characteristics (usable capacity vs. C rate) of various batteries.

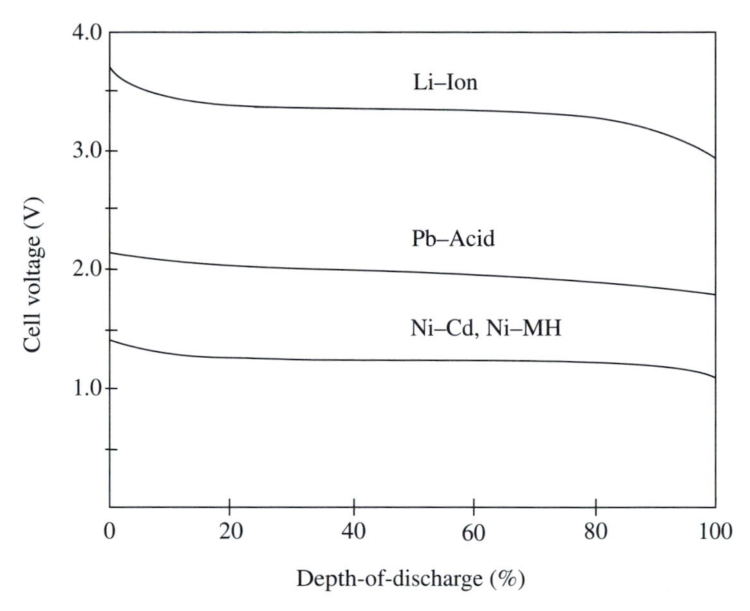

Fig. 7.4. Discharging characteristics (cell voltage vs. DOD) of various batteries.

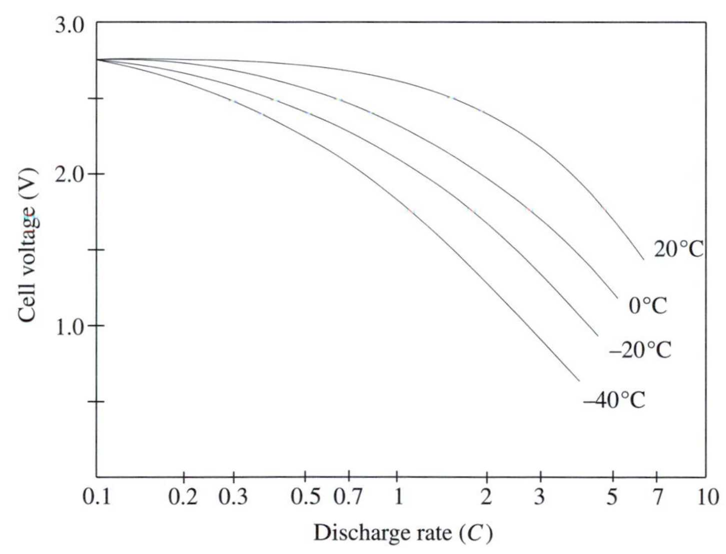

Fig. 7.5. Discharging characteristics (cell voltage vs. C rate) of a Pb–Acid battery at various temperatures.

reason is due to fact that when the temperature is lowered, the reduction in chemical activity and the increase in battery internal resistance are resulted. On the contrary, when the temperature is too high, chemical deterioration inside the

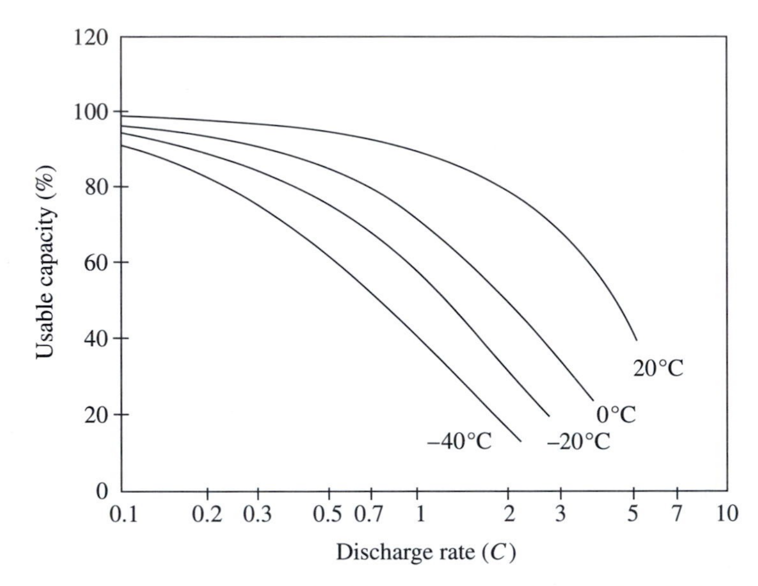

Fig. 7.6. Discharging characteristics (usable capacity vs. *C* rate) of a Pb–Acid battery at various temperatures.

cell and hence a loss of capacity may be resulted. Therefore, the thermal management of the battery environment is a principal task of each EV.

Ageing is also a parameter that affects the cell voltage and usable capacity of different batteries. In general, when the number of discharging cycles increases to certain extent, both the cell voltage and usable capacity decrease gradually and drop off rapidly near the end of cycle life. Compared to the effects of discharging rate and temperature, the ageing effect is relatively mild.

7.1.2 BATTERY CHARGING CHARACTERISTICS

When a battery is run out of its energy storage or its voltage reaches the cut-off voltage, it must stop discharging and needs to be recharged. When the battery is charged, its charging characteristics varies with the charging current, charging time and temperature. Some typical charging characteristics (cell voltage vs. input capacity) of those common EV batteries are shown in Fig. 7.7. It can be seen that different types of batteries have different profiles of charging characteristics.

In general, there are many methods to charge EV batteries because of their different charging characteristics. The key criterion for effective charging is to recharge the battery to its full capacity without causing extended overcharge or excessive temperature. Otherwise, the presence of over-voltage or overheating may result in permanently deteriorating the battery's performance and life, or even causing serious safety hazard. According to the selection of charging voltages and currents, the charging methods for EV batteries can roughly be classified as:

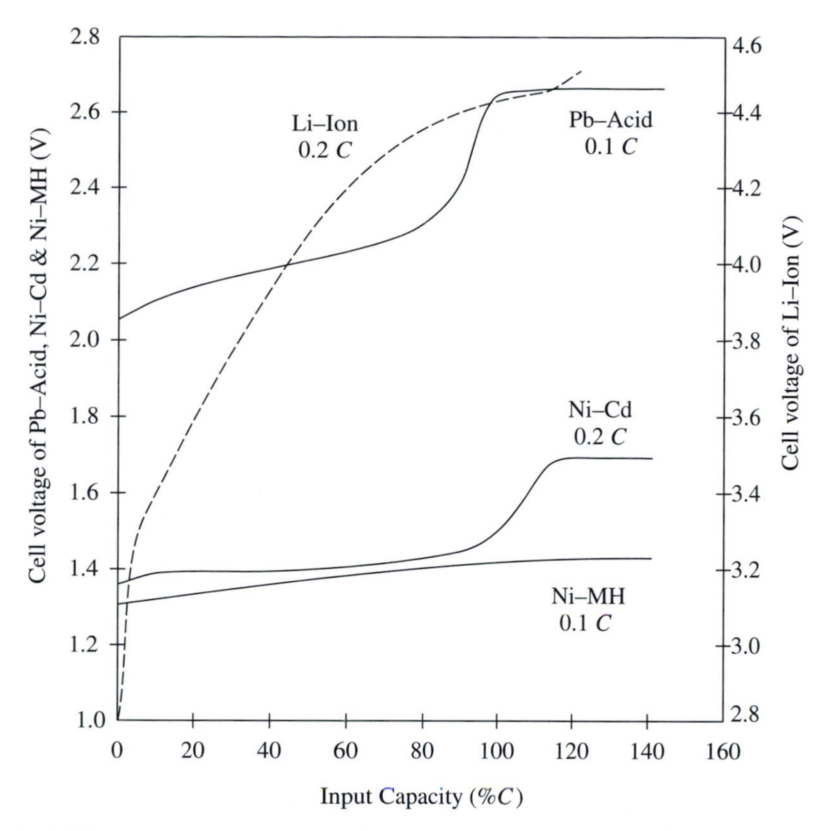

Fig. 7.7. Charging characteristics (cell voltage vs. input capacity) of various batteries.

- controlled voltage charging
- controlled current charging and
- controlled voltage and current charging.

At present, there are some general charging methods catering for a variety of batteries such as the constant-current low-rate charging or pulsating-current high-rate charging. However, these general methods cannot fulfil the stringent EV charging requirements—namely efficient and safe. In order to simultaneously achieve both high charge efficiency and high safety, the charging methods generally vary with the types of batteries (actually depending on their charging characteristics).

Pb–Acid battery

The Pb–Acid battery can generally be charged at any rate that does not produce excessive gassing, over-voltage or overheating. It can absorb a very high current during the early stage of charging, but should be limited to a safe current as the

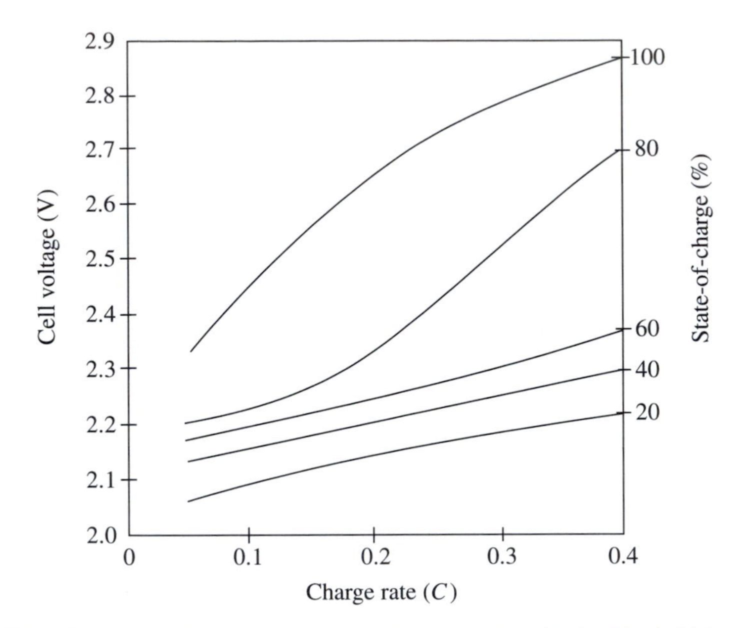

Fig. 7.8. Charging characteristics (cell voltage vs. C rate) of a Pb–Acid battery.

battery becomes charged. Figure 7.8 shows the relationship of cell voltage to the SOC and the charging rate. It indicates that a deeply discharged battery (such as 20% SOC) can absorb high charging rates with the cell voltage remaining relatively low and well below the gassing voltage (about 2.4 V). However, as the battery becomes charged, the cell voltage increases to excessive high values if the charging current is maintained at the high rate, leading to overcharge and gassing.

Recently, an advanced multiple-step charging method has been developed for the Pb–Acid battery applying to an EV (Chan and Chu, 1990). Its key feature is to minimise the possibility of overcharge and to improve the charge efficiency by using three steps of partial charging. As shown in Fig. 7.9, the charging procedure mainly includes four steps: namely the S1, S2, S3 and S4. The charging rate in the S1 is chosen at the $C/6$ rate before reaching the gassing voltage. Then, the charging current is cut off so that the cell voltage is allowed to decay until it reaches a predefined value, namely 2.2 V. Consequently, the S2 and S3 take a similar action which charges up to the gassing voltage and then decays down to the predefined voltage, except that their charging currents are respectively 75% and 50% of that in the S1. Finally, the charging current in the S4 is chosen at 25% of that in the S1, which continues to charge over the gassing voltage. The S4 will be terminated when there is no detectable rise in the cell voltage within 15 min. After completing the four main steps, pulse equalization will be activated. Based on the 25% charging current of the S1, the equalization process starts up when the cell voltage decays down to another predefined voltage, namely 2.13 V, and then stops at

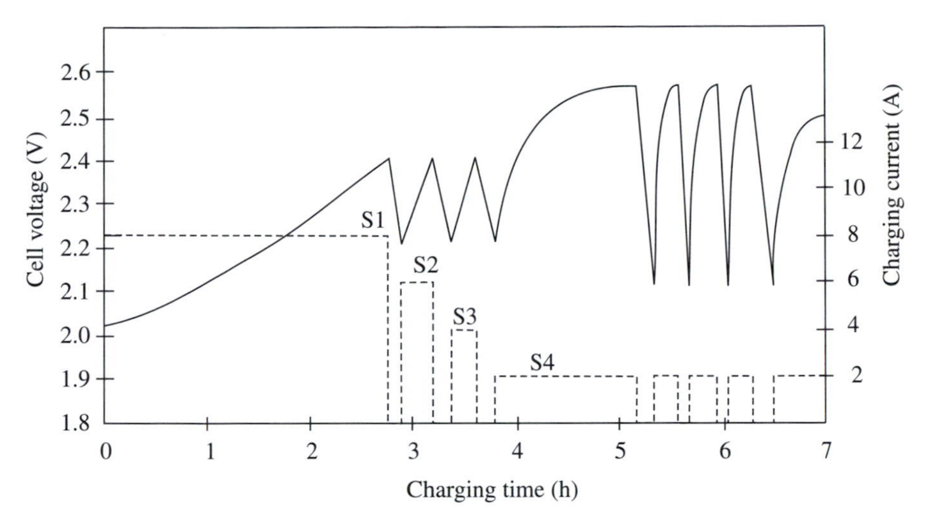

Fig. 7.9. Multiple-step charging method for a Pb–Acid battery.

which there is no detectable voltage rise for 10 min. The whole process will be repeated for three times to ensure that all battery cells can be fully charged, hence prolonging the battery cycle life. Finally, trickle charging is applied to compensate the battery self-discharge. This trickle charging current can be the same as that for equalization or the $C/100$ rate. It is activated whenever the open-circuit voltage drops below 2.13 V, and is stopped when there is no detectable voltage rise within 5 min.

Ni–Cd Battery

The Ni–Cd battery generally employs constant-current charging with voltage cut-off control because of its simplicity and effectiveness. Figure 7.10 shows its typical charging characteristics (cell voltage vs. input capacity) when the charging current is at the $C/10$ rate. It should be noted that if the battery is fully charged, there will be a significant rise in its cell voltage.

For commercially available Ni–Cd battery chargers, a commonly adopted charging method is to regularly apply about $1C$ current to charge the battery and then terminate when the cell voltage reaches a predefined upper voltage, namely 1.5 V. The charging process will be re-initiated whenever the open-circuit voltage decays down to a predefined lower voltage, namely 1.36 V. This simple on-off action continues at decreasing frequencies and decreasing on-time periods, thereby maintaining the battery in a float condition at 100% SOC. Due to the influence of temperature on the charging characteristics, both of these upper and lower voltages should be adjusted according to a temperature coefficient of $-4\,\mathrm{mV}/^\circ\mathrm{C}$. Figure 7.11 illustrates this simple and effective charging method for a Ni–Cd battery.

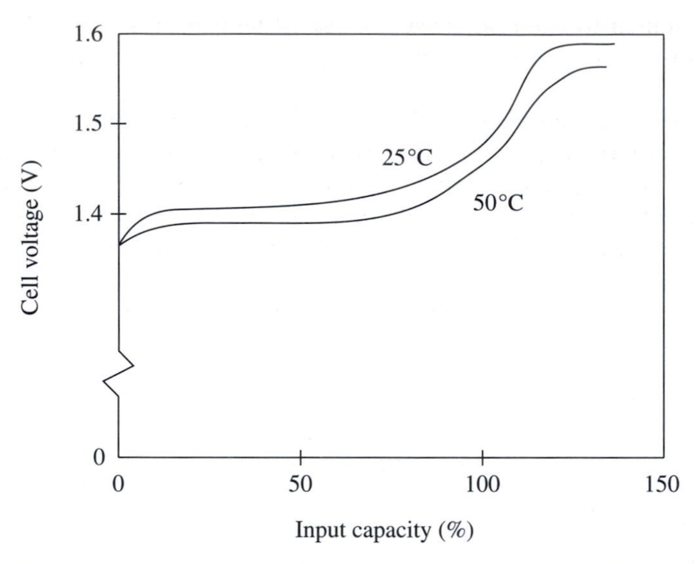

Fig. 7.10. Charging characteristics (cell voltage vs. input capacity) of a Ni–Cd battery.

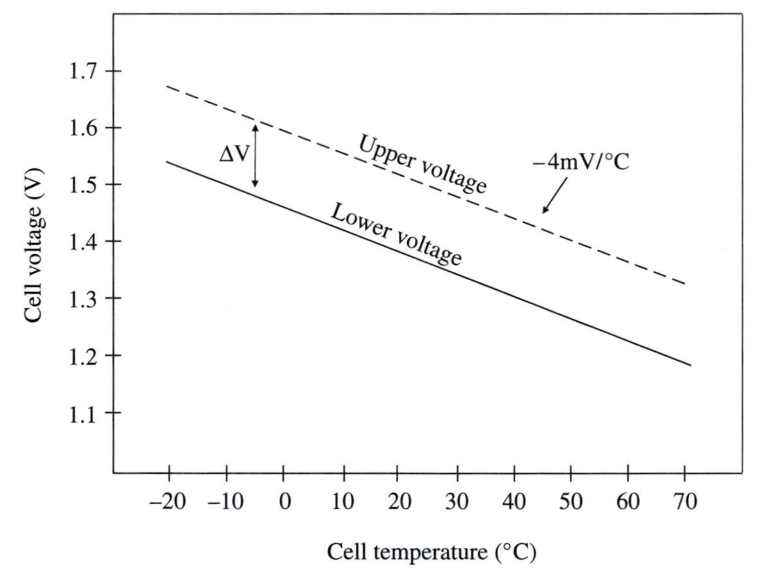

Fig. 7.11. On-off charging method for a Ni–Cd battery.

Ni–MH battery

Similar to the Ni–Cd, the Ni–MH battery generally adopts constant-current charging methods. However, it is sensitive to overcharge so that its charging rate should be limited to avoid an excessive rise of temperature. Its typical charging characteristics (cell voltage vs. input capacity) at different C rates are shown in Fig. 7.12.

Since proper charging control is highly desirable to avoid overcharging the Ni–MH battery, there are three major methods to terminate the charging process before overcharging, namely the cell voltage drop ($-\Delta V$) or plateau ($0\Delta V$), the temperature cut-off (TCO) and rate of temperature rise ($\Delta T/\Delta t$) as shown in Fig. 7.13. The method using the rate of temperature rise is preferable because it

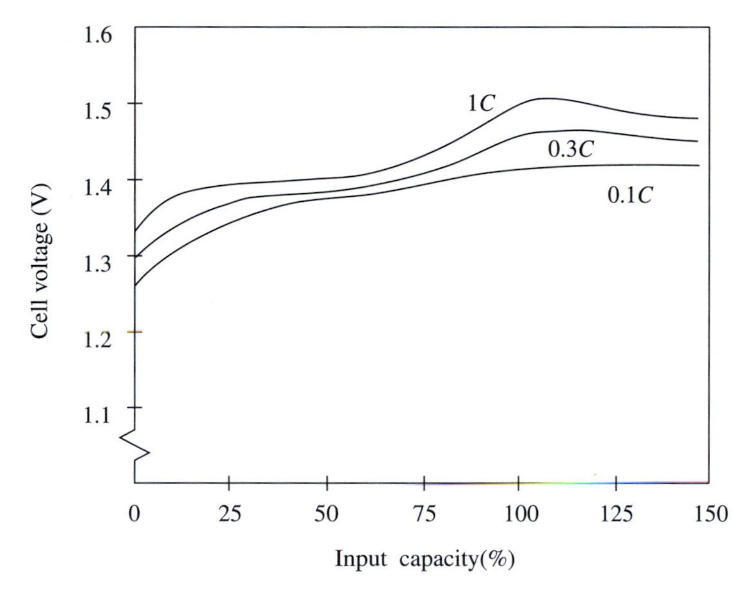

Fig. 7.12. Charging characteristics (cell voltage vs. input capacity) of a Ni–MH battery.

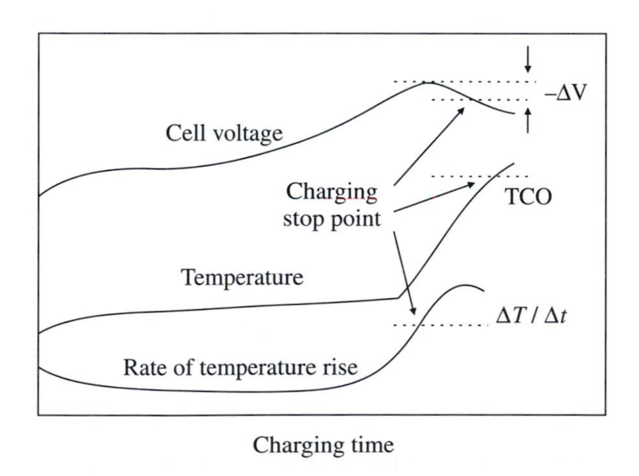

Fig. 7.13. Charging control methods for a Ni–MH battery.

can eliminate the influence of the ambient temperature and can also provide long cycle life. Typically, the rate of 1 °C/min is adopted for activating the termination. Nevertheless, all these methods can be used together to complement each other. For example, the detection of $\Delta T/\Delta t$ activating at 1 °C/min can be complemented by the TCO at 60 °C. After the termination, the Ni–MH battery can compensate its self-discharge by applying trickle charging at the $C/20$ rate.

Li–Ion battery

Since the application of the Li–Ion battery to EVs is still under development, the corresponding charging methods have not yet been well developed. Nevertheless, some methods have recently been proposed. A typical one is called the multistage constant-current charging method. In the initial stage, the charging current is maintained at its rated value until any one of the battery cells reaches a predefined voltage (normally close to the full-charge cell voltage). Subsequently, an electronic control circuit installed for each cell will bypass its charging current to avoid any possible overcharge, and will also activate another stage that adopts a lower value of constant charging current. This process repeats until the constant charging current is gradually reduced to the predefined minimum. After employing this minimum constant current to charge the battery for a predefined period of time, the whole charging process will be terminated.

7.1.3 BATTERY CHARGERS

Depending on whether the battery charger is installed inside or outside the EV, it is generally classified as on-board or off-board. The on-board charger is designed with a low charging rate, and is dedicated to charge the battery for a long period of time (typically 5–8 h). Due to the limitation of allowable payload and space of the EV, the on-board charger needs to be lightweight (typically less than 5 kg) and compact. Since both the charger and the BMS (which functions to monitor the battery's voltage, temperature and SOC) are inside the vehicle, they can easily communicate to each other based on the internal wiring network. The corresponding charging method is predefined to suit the battery used in the EV. On the other hand, the off-board charger is designed with a high charging rate, and has virtually no limitation on its weight and size. Since the off-board charger and the BMS are physically separated, they should have a reliable communication by wiring cables or wireless radios. Based on the information of the battery's type, voltage, temperature and SOC supplied by the BMS, the off-board charge will adopt a proper charging method to charge the battery without any excessive overcharge and overheating.

Based on the modes of energy transfer from the power supply to the EV, the battery chargers can be named as conductive or inductive chargers. The former is by plugging an ac power wire into the socket on the EV, whereas the latter is based on energy transfer from the power supply to the EV via magnetic induction coupling. Since the conductive charger takes the definite advantages of simplicity and high efficiency while the inductive charger has the distinct merits of easy to use and free from electric shock under all-weather conditions, they are most welcome for indoor charging and outdoor charging, respectively. It should be noted that there are no contradiction between the classification of on-board/off-board and conductive/inductive chargers. In principle, both on-board and off-board chargers can be based on conductive or inductive modes of energy transfer.

Conductive charger

The conductive charger for EVs takes the advantages of maturity, simplicity and low cost because it simply makes use of plugs and sockets to conduct electrical energy via physical metallic contacts. There are many circuit configurations suitable for the conductive charger. Figure 7.14 shows two basic charger circuits, with and without regulation. For practical application to EVs, we have recently developed a microprocessor based charger circuit as shown in Fig. 7.15 which adopts multiple-step constant-current charging control for the Pb–Acid battery (Chan and Chu, 1990). Similar schematics except supplied by three-phase power are applicable to the fast-charging conductive charger.

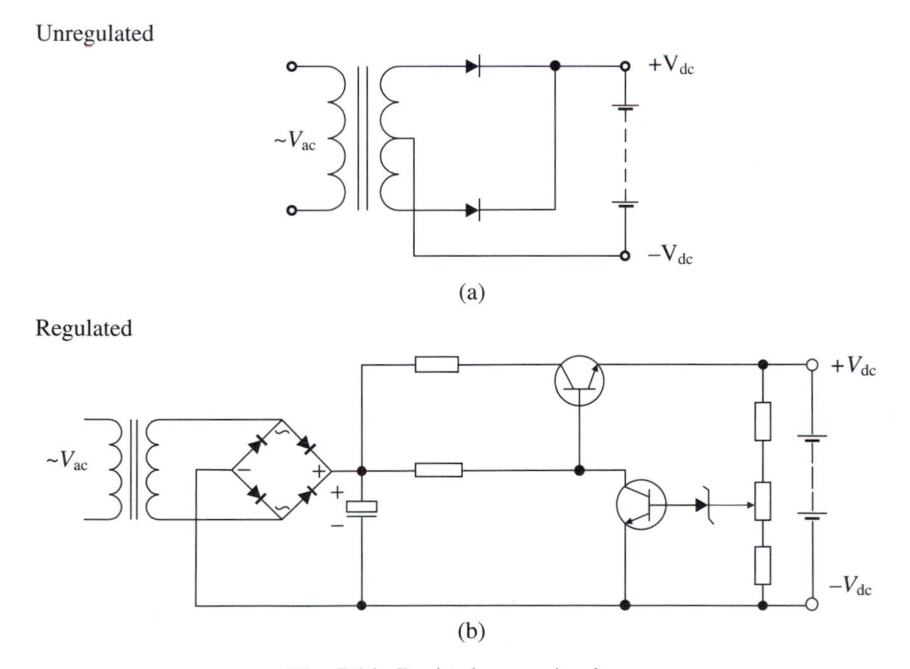

Fig. 7.14. Basic charger circuits.

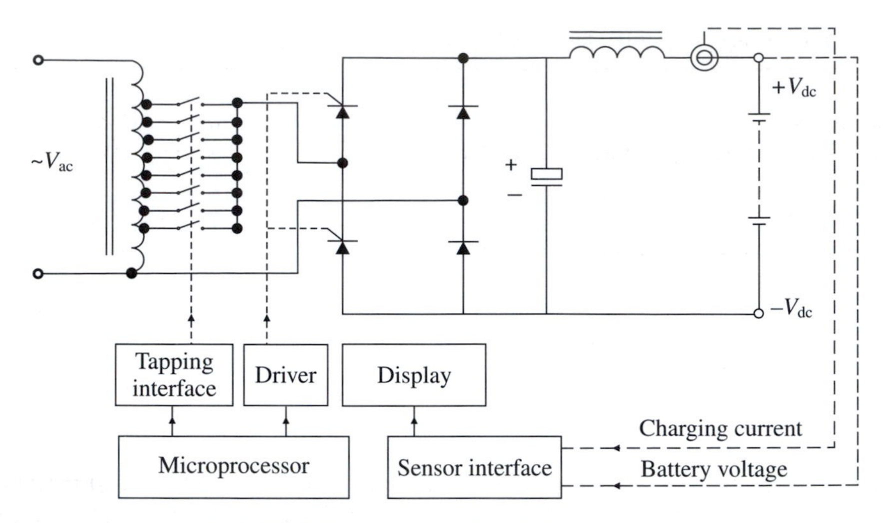

Fig. 7.15. Microprocessor based charger circuit.

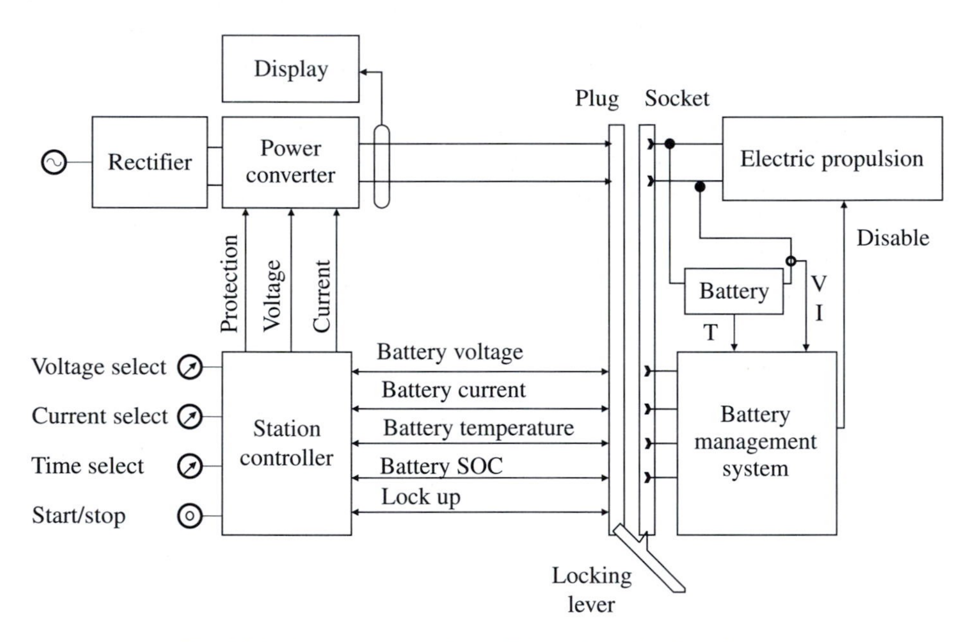

Fig. 7.16. Arrangement of an off-board conductive charger.

Figure 7.16 shows a typical arrangement of an off-board conductive charger for the EV. The off-board charger consists of a rectifier to convert ac input power into dc charging power, and a power converter to regulate the flow of the dc power. This dc power is fed to the EV battery via the charging cable with the plug that

Table 7.1 Standard power levels of conductive chargers

Level 1 Convenience Plugs into a common grounded wall outlet	1-phase 120 V ac, 15 A* ac 1.44 kW (max)
Level 2 Private/public Requires EV supply equipment installation	1-phase 208–240 V ac, 30–60 A ac 14.4 kW (max)
Level 3 Opportunity Requires commercial equipment installation	3-phase 208–600 V dc, 400 A dc 240 kW (max)**

*Receptacle rating (maximum continuous current of 12 A).
**Maximum allowed by standards.

mates with the socket on the EV. A locking lever is also equipped to assist insertion and extraction of the plug, as well as to activate a signal to confirm the lockup. In the absence of this signal, the charger will not allow any power transfer so as to ensure safety precaution. Based on the communication between the off-board charger and the on-board BMS, the charging dc power is on-line regulated by the power converter. The off-board charger also displays the charging voltage, charging current and charging energy as well as the desired charging fee (Nor, 1992).

According to the SAE J1772 standard, there are three power levels for EV conductive chargers. As listed in Table 7.1, these three charging power levels and their charging currents essentially satisfy all charging needs of various EVs (Friedman, 1997; Griffith and Gleason, 1997). At present, a number of manufacturers, such the Aero Vironment, EVI, Ford, Norvik and SCI Systems, are actively involved in the development of SAE J1772 compliant EV conductive chargers.

Inductive charger

Inductive charging allows electrical energy being transferred from chargers to EVs by induction. As shown in Fig. 7.17, the principle of inductive charging is based on the magnetic coupling between two windings of a high-frequency transformer. One of the windings is installed in the charger terminal while the other is embedded in the EV. Firstly, the main ac supply with a frequency of 50–60 Hz is rectified and converted to a high-frequency ac power of 80–300 kHz within the charger module, then the high-frequency ac power is transferred to the EV side by induction, and finally this high-frequency ac power is converted to dc power for battery charging. The whole process is free from any metallic contacts between the charger and the EV, hence providing a very convenient way for battery charging. This inductive charging approach has a distinct merit over the conductive one—inherently safe under all-weather operation (including rainy, snowy and dirty conditions). The main drawbacks are the high investment cost and inevitable induction losses.

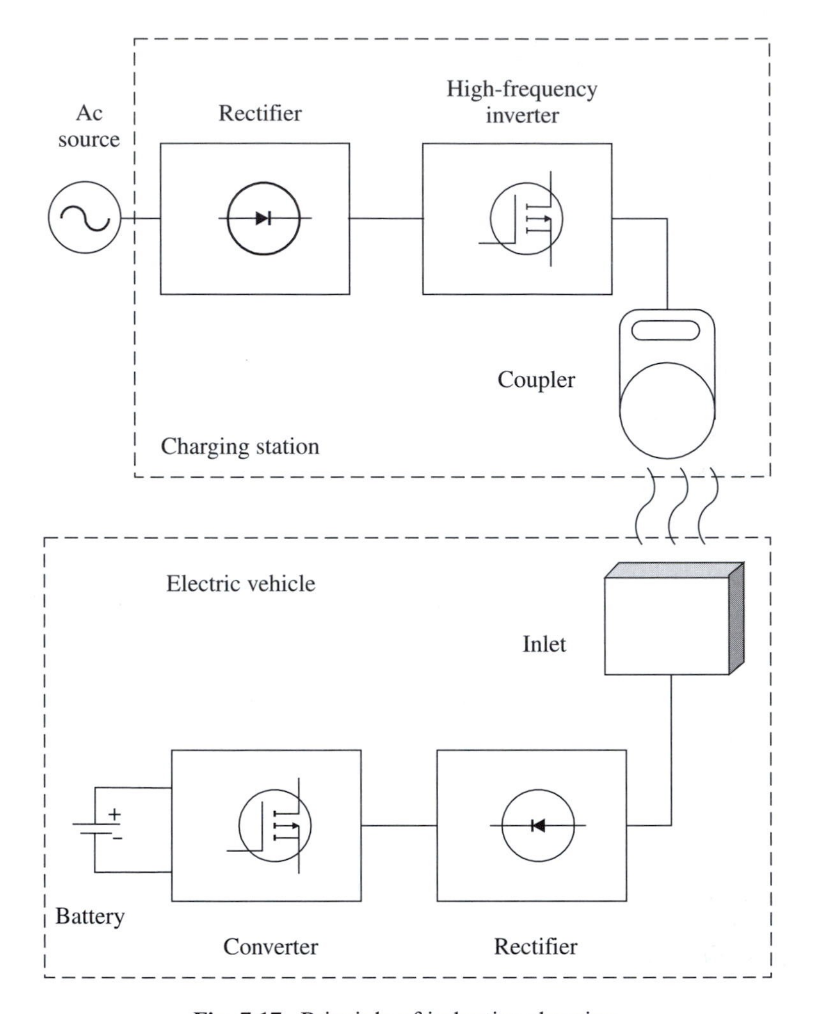

Fig. 7.17. Principle of inductive charging.

In 1995, the SAE published the SAE J1773 standard—Electric Vehicle Inductive Charge Coupling Recommended Practice. It specializes the core and winding for both the coupler and inlet. Manganese–zinc ferrite is recommended as the material for the cores, offering low core losses at high frequencies (80–300 kHz) and high level of saturation. Since the operating frequency is so high that the loss caused by the skin effect in the winding is substantial, the Litz wire is recommended as the winding for both the coupler and inlet. This Litz wire consists of strands and each strand is insulated so as to reduce the loss caused by the skin effect. In order to reduce the high switching losses of the high-frequency inverter, soft-switching techniques, such as zero-current switching (ZVS) and/or zero-voltage

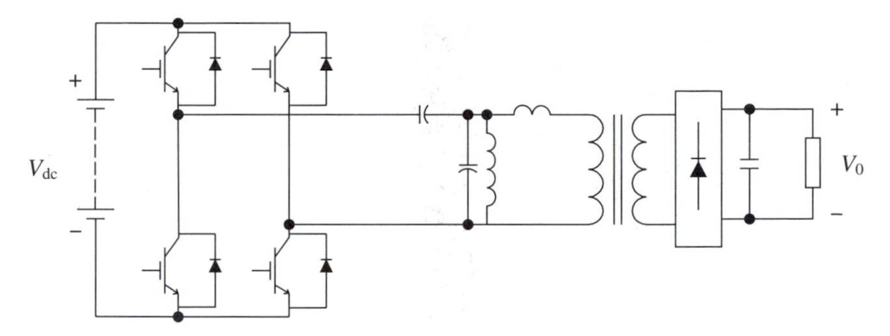

Fig. 7.18. Soft-switching power converter for inductive charging.

switching (ZVS), can be employed. Figure 7.18 shows a typical soft-switching circuit for inductive charging in which the ZCS series-resonant inverter circuit belongs to the charger and coupler while the conventional rectifier is inside the inlet. During inductive charging, information on the battery's voltage, current, temperature and SOC need to be transferred to the off-board charger. These communications are usually realised by radio-frequency (RF) or microwave signals because of the absence of electric connection between the charger and the vehicle.

Magne Charge has actively engaged in the development of EV inductive chargers, ranging from low power of 1 kW to high power of 120 kW (Kosowski, 1997). All of them adopt the same coupler and inlet as recommended by the SAE J1773 standard. Besides that, there is the automatically operated inductive charger for EVs. This automatic park-and-charge concept is to incorporate the inductive charger into the parking lot so that the EV driver needs no bothering about those cumbersome and dangerous charging cables. The use of this system is very easy and the charging process takes place automatically once the driver ensures that the EV is correctly parked. In France, this automatic inductive charger has recently been implemented and tested by the Praxitele consortium (including the INRIA, INRETS, Renault, CGEA, Dassault Automatismes et Télécommunications and EDF).

7.2 Battery indication and management

The EV battery is the key component to influence the whole EV operating behaviours, such as the driving range, acceleration rate and hill-climbing capability. Since the battery characteristics and behaviours are highly nonlinear, a dedicated BMS is highly necessary. The BMS is generally composed of some sensors (voltage, current and temperature), a microprocessor-based control unit and some I/O interfaces. The basic functions of the BMS are to monitor the battery's working behaviors (voltage, current and temperature), hence predicting its SOC and the corresponding residual driving range, and to manage the battery's

Table 7.2 Main tasks of a BMS

Tasks	Input sensing	Output control
Prevention of battery overcharge	Battery voltage, current and temperature	Battery charger
Avoidance of battery over-discharge	Battery voltage, current and temperature	Power converter of electric motor
Control of battery temperature	Battery temperature	Cooling or heating device
Balancing of module voltages and temperatures	Battery voltage and temperature	Battery balancing unit
Prediction of SOC and residual driving range	Battery voltage, current and temperature	Display unit
Battery diagnosis	Battery voltage, current and temperature	Off-line analysing unit (PC)

operation (avoiding over-discharge, overcharge, over-temperature and unbalance of module voltages) so as to maximize its storage capacity and cycle life. The main tasks of a BMS as well as the corresponding input sensing and output control are summarized in Table 7.2.

7.2.1 BATTERY INDICATING METHODS AND DEVICES

Similar to the fuel gauge of ICEVs, the SOC indicator functions to provide the EV driver accurate information on how much energy content remained in the battery. Hence, the driver can plan the future driving range before recharging. Theoretically, the SOC is defined as:

$$\mathrm{SOC} = \frac{C_\mathrm{r}}{C_\mathrm{t}} \times 100\%,$$

where C_r and C_t are respectively the residual and total usable coulometric capacity in Ah of the battery. In general, as shown in Table 7.3, C_t is significantly affected by the discharging rate or current I. Thus, practically, the SOC is defined as:

$$\mathrm{SOC}_I = \frac{C_{\mathrm{r}I}}{C_{\mathrm{t}I}} \times 100\%,$$

where SOC_I, $C_{\mathrm{r}I}$ and $C_{\mathrm{t}I}$ are the SOC, residual capacity and total usable capacity at a constant discharging current I, respectively.

A well-known relationship between the usable capacity and the discharging current is expressed by the Peukert equation:

$$C_{tI} = KI^{(1-n)},$$

Table 7.3 Effect of discharging rate on total usable capacity of batteries

Discharging rate (C)	C_t/C_∞
C	0.487
$C/3$	0.637
$C/5$	0.711
$C/10$	0.801
$C/20$	0.875
$C/40$	0.923
$C/60$	0.945
$C/100$	0.962
C/∞	1.000

where K and n are constants (Bode, 1997). These two constants can be determined by:

$$n = \frac{\log(t_2/t_1)}{\log(I_1/I_2)}$$

$$K = I_1^n t_1 = I_2^n t_2,$$

where I_1 and I_2 represent the high-rate and low-rate discharging currents with the discharging times of t_1 and t_2, respectively. For example, a 50-Ah exhibits different usable capacities at various discharging rates as listed in Table 7.4. Considering the maximum and minimum discharging rates ($1.5C$ and $C/10$) rate of discharge, we have $I_1 = 75$A, $I_2 = 5$A, $t_1 = 0.36$ h and $t_2 = 10.92h$. Thus, $n = 1.26$ and $K = 82.97$. Hence, the usable coulometric capacity of this battery at different discharging currents can be written as $C_{tI} = 82.97 I^{-0.26}$.

Besides the Peukert equation, there are other methods to predict the total usable capacity at different discharging rates or currents. A typical one is to employ a polynomial as described below:

$$C_{tI} = It$$

$$\log t = K_0 + K_1 \log I + K_2 \log^2 I + K_3 \log^3 I,$$

Table 7.4 Usable capacities at various discharging rates

Discharging rate (C)	Discharging current (A)	Discharging time (h)	Usable capacity (Ah)
0.1	5	10.92	54.6
0.2	10	4.73	47.3
0.4	20	1.97	39.4
1	50	0.67	33.5
1.5	75	0.36	27.0

where K_0, K_1, K_2 and K_3 are the coefficients to relate the discharging times t and the corresponding I. By adopting this method, the accuracy can be as high as 1% over a wide range of discharging rates.

Having formulated C_{tI} with respect to I, we need to know C_{rI} to deduce the SOC. There are many methods that have been developed to measure C_{rI} or to directly indicate the SOC. Those viable methods are the specific gravity, open-circuit voltage (OCV), constant-current voltage and coulometric capacity measurements:

(1) Specific gravity—it is applicable to those batteries that the specific gravity of the electrolyte depends on the concentration of the electrolyte, hence the SOC. The specific gravity measurement is generally based on a hydrometer. Although this method can yield a reasonable accuracy on the SOC estimation, it is impractical for continual battery operation nor the sealed type battery. Also, the specific gravity measurement needs a lengthy stabilisation period after charging or discharging because of the slow diffusion rate of the electrolyte.
(2) Open-circuit voltage (OCV)—it is a simple and convenient method, but is applicable to those batteries that the OCV significantly varies with the battery SOC. Figure 7.19 shows a typical linear relationship between the OCV and the

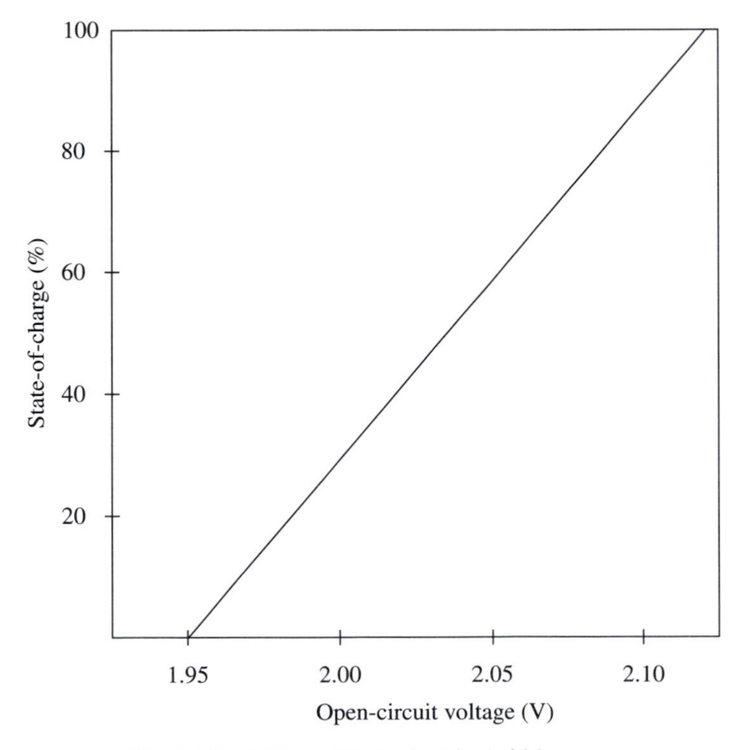

Fig. 7.19. SOC vs. OCV of a Pb–Acid battery.

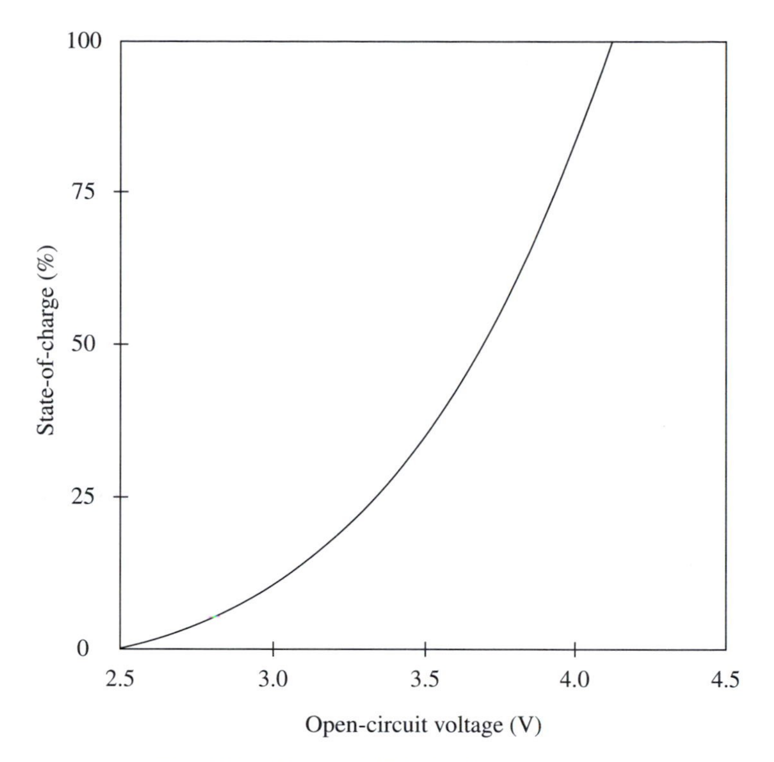

Fig. 7.20. SOC vs. OCV of a Li–Ion battery.

SOC of a Pb–Acid battery at the $C/10$ rate, whereas Fig. 7.20 is a nonlinear relationship between the OCV and the SOC of a Li-Ion battery at the $C/10$ rate. The key drawback of this method is due to the fact that the OCV generally needs a lengthy stabilization period (typically 12 h) after charging or discharging.

(3) Constant-current voltage—provided that the load current is constant, the load voltage is directly proportional to the OCV, hence estimating the battery SOC. However, this constant-current situation is impractical for variable-load applications such as EVs.
(4) Coulometric capacity—it simply counts the Ah that has been taken out or put into the battery. It can be applied to estimate the SOC of all batteries provided that the total usable coulometric capacity as represented by a Peukert equation is known. This method does provide reasonable accuracy for short-term estimation of the battery SOC; however, it suffers from the accumulation of errors over a long period of estimation.

Since the OCV method takes the merit of high accuracy under stabilized conditions while the coulometric capacity method possesses the merit of high accuracy under short-term operation, a combined approach is becoming attractive. This

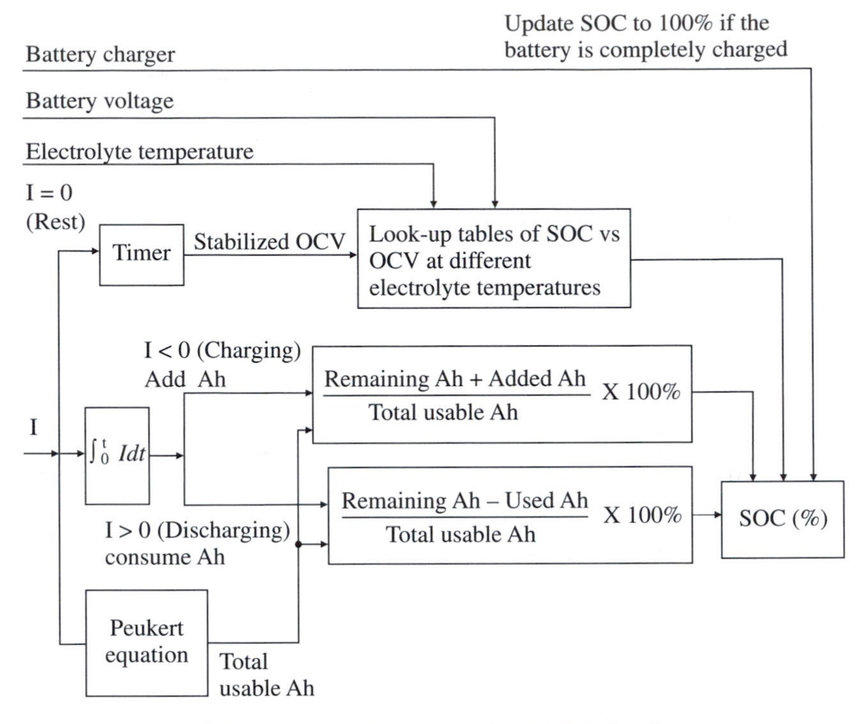

Fig. 7.21. Combined approach for SOC indication.

combined approach is based on the use of coulometric capacity method during operation and the corresponding cumulative error of the SOC is regularly compensated by the OCV method. Figure 7.21 illustrate the principle of this approach. Nevertheless, the regularity of compensation heavily depends on the stabilization period of the battery voltage. In fact, the OCV of most batteries desires a stabilization period of at least 12 h. Figure 7.22 shows the recovery characteristics of the OCV of a typical Pb–Acid battery under various SOC conditions after discharging. Recently, a prediction method has been developed which can estimate the OCV of a Pb–Acid battery within 5-min stabilization period after discharging (Aylor *et al.*, 1992). This 5-min prediction method essentially employs two logarithmic asymptotes, which are based on two measurements at the 1st and 5th min after stabilization, to approximate the recovery characteristic after discharging. However, it may not provide accurate prediction when there are numerous intermittent regenerative charging for practical EV applications.

Very recently, a new estimation method to directly determine the battery available capacity (BAC), rather than the SOC, has been purposely developed for EVs (Chan *et al.*, 2000a). This method can significantly improve the estimation accuracy of the BAC by using artificial neural networks and fuzzy logics, in which the variable discharge current regimes as well as the temperature and aging influences

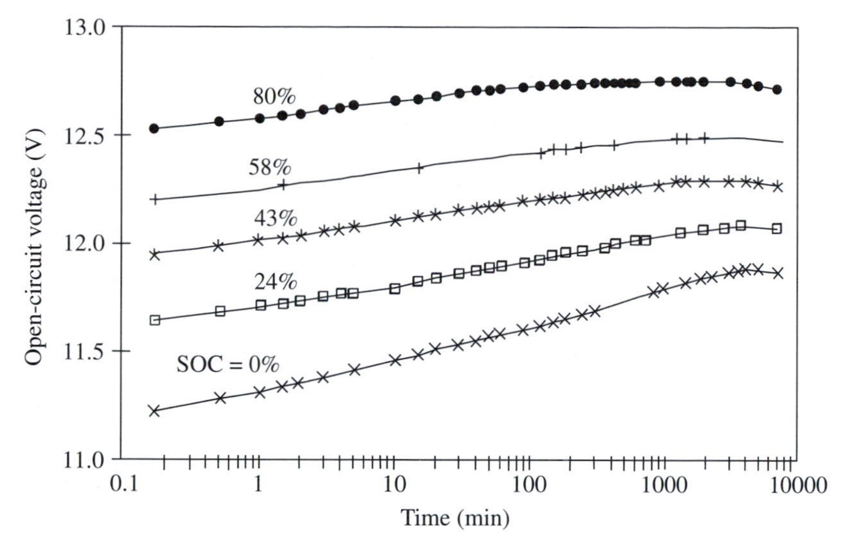

Fig. 7.22. OCV recovery characteristics of a Pb–Acid battery.

are taken into account. Also, the proposed method can readily be real-time implemented in EVs (Chan *et al.*, 2000b).

Besides the battery SOC indicator, there are a number of battery indicators making use of the knowledge of SOC. Particularly, the residual energy indicator and residual range indicator are desirable for the EV driver. Both the SOC and residual energy indicators work as the fuel gauge of the ICEV because they show the remaining coulometric and energy capacities of the EV battery. Increasingly, the residual range indicator can predict the remaining driving range before using up the energy stored in the EV. It should be noted that this residual range indicator can only provide a short-term estimation because the residual range heavily depends on the driving habit and road load.

Based on the SOC indication, the residual coulometric capacity can be easily calculated provided that the total usable capacity is known. Then, the residual energy of the battery can readily be obtained by multiplying the residual coulometric capacity to the average voltage (averaging the instantaneous voltage and the cut-off voltage). Hence, based on the available or pre-defined energy consumption parameter of the EV, the residual driving range can finally be estimated. In general, this energy consumption parameter is expressed as the travelling distance per consumed energy within the last 5 min. By on-line updating this parameter, the residual range indicator can offer a reasonably accurate short-term prediction.

The development of various EV battery indicators, namely the SOC indicator, residual energy indicator and residual range indicator, has been continually fuelled by the improvement of microelectronic devices such as microprocessors, microcontrollers and memories. For example, the University of Hong Kong has developed a battery indicating and monitoring system which includes the battery

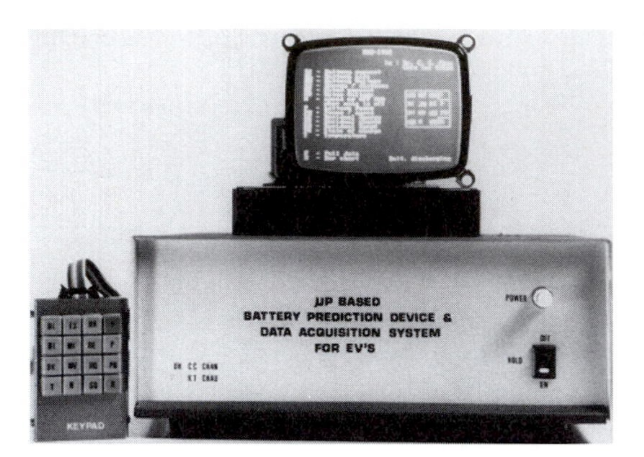

Fig. 7.23. Battery indicating and monitoring system.

SOC indication, residual energy indication, residual range prediction and data acquisition (Chan and Chu, 1988). Figure 7.23 shows its microprocessor-based hardware, monitor display and keypad interface. On the other hand, Cruising Equipment has developed a commercially available battery indicator, so-called the E-meter (Proctor, 1997). This E-meter can display four important parameters, namely the voltage, current, ampere-hours consumed (or kilowatt-hours consumed) and an estimation of how long the battery will last. Recently, Denso and Batech have also developed similar products for EVs.

7.2.2 BATTERY MANAGEMENT METHODS AND DEVICES

As mentioned before, the BMS handles both battery indication and management. Actually, they share the same sensory inputs (namely the battery voltage, current and temperature), but have different missions. The battery indication and its devices are to provide the battery SOC, residual energy and/or residual range prediction. On the other hand, the battery management and its devices are to provide battery protection and diagnostics. For example, some advanced batteries such as the Ni–MH and Li–Ion can be permanently damaged if they are over-charged or over-discharged due to cell–cell capacity deviations.

By on-line monitoring the battery voltage and current, proper battery management devices of the BMS function to prevent the battery from over-current discharge (the discharging current is higher than the maximum allowable value), under-voltage discharge (the battery voltage during discharging is lower than the cut-off voltage), over-current charge (the charging current is higher than the maximum allowable value), and over-voltage charge (the battery voltage during charging is higher than the gassing voltage). These threshold values generally vary with different battery types, battery models, discharging/charging currents, operating temperatures and ageing.

There are three possible ways to control the cell–cell voltage variation so that the difference in the cell–cell SOC can be minimized, namely the equalization charging,

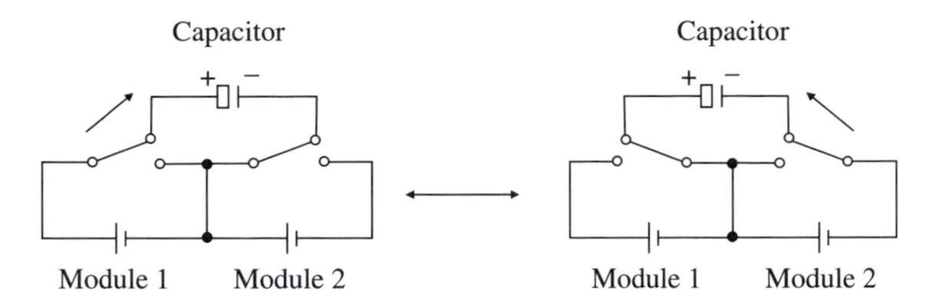

Fig. 7.24. Principle of a switched capacitor balancer.

charging current bypass circuit and switched capacitor balancer. Firstly, the equalization charging employs current pulses to charge up all battery cells close to full charge. This method can be applied to almost all kinds of batteries during normal charging. Secondly, the charging current bypass circuit is to divert the charging current once the corresponding battery module has been fully charged. This method is useful for fast charging, particularly important for those overcharge-sensitive batteries such as the Li–Ion. Thirdly, the switched capacitor balancer functions to balance the voltage of all battery modules (Shinpo and Suzuki, 1997). The principle of operation is illustrated in Fig. 7.24 in which two modules are used for exemplification. This balancer consists of capacitors connected in parallel with battery modules and switching circuits to change connections between capacitors and battery modules. If the module 1 has a higher voltage than the module 2, the capacitor will be connected to the module 1 so that the capacitor is charged to the voltage of the module 1. Then, this capacitor is disconnected with the module 1 and connected to the module 2. The module 2 will be charged until the capacitor voltage is equal to the voltage of the module 2. By repeating this action, the voltages of these two modules will be closed to each other and be balanced. This method is particularly useful to those batteries whose SOC are sensitive to the cell voltage. It should be noted that the first and second methods can be used only during the charging process, whereas the third method can be applied whenever necessary. Nevertheless, the third method is the most complicated one.

Since the battery electrical characteristics for both discharging and charging highly depend on the battery electrolyte temperature, a battery management device of the BMS functions to maintain all battery modules within the normal operating temperature range during discharging and charging periods. In general, an air cooling or warming system is chosen in such a way that the required energy consumption is minimum. Figure 7.25 shows a typical layout of battery modules in which the cooling or warming air goes into the gaps between battery modules and blows out of the apertures located on the bottom of the battery tray. Energy consumption can also be reduced by controlling the fan speed in accordance with the battery temperature (Asakura and Kanamaru, 1996).

The final role of the BMS is to acquire all battery operating data (during both charging and discharging), and hence to transfer these data to external devices such as the PC for off-line analyses and diagnoses. This feature offers the

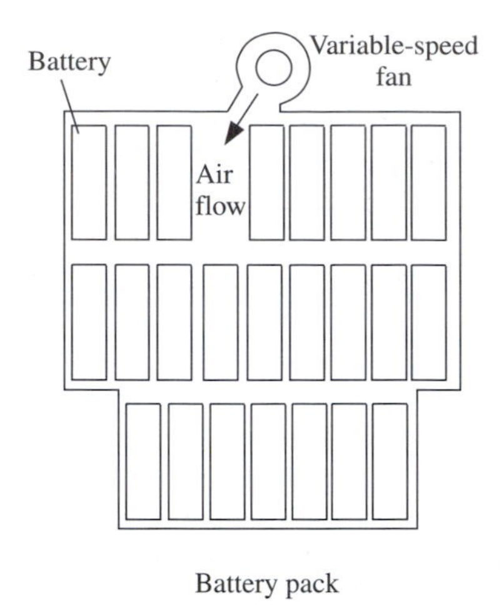

Fig. 7.25. Layout of battery modules.

possibility to build up a comprehensive database of various battery modules. Hence, continual research and development can be carried out to improve the battery design and modelling, especially the effects of temperature and ageing.

As shown in Fig. 7.26, AEVT/Amerigon has developed a commercial EMS which incorporates the BMS and relevant energy control in EVs (Chan *et al.*, 1994).

Fig. 7.26. Commercial EMS. (Photo courtesy of AEVT.)

Based on the sensory inputs including speeds, temperatures, currents and voltages of various EV subsystems, the EMS functions to predict the range for standard or user-specified driving cycles, suggest driving profile changes to use energy more efficiently, direct regenerative energy from braking to the battery pack, select the battery charging algorithm based on the battery SOC and cycle-life history, modulate climate control in response to current driving conditions, and adjust lighting brightness in response to the external light intensity.

7.3 Temperature control units

Similar to ICEVs, EVs need to offer a comfortable interior climate to their drivers. However, conventional temperature control units for ICEVs usually consume significant amount of electrical energy (typically 2–4 kW). This level of energy consumption may greatly reduce the driving range of EVs. In order to provide the drivers comfortable interior climate without sacrificing the EV driving range, dedicated temperature control units are being developed for EVs.

7.3.1 AIR CONDITIONERS

At present, conventional air-conditioners in ICEVs combine both cooling and heating capabilities. The corresponding cooling unit consists of an engine-driven compressor, an outer heat exchanger (condenser), an inner heat exchanger (evaporator) and an expansion valve, whereas the heating unit simply makes use of the waste heat of the engine's cooling water to heat the heater core. With a proper temperature control procedure, both comfortable inner climate and sufficient windshield defrosting can be provided. However, those conventional air-conditioners have relatively high energy consumption which may seriously affect the EV driving range. Even so, EVs have no heat source equivalent to the engine's cooling water, and the heating function has to be provided by the principle of air-conditioning. Therefore, the EV air-conditioners should be specifically designed which can offer both cooling and heating capabilities, and have high efficiency for both functions.

Taking into account the factors of EV operation, the EV air-conditioners should satisfy the following requirements:

- high efficiency
- compact size
- lightweight
- low cost
- low acoustic noise and
- workable in all-weather conditions.

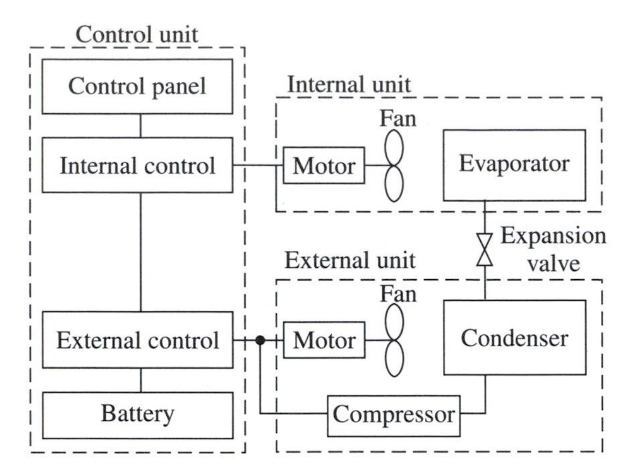

Fig. 7.27. Principle of an EV air-conditioner.

Figure 7.27 shows the schematic of a recently developed EV air-conditioner. It consists of a compressor, an internal heat exchanger (evaporator), an expansion valve, an external heat exchanger (condenser), an internal fan, an external fan and a control unit (Kanetuki and Arakawa, 1996). In order to achieve the above criteria, we need to adopt some advanced technologies which are described as follows:

(1) Heat exchangers—in ICEVs, frost does not form in the internal and external heat exchangers because the air-conditioner works only for cooling. Thus, these heat exchangers consist of fins and flat pipes with good thermal conductivity. However, in EVs, frost is likely to form on these heat exchangers when operating for heating purposes. To avoid this problem, the fin tube type heat exchangers should use round pipes to improve the defrosting efficiency. Moreover, both V-shaped slit fins and corrugated fins can be simultaneously adopted to increase thermal conduction.
(2) Fan motors—although the power level of both internal and external fans is only of about 100 W, a saving in their energy losses can definitely enhance the EV driving range. Thus, two variable-speed PM brushless motors driven by inverters are employed. To further enhance the overall compactness, single-chip low-power inverters can be directly incorporated into the motors.
(3) Compressor—by adopting the scroll compressor, the overall operational noise and vibration can be significantly reduced as compared to that of a rotary compressor, hence improving its reliability. Moreover, the horizontal arrangement of this scroll compressor can enhance easy installation.
(4) Compressor motor—the compressor motor provides the driving force for the scroll compressor. To realize high efficiency, the PM brushless motor driven by an inverter is employed, hence offering the overall efficiency over 80%. This motor usually runs at high speeds (typically over 7000 rpm).

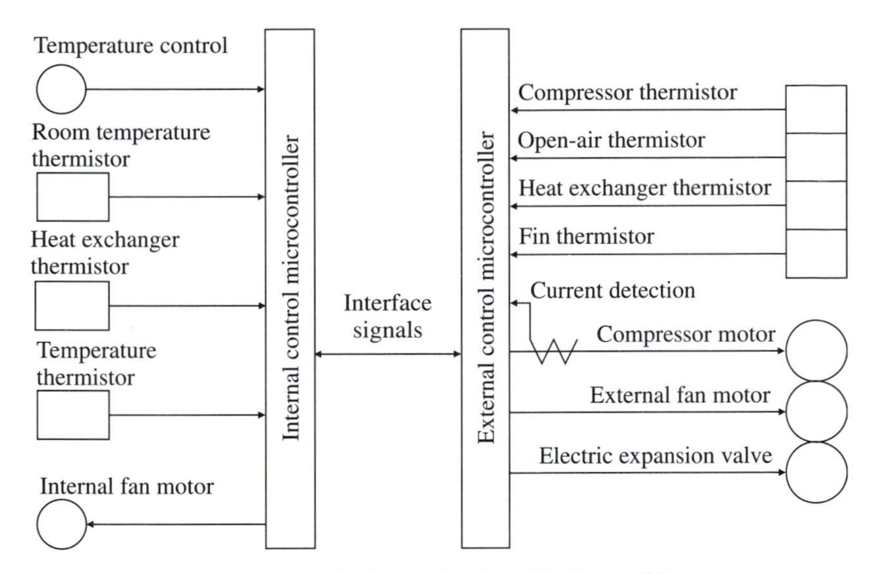

Fig. 7.28. Control schematic of an EV air-conditioner.

(5) Control unit—in order to conduct efficient temperature control of the whole air-conditioner, distributed microcontrollers are generally used for fan and compressor motor speed control, temperature detection, operating mode selection, data management, protection and display. Its control schematic is illustrated in Fig. 7.28.

(6) Refrigerant—the refrigerant of the air-conditioner adopts the HFC R-407C, which will not affect the ozone layer of the earth. Moreover, it is more efficient than those classical refrigerants in both cooling and heating capabilities.

This EV air-conditioner not only effectively realizes comfortable inner climate, but also the function of defrosting the windshield with high efficiency. Nevertheless, this kind of air-conditioners cannot supply enough heating capacity when EVs are operated in a cold weather (below 0 °C). Thus, an auxiliary electric heater can be installed and works together with the air conditioner to increase the total heating capacity when the ambient temperature may be down to −20 °C. On the other hand, some advanced thermodynamics technologies have been proposed to enable the EV air-conditioner efficiently workable down to −20 °C. For example, Denso has applied the gas injection technology to an EV air-conditioner which can increase its heating capacity by about 30% and the efficiency by about 20%.

7.3.2 THERMOELECTRIC VARIABLE TEMPERATURE SEATS

The basic drawback of air-conditioners for EV application is high energy consumption because it generally needs 2–4 kW for cooling or heating the whole vehicle compartment. Recently, one new idea has been proposed, namely cooling

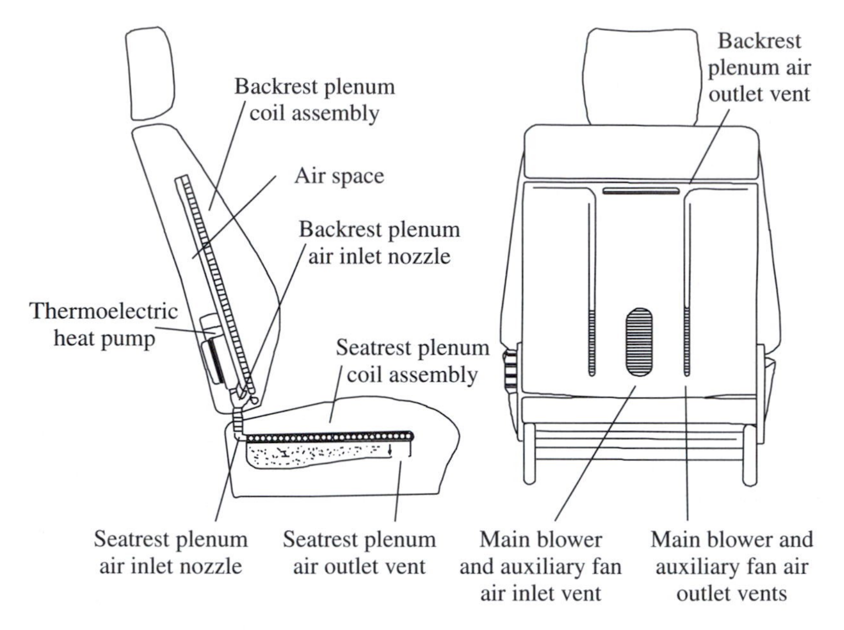

Fig. 7.29. Schematic of an EV variable temperature seat.

or heating the vehicle occupant directly, rather than the whole compartment. Figure 7.29 shows a variable temperature seat that is made available by Amerigon. It consists of a thermoelectric heat pump, blower and two plenum coil assemblies. The temperature effect is produced by a combination of conduction to the occupant through the seatrest and backrest, and through the convection of conditioned air escaping through the surface of the seat. Cooling and heating are provided by the thermoelectric heat pump and blower contained within the seat. A control module is located at the side of the seat which can adjust temperature, fan speed and air-flow direction. Recently, the HKU U2001 EV has equipped this thermoelectric variable temperature seat which consumes only 100 W per occupant (Chan *et al.*, 1994).

7.4 Power steering units

Vehicles are considered to perform three basic functions, namely run, stop and turn. Run is definitely its key function, and is governed by the propulsion system. In contrast to run, stop is governed by the braking system. Turn is equally important as run and stop, and is governed by the steering system. Initially, the development of power steering was to lighten the steering efforts for large-size vehicles such as trucks and buses. Then, this power steering has been applied to compact vehicles for easy steering, especially attractive to aged drivers. At present, power steering is almost a standard for modern ICEVs, which can greatly enhance the driveability.

Similar to ICEVs, EVs are generally equipped with power steering to enhance their driveability. Additionally, the corresponding power steering units need to be energy efficient. Among the available power steering units for EVs, there are two main types, namely electrohydraulic and electric.

7.4.1 ELECTROHYDRAULIC POWER STEERING

For conventional ICEVs, power steering is usually based on a hydraulic pump driven by the engine. Since the hydraulic pump always runs in proportion to the engine speed, there will be wastage of energy when the vehicle runs straightforward at high speeds. The corresponding energy consumption has been found to be about 3% of the total engine fuel. In order to minimize the energy consumption of power steering for EVs, the hydraulic pump of the power steering unit should be driven by an additional motor, rather than the propulsion motor. Figure 7.30 shows the electric-pump-type hydraulic (loosely termed electrohydraulic) power steering unit for EVs (Ijiri, 1996). It uses a highly efficient gear pump that is driven by an advanced motor such as the induction motor, PM brushless motor or SR motor. The steering gear is identical to the conventional hydraulic power steering unit. The inverter is usually supplied by an auxiliary battery. Based on this electrohydraulic power steering unit, the hydraulic pump works at low speed or stop when the vehicle travels in a normal straight line whereas runs at a high speed to provide the power steering effect when the vehicle turns a corner. At present, many modern EVs employ this electrohydraulic power steering unit (Zemke, 1997).

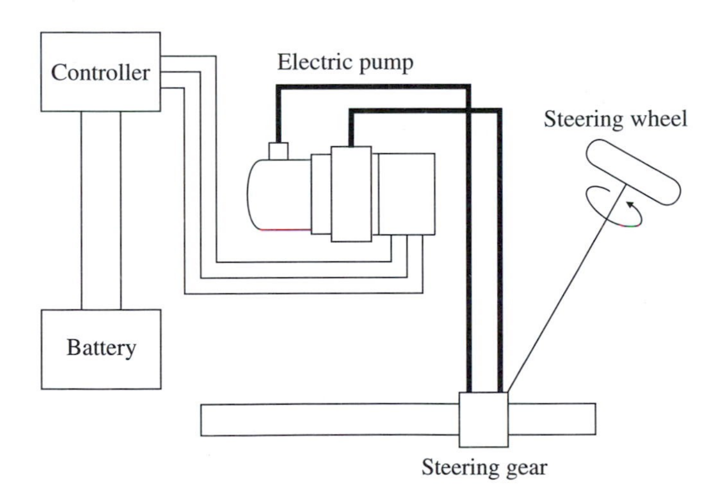

Fig. 7.30. Schematic of an electrohydraulic power steering unit.

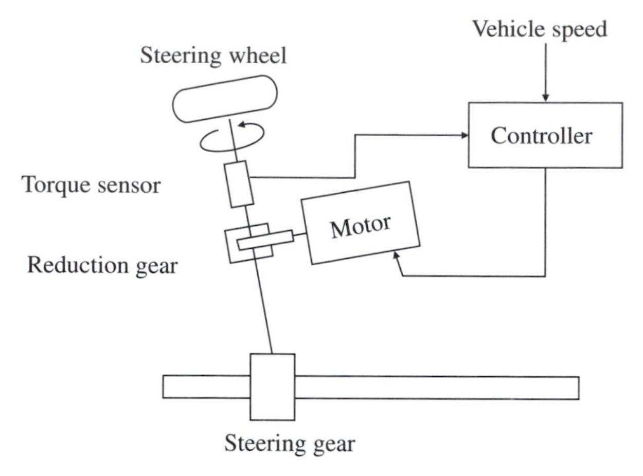

Fig. 7.31. Schematic of an electric power steering unit.

7.4.2 ELECTRIC POWER STEERING

By eliminating the hydraulic pump, the electric direct-driving (loosely termed electric) power steering unit directly employs an electric motor to generate the steering torque. Figure 7.31 shows this electric power steering unit which mainly consists of the electric motor, torque sensor, controller and reduction gear (Ijiri, 1996). In order to generate a suitable torque at the motor according to the desired steering effect, the highly reliable torque sensor functions to detect the steering effort. Then, based on the torque sensor signal and vehicle speed signal, the controller provides proper control to the motor. The reduction gear is used to amplify the motor torque and transmit it to the output shaft. The motor and its power converter are usually supplied by an auxiliary battery. Some latest EVs have been equipped with this electric power steering unit.

7.5 Auxiliary power supplies

The auxiliary power supply is basically composed of an auxiliary battery and a dc–dc power converter. Its function is to provide a stable power supply for all essential EV auxiliaries, even when the main energy source such as the main battery is fully discharged or malfunction. Modern EVs have many electric auxiliaries such as air-conditioners, audio accessories, driving control electronics, EMS electronics, horn, instruments, lamps, power steering units, power windows, window defrosters and windshield wipers. Figure 7.32 gives a comparison of auxiliaries in ICEVs and EVs. The key difference is that the ICEV auxiliary battery is charged by the alternator (usually named as dynamo) which is coupled to the engine, whereas the EV auxiliary battery is charged by the main battery via the dc–dc converter. Moreover, most auxiliaries of ICEVs work at only one

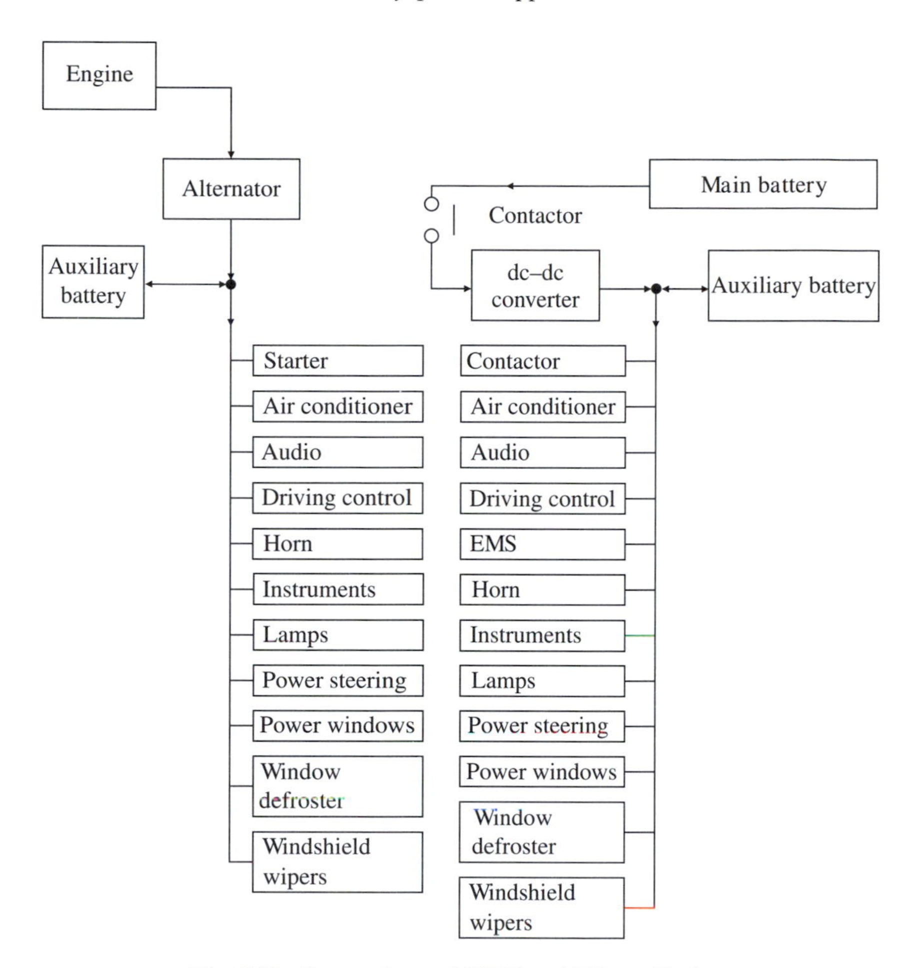

Fig. 7.32. Comparison of ICEV and EV auxiliaries.

voltage level, namely 12 V, whereas the auxiliaries of EVs may be designed to work at various voltage levels, such as 24, 48 and 120 V, so that the corresponding power consumption can be minimized. Therefore, the auxiliary power supply for EVs is generally more complex than that of ICEVs (Kim and Goong, 1996).

Table 7.5 summarizes the major types of EV auxiliaries and their typical power consumption. It is obvious that the air-conditioner plays the most important role among all EV auxiliaries. In fact, it spends about 60–75% of the total power consumption of all EV auxiliaries. In order to maximize its efficiency, the operating voltage is usually increased up to 120 V. Furthermore, in order to avoid short-time dry-off of the auxiliary battery due to those EV auxiliaries with high power consumption (namely the air-conditioner, power steering and window defroster), they should be operated only when the contactor is closed so that the necessary power can be directly drawn from the main battery.

Table 7.5 Power consumption of EV auxiliaries

Subsystem	Operation mode	Power (W)
Air-conditioner	Continuous	2000–4000
Audio	Continuous	20
Contactors	Continuous	20
Driving control	Continuous	150
Energy management system	Continuous	150
Headlamps and tail lamps	Continuous	120
Horn	Intermittent	10
Instruments	Continuous	30
Parking, turn and interior lamps	Intermittent	50
Power steering	Continuous	400
Power windows	Intermittent	80
Window defroster	Continuous	250
Windshield wipers	Continuous	40

7.5.1 AUXILIARY BATTERY

Besides the air-conditioner, power steering and window defroster, the total power consumption of other auxiliaries is about 700 W. Thus, the auxiliary battery should have the usable energy capacity to afford such power consumption to ensure driving safety and reliability, even when the main battery is fully discharged or the dc–dc converter is malfunction. The corresponding capacity is usually determined as the energy storage for about 1-h operation at given emergency loads. The other benefits of the auxiliary battery are to prevent the EV auxiliaries suffering from the voltage fluctuation and EMI contamination in the main battery due to electric propulsion.

7.5.2 DC–DC CONVERTERS

Figure 7.33 shows two flyback dc–dc converters (non-isolated and isolated) which are suitable for EV auxiliary power supplies. These dc–dc converters generally work with high efficiency (typically 85–95%) and are commercially available. The non-isolated type takes the advantages of simplicity and low cost, whereas the isolated type takes the merit of isolation between the high voltage level of the main battery and the low voltage level of the EV auxiliaries.

The dc–dc converter for EV auxiliary power supplies has two main functions—charging the auxiliary battery to a fully charged state, and supplying electrical energy to EV auxiliaries especially those with high power consumption. Figure 7.34 shows a typical load cycle of those EV auxiliaries and the corresponding optimal capacity of the dc–dc converter. This optimal capacity denotes the condition that the battery charging process can balance the battery discharging process, and the auxiliary battery is always in its fully-charged state. If we adopt higher capacities, the battery charging process will dominate the discharging

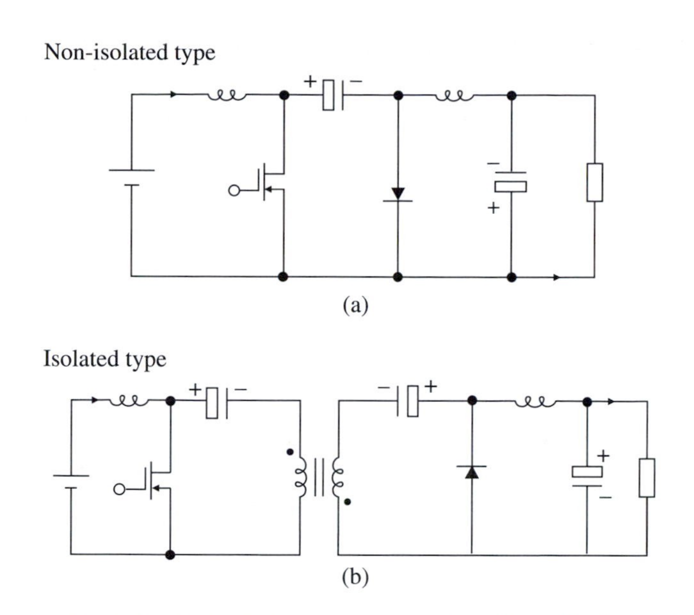

Fig. 7.33. Flyback dc–dc converters for EV auxiliaries.

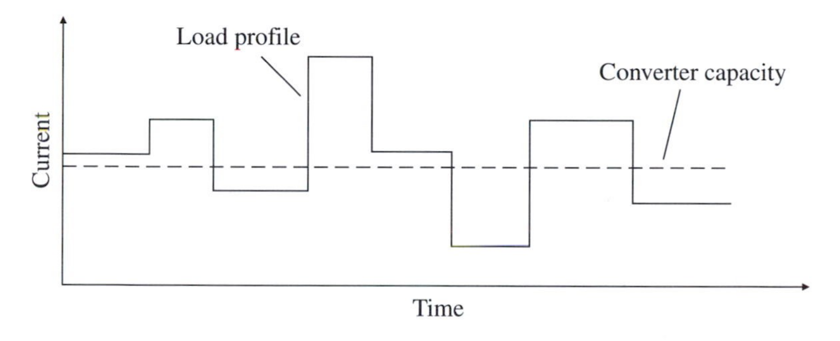

Fig. 7.34. Optimal dc–dc converter capacity.

process, hence either over-sizing the dc–dc converter or overcharging the auxiliary battery. On the contrary, if we adopt lower capacities, the battery discharging process will dominate the charging process, leading to cause the auxiliary battery losing the fully charged state for emergency use (Kim and Goong, 1996).

7.6 Navigation systems

In modern ICEVs, navigation systems are becoming attractive which enable the drivers to find the shortest route to destination and/or avoid the traffic-jam areas. Besides that, navigation systems for EVs can work with the EMS to recommend the most energy efficient route to destination, the residual range prediction on the basis of traffic conditions and the location of nearby recharging stations for extended trips. In general, there are two major types of navigation systems that have been developed for EVs, namely local based and global based.

7.6.1 LOCAL NAVIGATION

A recently developed local navigation system for EVs is the audio navigation system that was made available by Amerigon (Chan *et al.*, 1994). It is based on a low-cost CD-ROM driver which utilizes a voice interface. The driver's voice is used for input while speech stored on the CD-ROM is used for output. The driver is prompted to spell his destination and current location. Based on the local map and information stored in the CD-ROM, the system can calculate the shortest route and then verbally instruct the driver to follow this route. This navigation system takes the definite advantage that driving interference is kept to a minimum level as the driver does not have to take his eyes off the road to look at a video screen or take his hands off the steering wheel to press a button for further directions. The system block diagram is shown in Fig. 7.35. At present, the HKU U2001 EV has adopted this system for navigation.

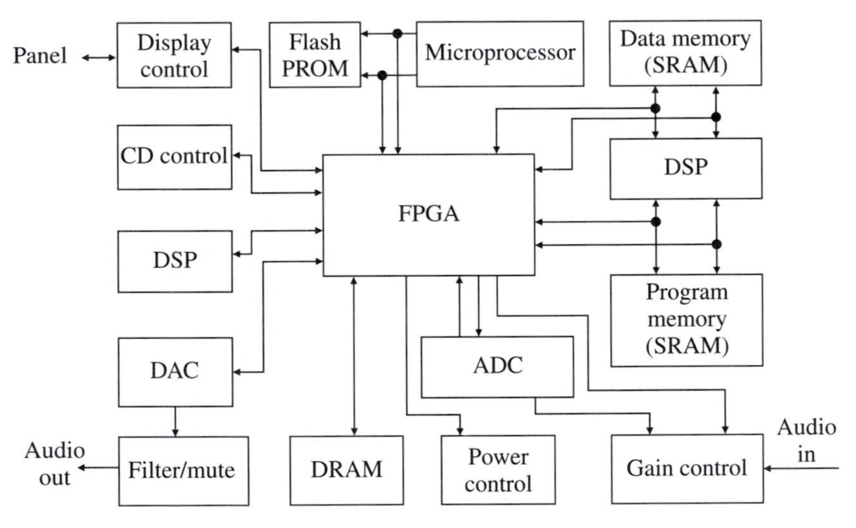

Fig. 7.35. Block diagram of an audio navigation system.

7.6.2 GLOBAL NAVIGATION

Different from the previous local navigation system that the vehicle position is roughly inputted by the driver, the global navigation system adopts the global position system (GPS) to automatically track the exact vehicle position. This GPS usually employs four satellites signals to compute four desired quantities, namely the vehicle position in three dimensions and the time. According to the exact vehicle position, the map-matching technology and the current traffic information, the global navigation system not only recommends the shortest and/or most energy efficient route with minimum traffic jam, but also works with the EMS to accurately predict the residual driving range for EVs. With the advent of this navigation system, the key shortcoming of EVs, namely the relatively short driving range, can be alleviated.

This GPS navigation system generally consists of the GPS receiver, GPS antenna, navigation computer, CD-ROM driver, display screen and audio speaker. It allows the driver to program the best route by inputting the destination such as hotels, airports, hospitals, tourist attractions, public charging stations and even individual street addresses. Once the route is set, it provides both visual display and audio prompts to guide the driver, turn by turn, to the destination. The system size is designed and manufactured to be very compact and can be easily installed. At present, a number of hi-tech companies such as Rockwell, Philips and North Valley Land & Timber have released their GPS navigation systems. It should be noted that these GPS navigation systems are applicable to both ICEVs and EVs although some special features such as the coordination with the EMS of EVs may not be available.

7.7 Regenerative braking systems

Regenerative braking is unique to EVs, which enables the vehicle's kinetic energy converting back to electrical energy during braking (deceleration or downhill running). The converted electrical energy will be stored in those receptive EV energy sources such as batteries, ultracapacitors and ultrahigh-speed flywheels to extend the driving range. If these receptive sources have been fully charged up, regenerative braking can no longer be applied and the desired braking effort can only be provided by the conventional hydraulic braking system. In fact, almost all modern EVs are equipped with the regenerative-hydraulic hybrid braking system which can save the braking energy, recover the kinetic energy and offer the conventional braking performances to the driver.

In general, the regenerative braking system is activated when the EV is decelerated for speed reduction, the accelerator pedal is released for coasting at highways, or the brake pedal is pressed for stopping. During normal speed reduction, the regenerative braking torque is usually kept at its maximum capability. When the EV is coasting at high speeds, its propulsion motor generally operates in the constant-power mode so that the motor torque is inversely proportional to the motor speed or vehicle speed. Thus, the higher the motor speed for constant-power operation, the lower the capability of the regenerative braking torque will be resulted. On the other hand, when the brake pedal is pressed, the propulsion motor usually operates at low speeds. Since the kinetic energy of the EV at such low speeds is generally insufficient for the propulsion motor to produce the

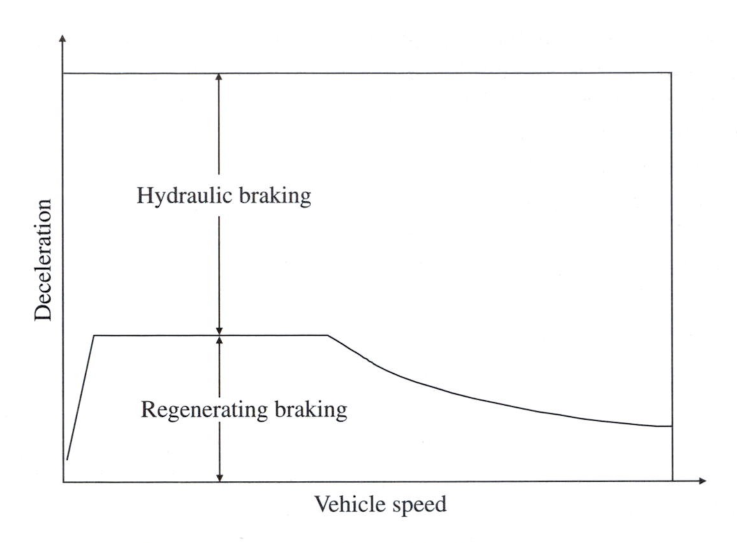

Fig. 7.36. Regenerative and hydraulic braking.

maximum braking torque, the corresponding regenerative braking torque capability will proportionally decrease with the reduction in the vehicle speed. Moreover, as shown in Fig. 7.36, the regenerative braking torque is in general insufficient to offer the same deceleration rate as available in conventional ICEVs. So, the hydraulic braking system generally coexists with the regenerative braking system for EVs. It should be noted that the hydraulic braking torque is applied only when the regenerative braking torque has already been maximized and it is still not large enough to fulfil the desired braking command.

7.7.1 SYSTEM CONFIGURATION

Since the regenerative-hydraulic hybrid braking system is unique to EVs or absent in conventional ICEVs, the coordination between regenerative braking and hydraulic braking is a key issue (Ogura *et al.*, 1997). Moreover, some special requirements should also be taken into consideration:

(1) In order to keep the EV driver having a smooth brake feel, the hydraulic braking torque has to be controllable according to the changes of the regenerative braking torque so that the total braking torque is the value expected by the driver. Also, the control of hydraulic braking should not affect the brake pedal stroke and therefore no abnormal feel is experienced by the driver.
(2) Since there is no engine to drive the pump for the production of hydraulic braking torque in EVs, an electric pump assisted hydraulic booster is usually needed. Instead of directly transmitting the hydraulic pressure generated by the driver's brake pedal operation to the wheel cylinders, the hydraulic brak-

ing torque is electrically controlled. Therefore, this regenerative-hydraulic hybrid braking system should be guaranteed with a fail-safe mechanism. To improve the reliability and satisfy the safety standard, a dual circuit arrangement is generally adopted. In case one circuit fails, the other circuit must be able to provide the necessary braking function.

(3) In order to stably brake the vehicle, the braking force distribution between the front and rear wheels has to be well balanced. Moreover, the maximum braking torque to the front and rear wheels should be kept lower than the maximum allowable value (depending on the rolling resistance coefficient) to prevent from skidding.

In order to realize the above requirements, the regenerative-hydraulic hybrid braking system is configured as shown in Fig. 7.37. By employing the electric pump, the hydraulic booster produces the desired hydraulic braking pressure which is activated by the driver's brake pedal operation. The brake control is used to collaborate with the motor control to deduce the regenerative braking torque as well as the desired hydraulic braking torque on the front and rear wheels. During regenerative braking, the regenerative control recovers the regenerative energy for battery charging. The ABS control and its valve are the same as that in conventional ICEVs, and function to prevent the vehicle from skidding.

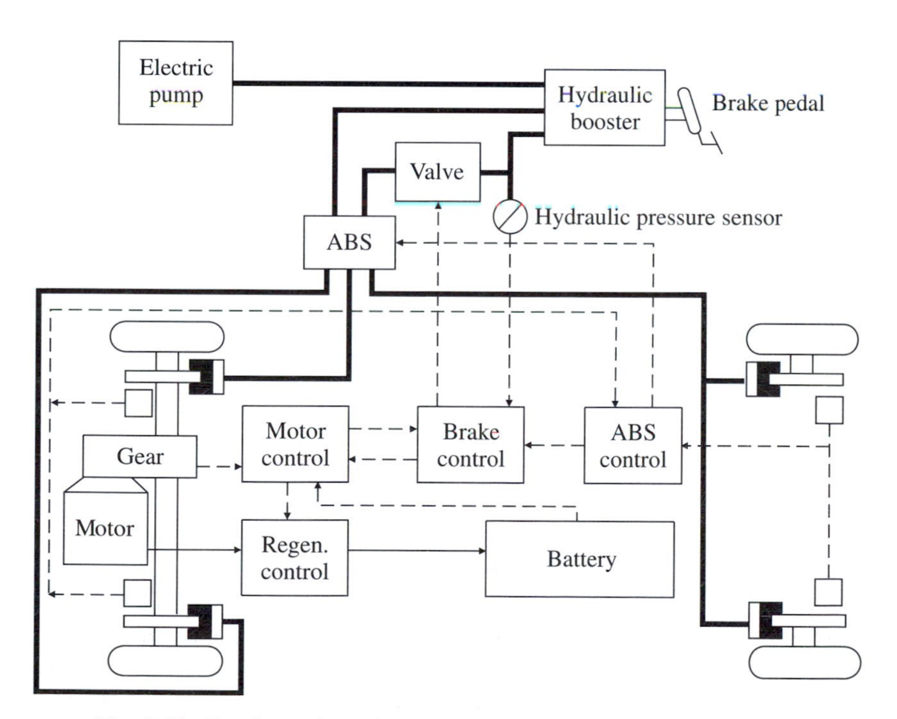

Fig. 7.37. Configuration of a regenerative-hydraulic braking system.

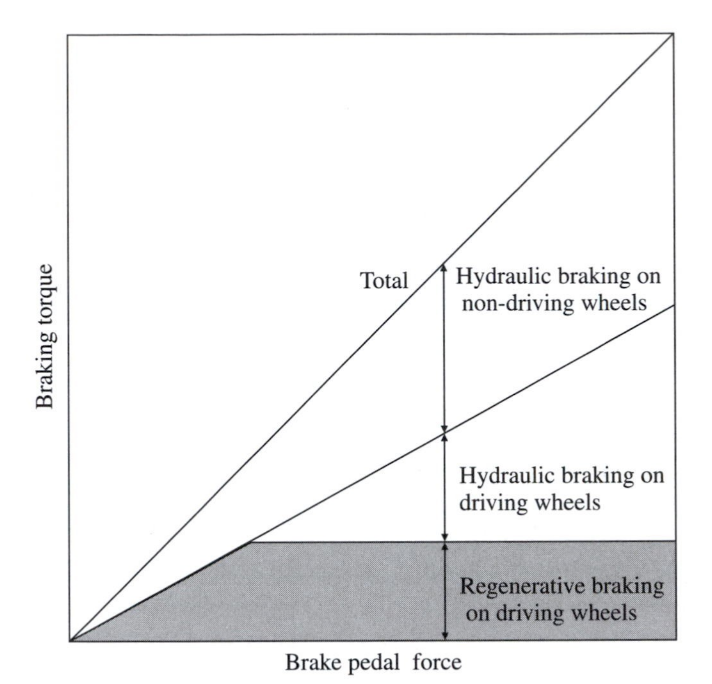

Fig. 7.38. Distribution of regenerative and hydraulic braking torques.

7.7.2 BRAKING CONTROL

As mentioned before, the total braking torque is the sum of the regenerative braking torque and the hydraulic braking torque. The control of their distributions is depicted in Fig. 7.38, which aims to provide the driver the same braking feel as that in conventional ICEVs while maintaining maximum regenerative braking. During low brake pedal force, only the regenerative braking torque is applied on the driving wheels, and is proportional to the pedal pressing force. The braking torque on the non-driving wheels is always due to the hydraulic braking which is also proportional to the pedal pressing force. When the pedal force is beyond a certain limit, the maximum regenerative braking torque is applied on the driving wheels, and the hydraulic braking torque is simultaneously applied on the driving wheels to top up the desired braking torque. Hence, the maximum regenerative braking torque can be kept constant to fully recover the kinetic energy.

During the whole regenerative braking process, the kinetic energy cannot be fully converted into the electrical energy for battery charging. The corresponding losses along the regenerative energy flow include the aerodynamic loss, rolling resistance loss, braking system loss, motor loss, device loss and charging loss. Nevertheless, modern EVs can generally benefit over 20% energy saving when employing regenerative braking.

References

Asakura, Y. and Kanamaru, K. (1996). RAV4-EV and its adaptability to the real market. *Proceedings of the 13th International Electric Vehicle Symposium*, 1–8.

Aylor, J.H., Thieme, A. and Johnso, B.W. (1992). A battery state-of-charge indicator for electric wheelchairs. *IEEE Transactions on Industrial Electronics*, **39**, 398–409.

Bode, H. (1977). *Lead Acid Batteries*. Wiley-Interscience, New York.

Chan, C.C. and Chau, K.T. (1997). An overview of power electronics in electric vehicles. *IEEE Transactions on Industrial Electronics*, **44**, 3–13.

Chan, C.C. and Chu, K.C. (1988). Intelligent battery management system. *Proceedings of the 9th International Electric Vehicle Symposium*, No. 88-052, 1–8.

Chan, C.C. and Chu, K.C. (1990). A microprocessor-based intelligent battery charger for electric vehicle lead acid batteries. *Proceedings of the 10th International Electric Vehicle Symposium*, 456–66.

Chan, C.C., Lo, E.W.C. and Shen, W.X. (2000a). The calculation approach of the available capacity of batteries in electric vehicles. *Proceedings of the 17th International Electric Vehicle Symposium*, CD-ROM.

Chan, C.C., Lo, E.W.C. and Shen, W.X. (2000b). Available capacity computation model based on artificial neural network for lead–acid batteries in electric vehicles. *Journal of Power Sources*, **87**, 201–204.

Chan, C.C., Zhan, Y.J., Bell, L., Yu, Z. and Araki, Y. (1994). The development of advanced electric vehicle U2001. *Proceedings of the 12th International Electric Vehicle Symposium*, 282–91.

Friedman, M. (1997). Conductive charging commercialization. *Proceedings of the 14th International Electric Vehicle Symposium*, CD-ROM.

Griffith, P. and Gleason, G. (1997). Demonstration of 300-kW rapid charger in transit bus application. *Proceedings of the 14th International Electric Vehicle Symposium*, CD-ROM.

Http://www.amerigon.com/.

Http://www.batech-jp.com/.

Http://www.carin.com/.

Http://www.garmin.com/.

Http://www.lara.prd.fr/.

Http://www.magellandis.com/.

Http://www.magnecharge.com/.

Http://www.sae.org/.

Ijiri, W. (1996). Power steering system for electric vehicle. *Proceedings of the 13th International Electric Vehicle Symposium*, 156–62.

Inductive Charge System for Electric Vehicles. Densei, 1998.

Inductive Chargers for Electric Vehicles. Boston Edison, 1998.

Kanetuki, M. and Arakawa, N. (1996). Development of compact high efficiency air conditioner for electric vehicle. *Proceedings of the 13th International Electric Vehicle Symposium*, 636–42.

Kim, C. and Goong, E.N. (1996). Capacity optimization of auxiliary power system for EVs. *Proceedings of the 13th International Electric Vehicle Symposium*, 149–55.

Kosowski, M.G. (1997). Technology plan for inductive high-power charging. *Proceedings of the 14th International Electric Vehicle Symposium*, CD-ROM.

Linden, D. (1995). *Handbook of Batteries*, (2nd edition). McGraw-Hill, New York.

Nor, J.K. (1992). Charging station for electric vehicles. *Proceedings of the 11th International Electric Vehicle Symposium*, 9.03, 1–11.

Ogura, M., Aoki, Y. and Mathison, S. (1997). The Honda EV plus regenerative braking system. *Proceedings of the 14th International Electric Vehicle Symposium*, CD-ROM.

Proctor, R.L. (1997). Assessing and increasing energy efficiency by using accurate state-of-charge instrumentation and data collection with GPS. *Proceedings of the 14th International Electric Vehicle Symposium*, CD-ROM.

SAE J1772—Electric Vehicle Conductive Charge Coupler Recommended Practice. Society of Automotive Engineers, 1996.

SAE J1773—Electric Vehicle Inductive Charge Coupling Recommended Practice. Society of Automotive Engineers, 1995.

Shinpo, T. and Suzuki, H. (1997). Development of battery management system for electric vehicle. *Proceedings of the 14th International Electric Vehicle Symposium*, CD-ROM.

Zemke, B.E. (1997). The technology inside the systems of General Motors EV1. *Proceedings of the 14th International Electric Vehicle Symposium*, CD-ROM.

8 EV simulation

Computer simulation has been widely accepted to be a powerful tool in different fields of research. Its key role is to provide virtual analysis of physical system behaviour before performing any expensive prototyping or time-consuming experimentation. Increasingly, it enables engineering designers to evaluate their designs thoroughly, and to identify some problems that are not easy to observe by measurement or experimentation. This tool can save the development cost in the order of millions of dollars that might otherwise be spent to correct flaws in later stage of engineering and manufacturing (Chau, 1996).

EVs are a complex system that integrates a number of subsystems such as the vehicle body, electric propulsion, energy source and energy management. The technologies involved are diversified and multidisciplinary, including electrical and electronic engineering, mechanical and automotive engineering, and chemical engineering (Chan, 1993). Some key technologies are listed below:

- lightweight, rigid, low aerodynamic and low rolling resistance chassis and body technology;
- high power density and high efficiency motor drive technology;
- high specific energy, high specific power, long cycle life and low cost energy storage/generation technology;
- intelligent energy management system;
- high efficiency and high reliability energy refuelling/recharging technology.

Because of these multidisciplinary and fast-changing EV technologies, the design process of EVs should be flexible, prompt and economical. Computer simulation not only facilitates this design process, but also enhances the system optimization of EVs (Chan, 2000; Wong, 2000).

8.1 System level simulation

The design of EVs is mainly composed of four main disciplines, namely the body design, energy source design, electric propulsion design and auxiliary design. The vehicle body design mainly involves the body structure, frame, bumpers and suspension. The energy source design typically involves the batteries and charger. The electric propulsion design mainly involves the electronic controller, power converter, electric motor, transmission and wheels. The auxiliary design mainly involves the brakes, steering, auxiliary power supply, temperature control and energy management. These subsystems are closely linked together and all the interactions among them are interrelated. Figure 8.1 shows a quality function

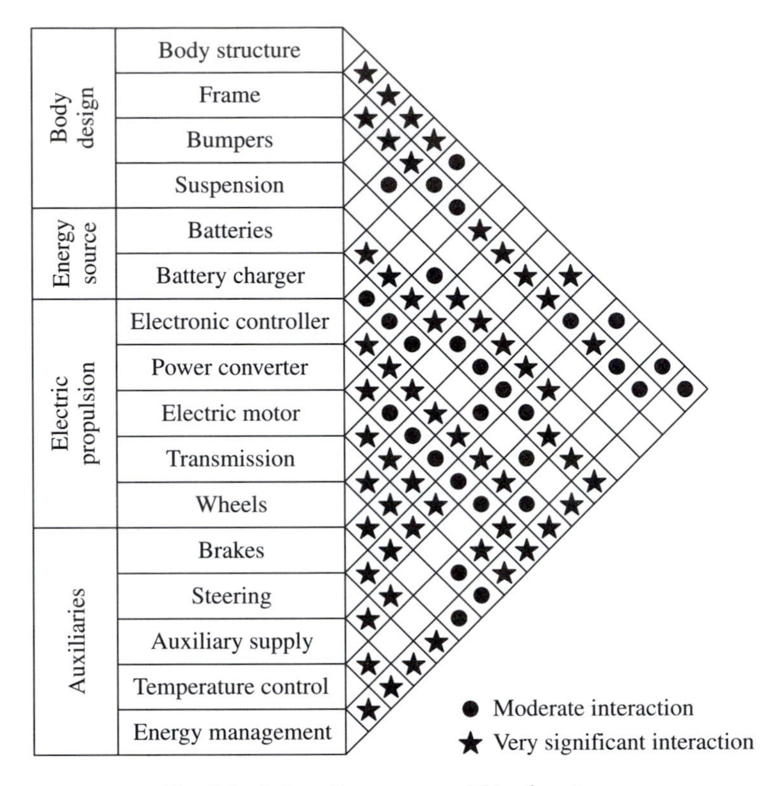

Fig. 8.1. Interactions among EV subsystems.

diagram illustrating the overall and significance of interactions among various EV subsystems. Those very significant interactions denote that they may affect the vehicle cost, performance, safety or other important objectives.

Due to the multidisciplinary nature of EVs, the simulation of EVs is a kind of mixed-technology process. Within the whole system, different EV subsystems are linked together by using mixed-technology linkages, namely the electrical link, electronic link, mechanical link and thermal link. For both electrical and electronic links, they can be further represented as analogue and digital signals. Figure 8.2 illustrates a typical mixed-technology mixed-signal representation of the whole EV system.

Having identified that there are different degrees of interactions among various EV subsystems and these subsystems are connected by mixed-technology mixed-signal linkages, it is hardly possible to perform the EV design and optimization based on analytical formulae. In fact, the whole mixed-technology mixed-signal simulation should be performed in the system level so that the trade-offs among the subsystem criteria can be carried out simultaneously to identify the preferred EV system integration.

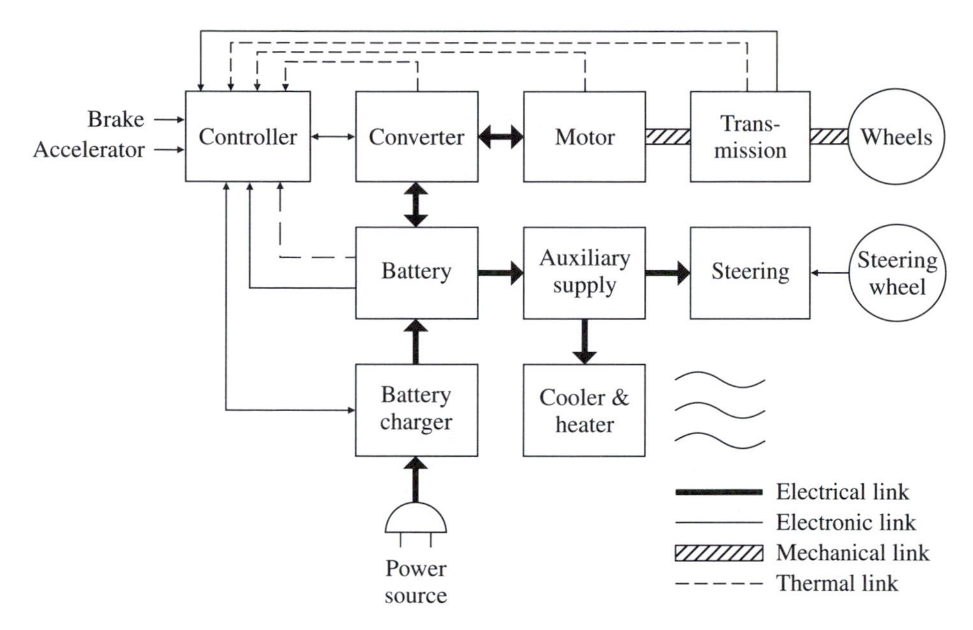

Fig. 8.2. Mixed-technology mixed-signal representation of an EV system.

8.2 EV simulator

MATLAB is a popular programming environment among researchers and engineers. It consists of a series of built-in functions and advanced toolboxes which enable users to perform computer simulation efficiently. As the m-files written for MATLAB are not complied into binary form, users are able to visualize the inner structure of the simulation programme and have the chance to realize how the programme is implemented. The MATLAB version for Windows also enables the use of graphical controls, facilitating interactive and user-friendly simulation.

An EV simulator, namely EVSIM, is newly developed by the authors and researchers at the International Research Centre for Electric Vehicles in the University of Hong Kong for the system level design, simulation and optimization of EVs (Chau and Chan, 1998). Based on MATLAB for Windows, this simulator is basically a mathematical tool that can solve a series of equations representing those subsystems and their interconnections within the whole EV system. To facilitate the pre-processing for data input and post-processing for data output, the simulator provides an interactive graphical user interface. So, rather than programming the tedious software, inputting the lengthy data or evaluating the massive output data, the EV designer can interactively perform data input, system simulation and output presentation. It also allows for graphically displaying and comparing various simulation results simultaneously. Increasingly, by using iterative simulation, optimization of various system performance criteria can be achieved. Of course, an on-line help menu should be provided to enhance the user-friendliness.

Based on the EVSIM, an intelligent EV virtual-reality (EVVR) system can be developed, which has the following features:

- combined software and hardware environment
- integration of existing modelling
- top-down and bottom-up approaches
- network-oriented platform-independent computing
- advanced visualization
- motor design and analysis
- controller design and analysis and
- energy management design and analysis.

8.2.1 SIMULATOR FEATURES

This EVSIM has a modular programming structure and is programmed as m-files. As shown in Fig. 8.3, the Main Menu consists of various subprogrammes which are graphically represented by pushbuttons. Hence, users can easily start the simulation with default values, modify the input values, evaluate the output results, perform particular optimizations or even quit the simulation by clicking

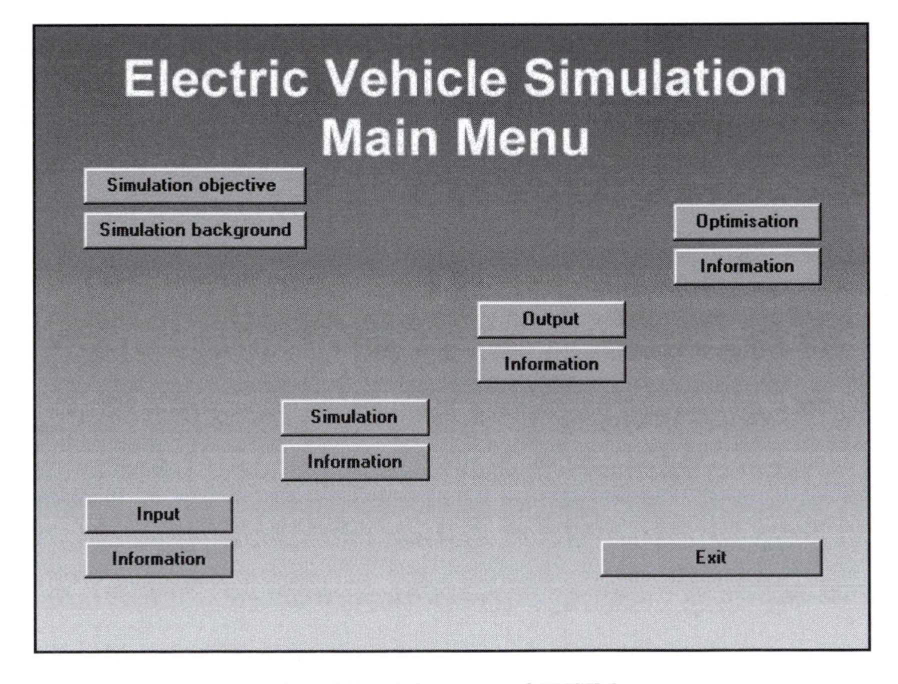

Fig. 8.3. Main Menu of EVSIM.

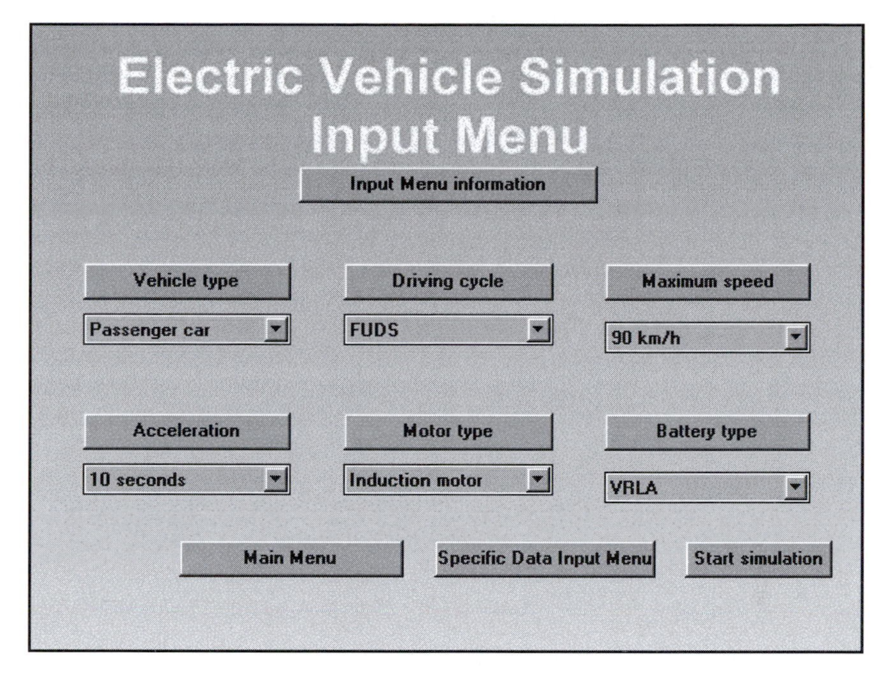

Fig. 8.4. Input Menu of EVSIM.

on those pushbuttons. The organization of the simulator is hierarchical. From the Main Menu, users can select to enter the Input Menu at which various input parameters are readily modified. As shown in Fig. 8.4, the following parameters can be chosen interactively:

- Vehicle type—it includes three main types of EVs, namely the passenger car, van and bus. These EVs have different default values of the vehicle payload, aerodynamic drag coefficient, transmission ratio, curb weight, frontal area and tire radius.
- Driving cycle—it is a standard or user-specified driving cycle for EVs. Those standard driving cycles include the FUDS, FHDS, ECE Urban, SAE J227a-C and Japan 10.15 Mode. The user-specified driving cycles can be freely designed to account for different driving habits in different countries.
- Expected performances—they refer to the maximum vehicle speed that is expected to be achieved and the minimum time to accelerate from zero to this maximum speed. These performance criteria are similar to those for conventional ICEVs.
- Motor type—it is the type of motors for electric propulsion in EVs. It can be the dc motor, induction motor, PM brushless motor or switched reluctance motor. Different types of motors have different default values of torque-speed characteristics, power densities and efficiency maps.

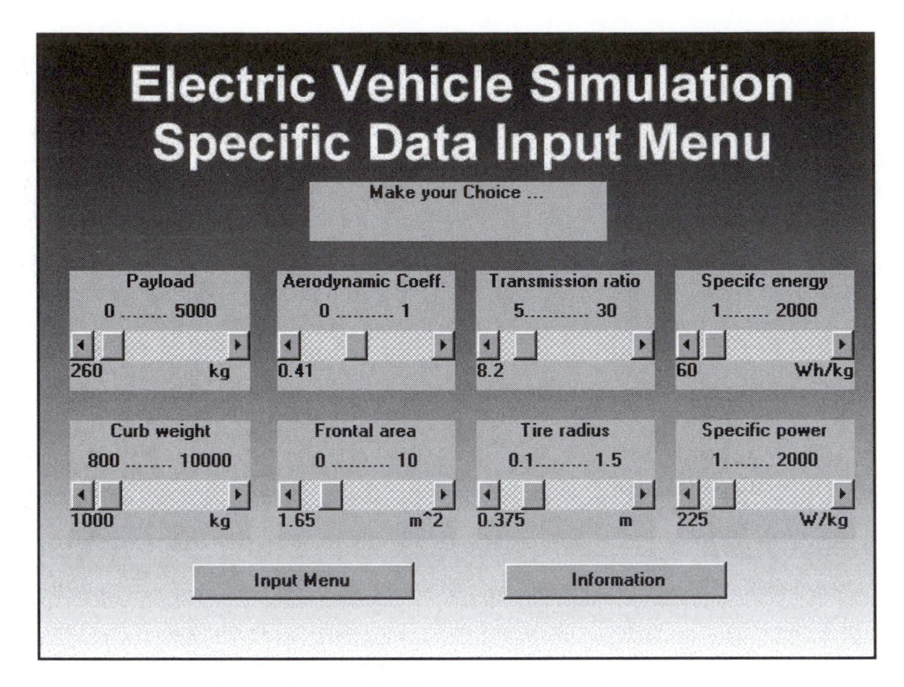

Fig. 8.5. Specific Data Input Menu of EVSIM.

- Battery type—it is the type of batteries for energy storage in EVs. It includes those viable EV batteries, namely the Pb–Acid, Ni–Cd, Ni–Zn, Ni–MH, Zn/Air, Al/Air, Na/S, Na/$NiCl_2$, Li–Polymer and Li–Ion. Different types of batteries have different default values of charging/discharging characteristics, specific energies, specific powers and efficiency maps.

Apart from using the default values, users can further select the Specific Data Input Menu to alter the vehicle payload, aerodynamic drag coefficient, transmission ratio, curb weight, frontal area and tire radius. As shown in Fig. 8.5, users can simply tune a particular parameter by altering the corresponding position of a slider, or directly type the desired value in the editable text box.

From the Main Menu, users can select to enter the Output Menu at which various output parameters are readily evaluated. Similar to the Input Menu, the Output Menu is also interactive and user-friendly. As shown in Fig. 8.6, the following parameters can be chosen for evaluation:

- Road load—it shows the selected driving cycle for the simulation and the corresponding road load characteristic desired by the EV.
- Wheel load—it shows the efficiency map of the selected transmission and the corresponding wheel speed, wheel torque and wheel power characteristics.
- Motor load—it shows the efficiency map of the selected motor and the corresponding motor speed, motor torque and motor power characteristics.

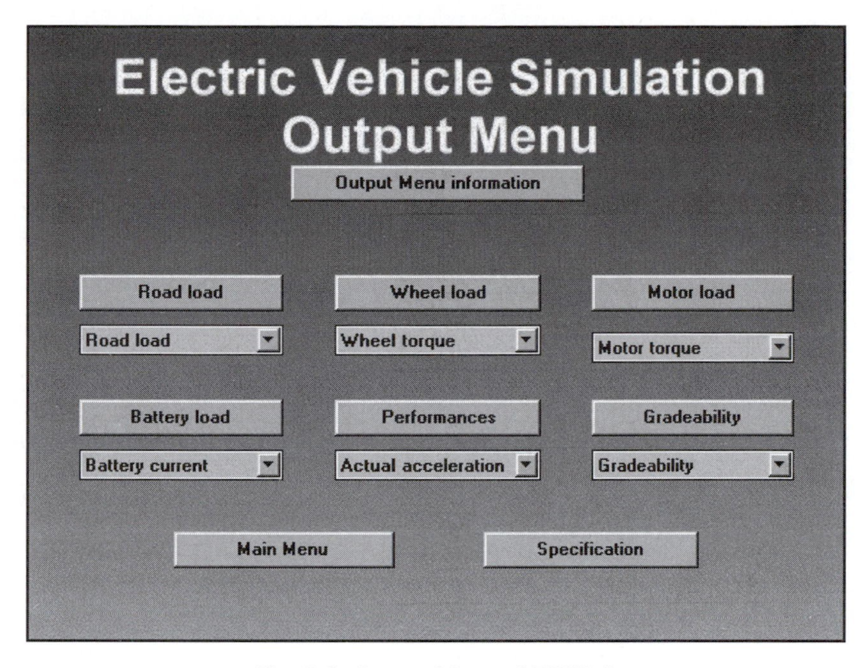

Fig. 8.6. Output Menu of EVSIM.

- Battery load—it shows the efficiency map of the selected battery and the corresponding battery current and battery energy consumption.
- Actual performances—they refer to the actual maximum speed and acceleration rate that can be achieved by the EV.
- Gradeability—it shows the maximum slope of the hill that the EV can climb up at 20 km/h.
- Specification—it summarizes all important output parameters of the EV.

For advanced EV designers, they may like to optimize the EV system based on certain criteria. As shown in Fig. 8.7, the EVSIM provides three pre-defined optimization schemes:

- Transmission ratio—it optimizes the transmission ratio in such a way that all operating points are within the allowable torque-speed envelop and the overall system efficiency is maximum. Hence, the total driving range can be maximized.
- System voltage—it optimizes the system voltage level in such a way that the corresponding weight ratio of any energy sources with respect to the total vehicle weight can provide the maximum driving range while all expected performance can be achieved.
- Hybridization ratio—it optimizes the weight/volume ratio between any two energy sources so that the specific energy and specific power requirements can be decoupled. Hence, both the driving range and acceleration rate can be optimized.

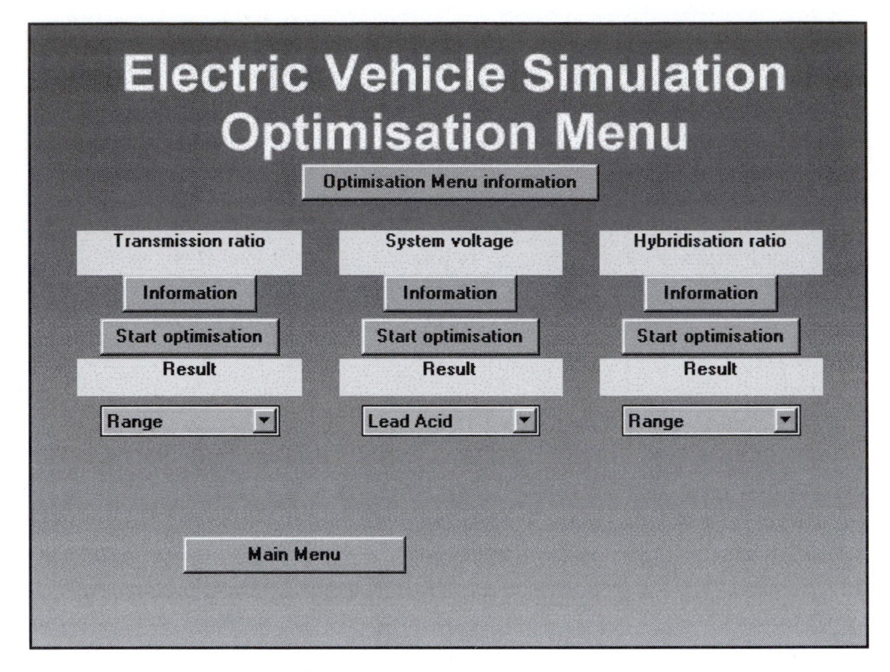

Fig. 8.7. Optimization Menu of EVSIM.

8.2.2 SIMULATOR MODULES

Since the EVSIM has a modular programming structure, besides those for data input, data output and graphical interface, this simulator consists of eight modules, namely driving cycle module, vehicle module, transmission module, motor module, motor performance module, controller module, energy source module and vehicle performance module. Figure 8.8 shows the flowchart of the simulation process. Given a standard or user-specified driving cycle, the corresponding road load, transmission load, motor load, controller load and energy source load are simulated in sequence. Actually, this sequence is a backward flow of the realistic process.

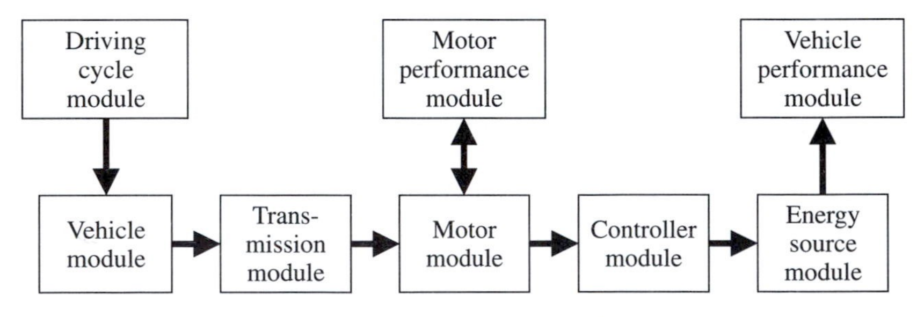

Fig. 8.8. Simulation flowchart of EVSIM.

(1) Driving cycle module—it provides the time-varying EV driving speed which works as the reference or entry of the whole simulation process. Inside the module, all standard and user-defined driving cycles are stored as discrete data in look-up tables, and any values between those discrete data are generated by interpolation.
(2) Vehicle module—the entries of this module are the selected driving cycle and basic vehicle parameters. Based on them, all force parameters including the aerodynamic drag force, rolling resistance force and climbing force are calculated inside the module. Hence, the total road load of the EV can be generated. All the necessary formulae used by this module has been derived in Chapter 3.
(3) Transmission module—it is a lumped module which represents the overall mechanical transmission including the intermediate gearing, final gearing and differential device. The entries of this module are the road load and the pre-defined transmission efficiency map on the plane of torque vs. speed. Since this map is stored as a discrete table, any values between those discrete data are determined by interpolation. Hence, the necessary driving torque and power produced by the motor can be deduced.
(4) Motor module—the entries of this module are the desired motor speed, torque and power as well as the pre-defined torque-speed characteristics, overloading capability and efficiency map of various EV motors. Inside the module, the necessary rated motor speed and power will be calculated and fed into the motor performance module. After finalizing the rated motor speed and power, the corresponding input current and power flowing into the motor can be deduced. It should be noted that this module actually represents both the motoring and regenerative modes, depending on driving and braking (Chau, Chan and Wong, 1998).
(5) Motor performance module—the entries of this module are the rated motor speed and power outputted by the motor module. Hence, the corresponding acceleration rate and maximum vehicle speed can be calculated, which should be better than the expected acceleration rate and maximum speed set by the designer in the Input Menu. Otherwise, the rated motor speed and power are increased in such a way that the expected performances can be achieved.
(6) Controller module—based on the entries of the necessary input current and power of the motor, the desired dc current and power that should be provided by the energy source are calculated in the module. No matter it is a dc chopper for a dc motor or an inverter for an ac motor, this module can operate in bidirectional modes, namely the positive and negative current/power for the motoring and regenerative operation, respectively (Chan and Chau, 1997).
(7) Energy source module—this module consists of all viable EV energy sources, including batteries, fuel cells, capacitors and flywheels. For those EV batteries, a well established battery model is used. Hence, the corresponding usable capacity is dictated by the pre-defined efficiency maps and the charging/discharging characteristics. For those fuel cells, a simple model containing the specific energy, specific power and constant efficiency is adopted. It should

be noted that fuel cells themselves cannot allow for accepting electrical energy regenerated by EVs during braking or downhill. Similarly, a simple model containing only the specific energy, specific power and constant efficiency is employed for capacitors and flywheels (Chau *et al.*, 1999).

(8) Vehicle performance module—the whole vehicle performances, including the acceleration rate, maximum speed, driving range per charge, gradeability and energy consumption, are calculated and summarized in this module.

8.2.3 PERFORMANCE EVALUATION

Computer simulations have been performed to evaluate performances of various EVs. These simulations are driven by various driving cycles. Based on a typical passenger EV operating at the FUDS, the desired torque-speed operating points as well as the rated and maximum torque-speed characteristics of an EV motor are shown in Fig. 8.9. We should note that the majority of operating points should be embraced by the motor's rated characteristic, while all operating points must be embraced by its maximum characteristic. Moreover, the discharging current profile of the corresponding battery is shown in Fig. 8.10, indicating that the current fluctuation is high and there are negative discharging currents (regenerative charging currents).

By using different types of EVs (passenger car, van and bus) and different profiles of driving cycles (FUDS, FHDS, J227a-C, ECE Urban, Japan 10.15

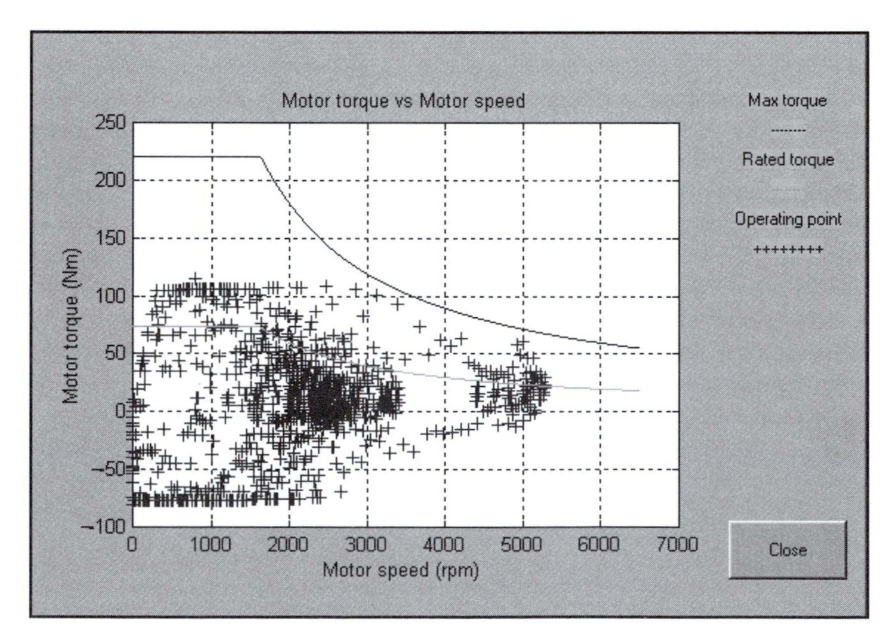

Fig. 8.9. EV motor torque-speed operating points and characteristics.

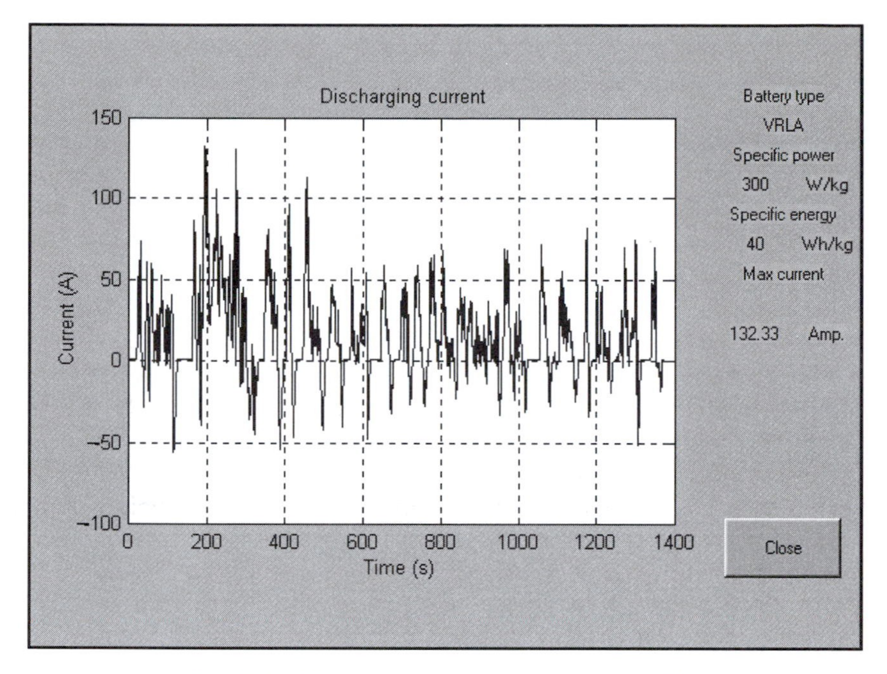

Fig. 8.10. EV battery discharging current profile.

Mode), we carry out a series of simulations. Thus, the desired rated power and maximum power of the EV motor, and hence the fuel economy, acceleration rate and gradeability of the EV are shown in Table 8.1. It should be noted that the resulting motor specifications and vehicle performances vary with different driving cycles for the same type of EV.

Another useful application of this simulator is to evaluate different EV subsystems or components, such as EV motors, transmissions and energy sources. Table 8.2 illustrates a simulation study on the selection of EV motors for the improvement of fuel economy. It is obvious that there is a significant improvement in the fuel economy due to the replacement of the induction motor by the PM brushless motor. The reason is due to the fact that the high efficiency region of the PM brushless motor is much wider than that of the induction motor, and covers most of the operating points.

8.3 System optimization

Due to high complexity, large flexibility and great diversity of EV systems, there are much room for designers to optimize EV configurations. Optimization of EV system configurations plays a significant role in EV development. Among many optimization schemes, three of them are discussed in this section, namely the transmission

Table 8.1 Performance of different EVs under different driving cycles

	FUDS	FHDS	J227a-C	ECE Urban	Japan 10.15
Passenger car					
Motor rated power (kW)	12.84	10.54	4.74	12.83	6.89
Motor max. power (kW)	38.52	31.63	14.21	38.48	20.67
Fuel economy (km/kWh)	5.19	4.57	5.35	5.85	6.03
Acceleration(0–50 km/h) (s)	5.64	7.20	14.10	5.64	8.40
Gradeability (%)	24.38	17.38	8.89	24.35	15.47
Van					
Motor rated power (kW)	46.10	38.49	14.97	40.78	23.04
Motor max. power (kW)	138.28	115.45	44.92	122.33	69.12
Fuel economy (km/kWh)	1.39	1.04	1.56	1.70	1.75
Acceleration (0–50km/h) (s)	4.93	6.19	15.22	5.28	9.31
Gradeability (%)	28.89	20.92	9.21	28.73	15.11
Bus					
Motor rated power (kW)	148.91	121.78	78.18	150.67	150.67
Motor max. power (kW)	446.74	365.34	234.54	452.00	452.00
Fuel economy (km/kWh)	0.49	0.50	0.47	0.53	0.56
Acceleration (0–50 km/h) (s)	5.89	7.12	11.17	5.40	5.40
Gradeability (%)	22.83	18.23	11.07	28.41	28.41

Table 8.2 Fuel economy of different EVs using different EV motors

	Fuel economy (km/kWh)		
	Induction motor	PM brushless motor	Improvement (%)
Passenger car			
FUDS	5.19	6.18	18.30
FHDS	4.57	5.07	10.94
J227a-C	5.35	5.94	11.03
ECE Urban	5.85	6.95	18.80
Japan 10.15	6.03	7.07	17.25
Van			
FUDS	1.39	1.65	18.71
FHDS	1.04	1.16	11.54
J227a-C	1.56	1.74	11.54
ECE Urban	1.70	2.01	18.24
Japan 10.15	1.75	2.01	14.86
Bus			
FUDS	0.49	0.59	20.41
FHDS	0.50	0.55	10.00
J227a-C	0.47	0.53	12.77
ECE Urban	0.53	0.63	18.87
Japan 10.15	0.56	0.65	16.70

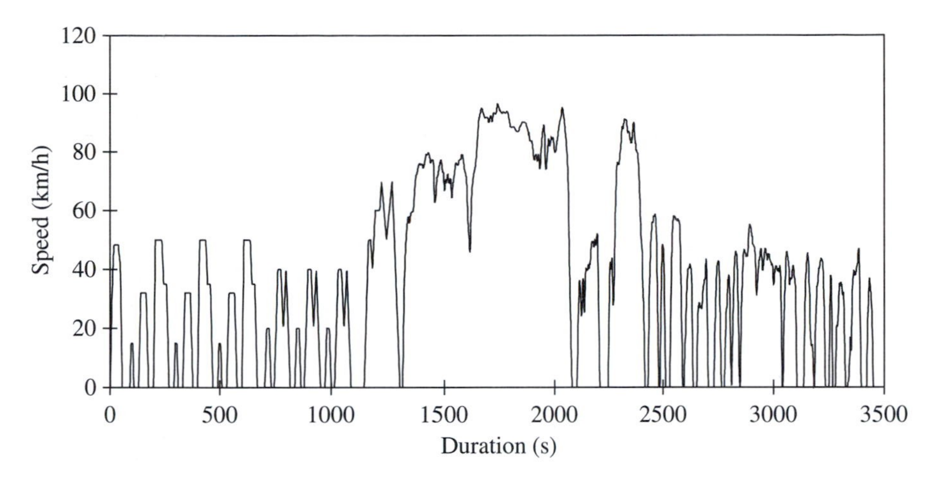

Fig. 8.11. Combined reference cycle for EV optimization.

ratio optimization, system voltage optimization and hybridization ratio optimization. As mentioned in the previous section, the system performances vary with the driving cycles. To minimize the effect of such variations, a combined reference cycle, incorporating the FUDS, FHDS, J227a-C, ECE Urban and Japan 10.15 Mode, is derived for optimization. Figure 8.11 shows this combined reference cycle.

8.3.1 TRANSMISSION RATIO

The transmission system of EVs is different from that of ICEVs. In a conventional ICEV, there is no alternative to the use of gear-changing transmission. In an EV, a fixed-gear transmission system is generally employed to eliminate the bulky gearbox and clutch. However, the fixed-gear transmission ratio depends on many factors, such as the torque-speed characteristics, acceleration rate, maximum speed and fuel economy. In fact, it greatly affects the specifications, cost and driveability of the whole EV. Therefore, the selection of this ratio should be based on computer optimization.

For passenger EVs, the fixed-gear transmission ratio generally has a range from 4.3 to 9.5, where the lower and upper limits are governed by the desired acceleration rate and maximum speed, respectively. The corresponding normal ranges for vans and buses are 4.9–12.1 and 6.8–15.2, respectively. Based on a constraint that the desired maximum speed of a passenger EV is at least 100 km/h, the fuel economy vs. acceleration rate (necessary time to accelerate from zero to 100 km/h) under different transmission ratios ranging from 4.3 to 9.5 is plotted in Fig. 8.12. When the ratio is increased from 4.3 to 6.1, the acceleration time is shortened and the fuel economy is improved. However, when this ratio is further increased up to

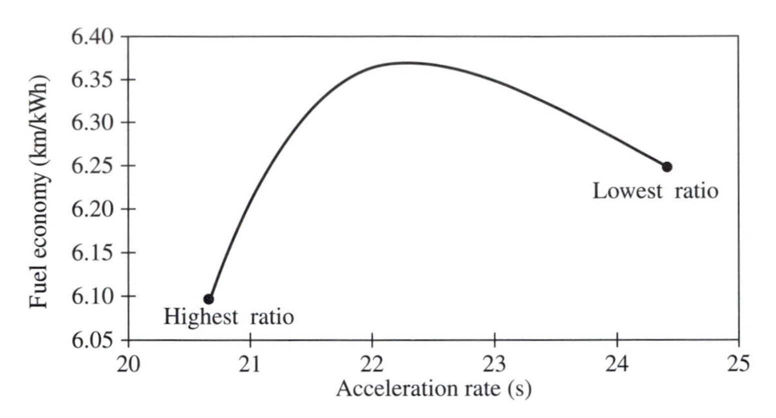

Fig. 8.12. Fuel economy vs. acceleration rate at different transmission ratios.

Table 8.3 Optimal transmission ratios of different EVs

	Optimal ratio	Fuel economy (km/kWh)	Acceleration rate (s)	Maximum speed (km/h)
Passenger car	6.06	6.37	22.28	151.64
Van	7.59	1.60	20.80	153.98
Bus	9.29	0.64	21.18	150.96

9.5, the fuel economy becomes worse even though the acceleration time is further shortened. In case the fuel economy is a key parameter to be optimized, the optimal transmission ratio should be selected as 6.1. Table 8.3 shows the optimal transmission ratio of those typical EVs (passenger cars, vans and buses) and hence the corresponding fuel economy, acceleration rate and maximum speed.

8.3.2 SYSTEM VOLTAGE

So far, there is no standard system voltage level for different types of EVs. In fact, there is a significant difference among the system voltage of existing EVs. This system voltage affects many aspects of the EV design and performance. There are many contradictory effects on the selection of this system voltage—the higher the voltage level, the more the number of battery packs, the more the connections between battery modules (bad connections may cause system failure), the heavier the battery weight, the larger the total battery capacity, the higher the motor voltage rating and the lower the inverter current rating. The key factor is the battery-to-vehicle weight ratio. For passenger EVs, this battery-to-vehicle weight ratio generally has a range from 10 to 40%, where the lower and upper limits are dictated by the desired driving range and acceleration rate, respectively. Based on the combined reference cycle, the acceleration rate (necessary time to accelerate from zero to 100 km/h) with respect to the driving range (number of combined

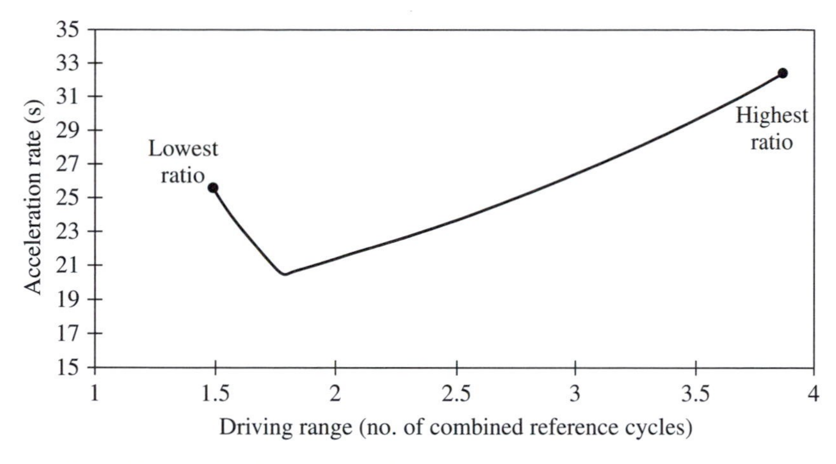

Fig. 8.13. Acceleration rate vs. driving range at different weight ratios.

reference cycles) under different battery-to-vehicle weight ratios ranging from 10 to 40% is plotted in Fig. 8.13. In case the acceleration rate is the key parameter to be optimized, the optimal battery-to-vehicle weight ratio should be selected as 22%. Ranging from 10 to 22%, the available power from the battery is less than the motor output power capability so that both acceleration rate and driving range improve with the increase of the weight ratio. Beyond 22%, the motor output power capability has already been fully utilized and any additional weight ratio will degrade the acceleration rate though the driving range can be further increased. In case the driving range is the key parameter to be optimized while the acceleration rate is a constraint only, the weight ratio can be maximized in such a way that the acceleration rate is marginally satisfied.

8.3.3 HYBRIDIZATION RATIO

As discussed in Chapter 6, many energy sources (batteries, fuel cells, ultracapacitors and ultrahigh-speed flywheels) have been proposed and developed for EVs. Since none of them can individually fulfil all demands of EVs to enable them competing with ICEVs, the concept of multiple energy sources, so-called the hybridization of energy sources, has also been introduced. In this section, the hybridization ratio of any two energy sources is optimized in such a way that the objective function of those important operating performances, namely the acceleration rate, driving range and fuel economy, can be maximized.

In the battery & battery hybrids, one battery provides high specific energy while another battery offers high specific power. Taking into account the maturity and cost, the Zn/Air & VRLA hybrid seems to be a natural choice. It combines the merits of the 230 Wh/kg of Zn/Air for long driving range and the 300 W/kg of VRLA for acceleration and hill climbing. This choice also overcomes the incapability of the mechanically rechargeable Zn/Air which cannot accept the precious

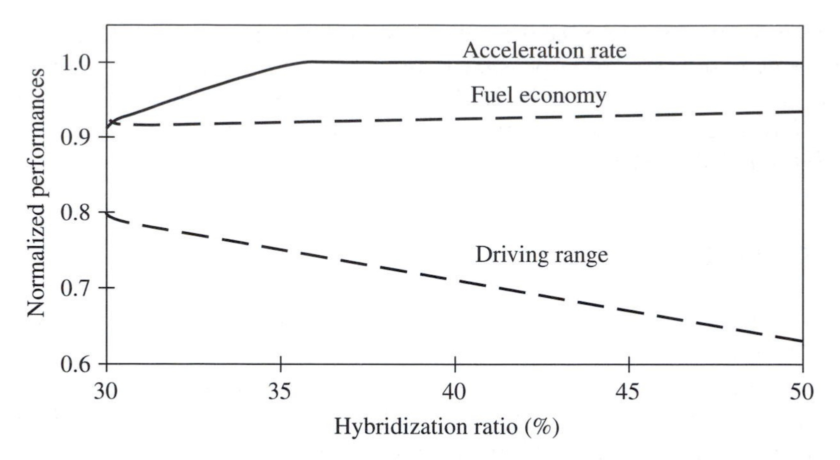

Fig. 8.14. Normalized performances vs. hybridization ratio.

regenerative energy during braking or downhill. Based on the EV simulator, the normalized operating performances with respect to the weight ratio of VRLA to Zn/Air (so-called the hybridization ratio) for a passenger EV is shown in Fig. 8.14, in which the references for normalization are based on using the long-term battery defined by the USABC. It can be observed that both the acceleration rate and fuel economy increase with the hybridization ratio, but suffering from a reduction in the driving range. The saturation phenomenon of the acceleration rate is due to the fact that the VRLA battery component is sufficient enough to fully satisfy the need of the maximum motor output power. In case we select the objective function as a simple summation of those operating performances, its variation with respect to the hybridization ratio is shown in Fig. 8.15. Thus, the corresponding optimal hybridization ratio for a passenger EV should be selected as 35%.

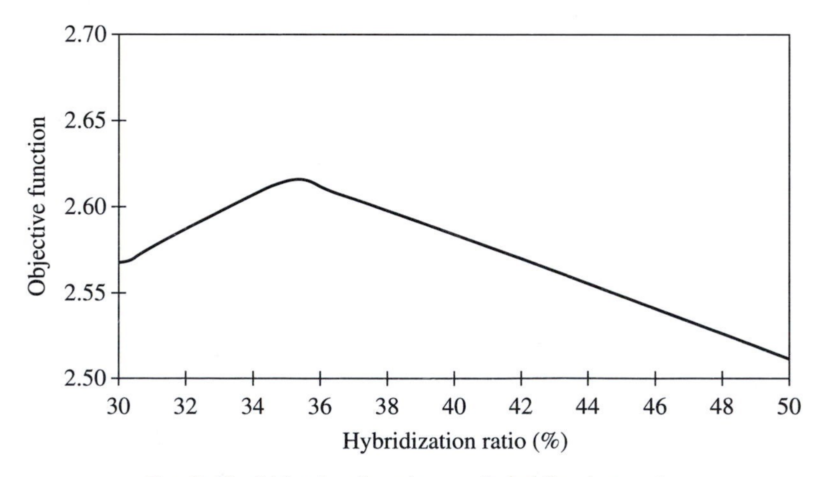

Fig. 8.15. Objective function vs. hybridization ratio.

Similarly, we can determine the optimal hybridization ratio of different types of hybrid energy sources. Other possible hybrids that are considered to be practically viable include Zn/Air & Ni–MH, Zn/Air & Li–Ion, VRLA & ultracapacitor, Ni–MH & ultracapacitor, Li–Ion & ultracapacitor, SPFC & VRLA, SPFC & Ni–MH, and SPFC & Li–Ion.

8.4 Case study

Based on the EV simulator, many practical applications can be performed. For exemplification, we carry out a case study on the implementation of electric light bus transportation in a specific urban area. This transportation system consists of five light bus routes with both flat and hilly regions as shown in Fig. 8.16. The route length, journey time, waiting time and working hours of these routes are listed in Table 8.4. The specifications of such a light bus are tabulated in Table 8.5.

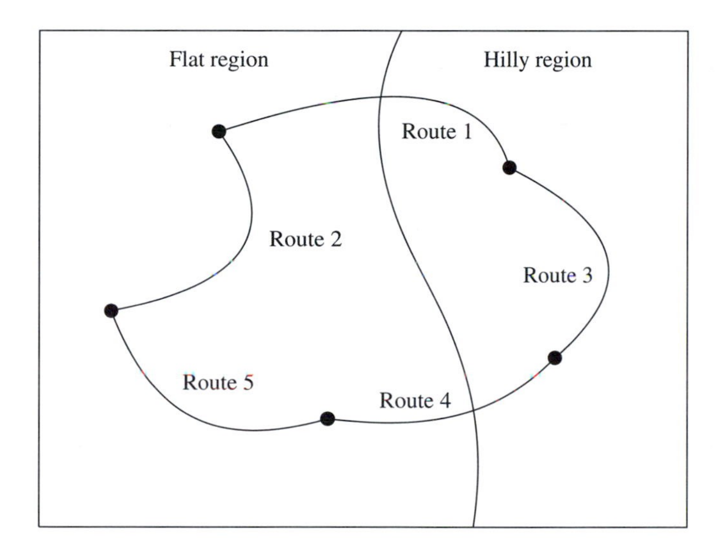

Fig. 8.16. Route distribution.

Table 8.4 Route schedule

Route	Length (km)	Journey time (min)	Waiting time (min)	Working hours
1	8.1	24	8.5	0645–2400
2	11.2	40	13	0630–2100
3	3.1	30	8	0640–2330
4	2.5	24	8	0635–2335
5	8.6	30	9	0000–2400

Table 8.5 Light bus specifications

Seater number	16
Overall length (mm)	6000–6200
Overall width (mm)	2070
Overall height (mm)	2580
Motor rated speed (rpm)	3000
Motor rated power (kW)	35
Motor rated voltage (V)	288
Curb weight (kg)	3200–3400
Payload capacity (kg)	810–1010
Maximum speed (km/h)	80
Acceleration rate (0–48km/h) (s)	7
Gradeability (%)	10

Since the driving profile of these light buses mainly depends on the road traffic conditions and the road grades, the pattern of the Japan 10.15 Mode driving cycle is employed to represent the urban driving profile with zero or near-zero road grade while two modified patterns are used to describe those profiles with significant road grades. These three driving patterns, namely A, B and C, are shown in Figs. 8.17, 8.18 and 8.19, respectively. Thus, the pattern A is adopted when the road grade lies between −6.7 and 5%, the pattern B is used when the grade is more negative than −6.7% or between 5 and 10%, and the pattern C is for the grade greater than 10%.

Although the passenger demand generally varies throughout the whole day, it needs to be modelled. Figures 8.20 and 8.21 show the models of this passenger demand, respectively for the forward and reverse directions from one terminal to another. In the normal operating period, the passenger capacity ratio is 0.5,

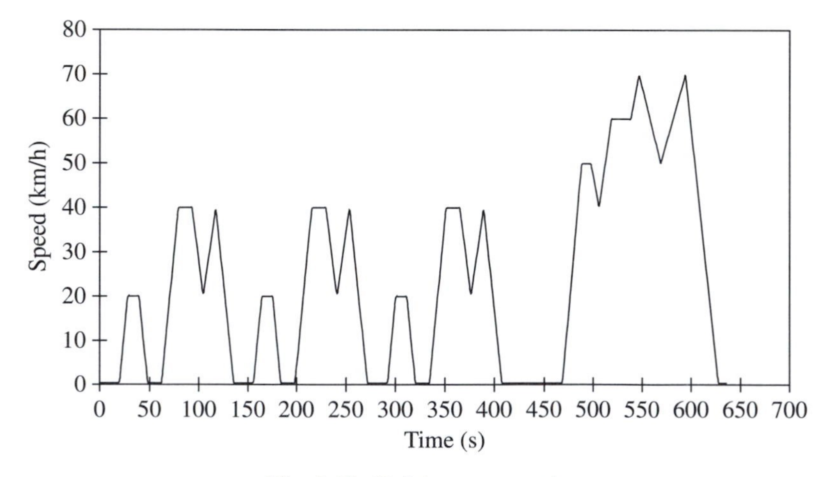

Fig. 8.17. Driving pattern A.

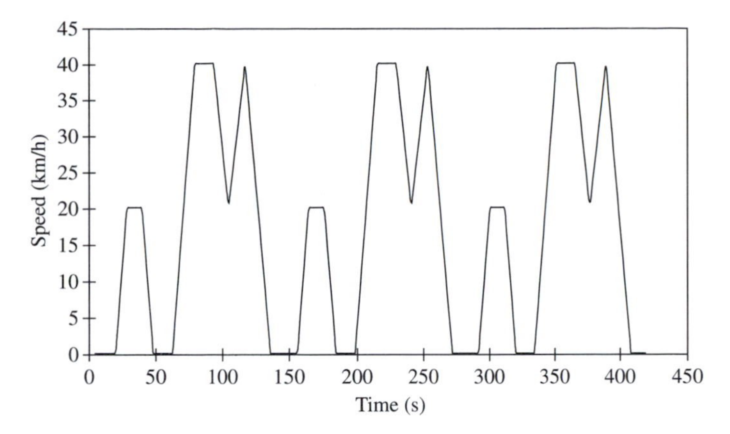

Fig. 8.18. Driving pattern B.

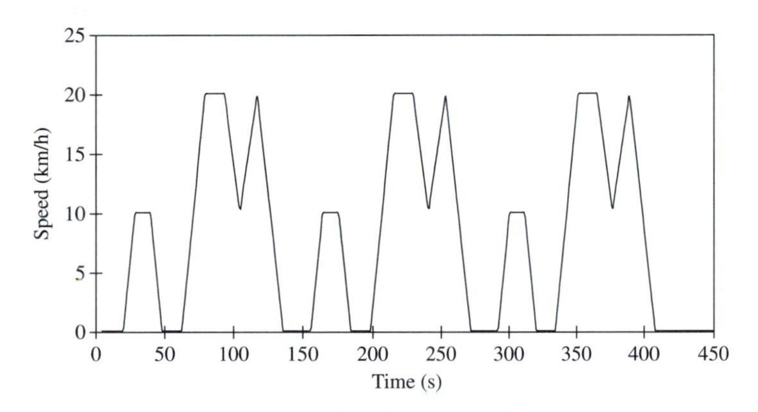

Fig. 8.19. Driving pattern C.

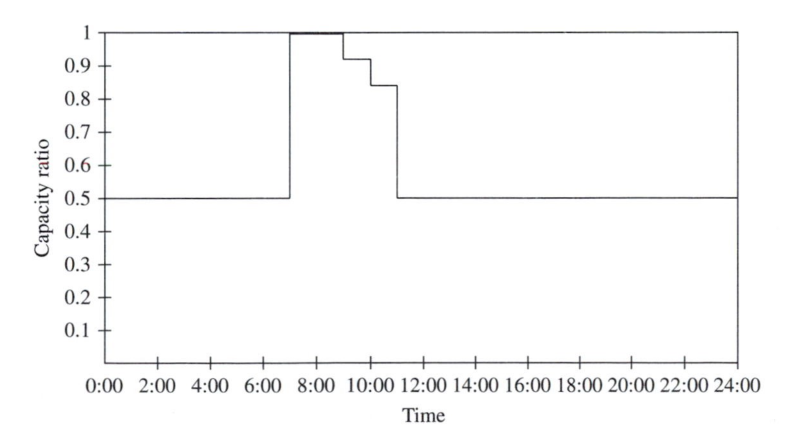

Fig. 8.20. Passenger demand in forward direction.

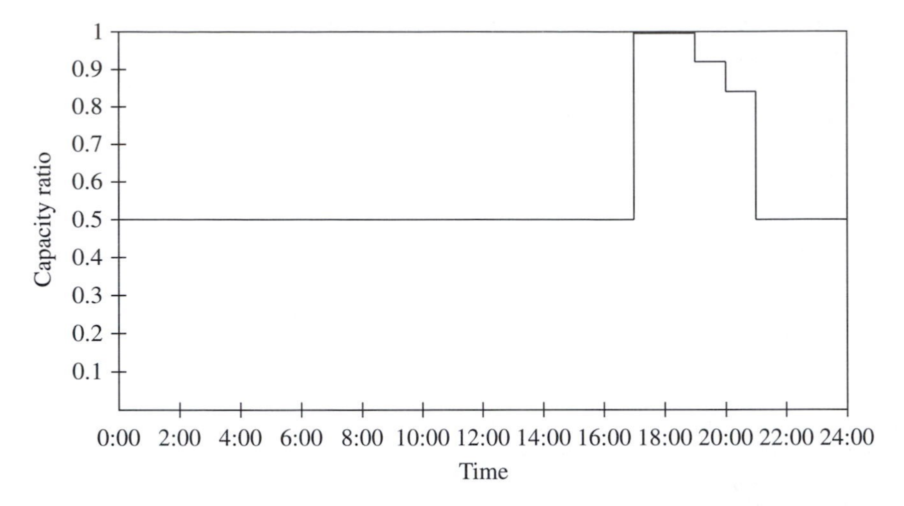

Fig. 8.21. Passenger demand in reverse direction.

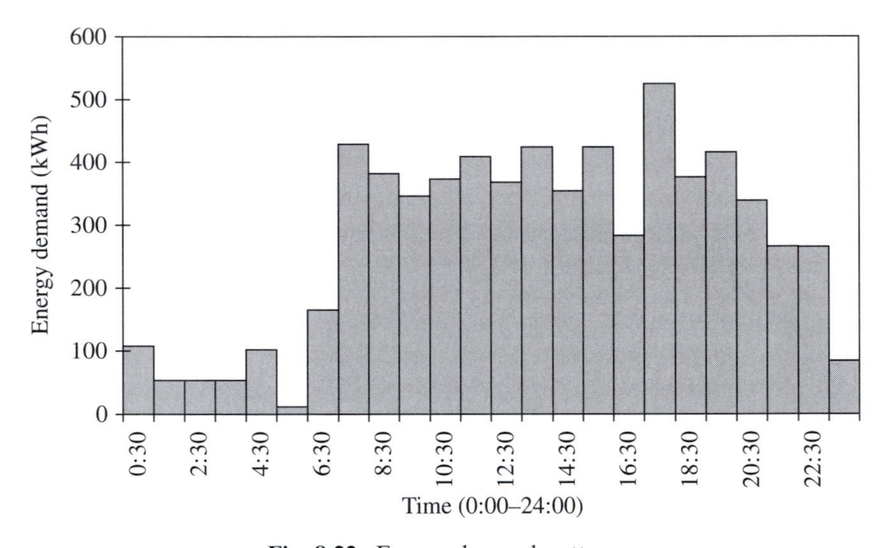

Fig. 8.22. Energy demand pattern.

namely eight passengers. The peak hours in the forward direction begin from 7 am with the unity capacity ratio and end at 11 am with the normal capacity ratio, while the peak hours in the reverse direction are from 5 pm to 9 pm with a similar staircase pattern.

By using the EV simulator, all the operating performances such as the desired motor ratings and battery ratings are readily determined. Furthermore, we can determine the total energy demand due to this transportation system. Figure 8.22

shows the corresponding energy demand pattern, hence its total daily demand is 6.6 MWh. It should be noted that this demand pattern does not represent the desired energy instantaneously drawn from our power system. Actually, it indicates the total battery charging energy that can be used for power system load levelling.

References

Chan, C.C. and Chau, K.T. (1997). An overview of power electronics in electric vehicles. *IEEE Transactions on Industrial Electronics*, **44**, 3–13.

Chan, C.C. (1993). An overview of electric vehicle technology. *Proceeding of IEEE*, **81**, 1201–13.

Chan, T.W.D. (2000). *System Level Simulation for Electric Vehicles*. M.Phil. thesis, The University of Hong Kong, Hong Kong.

Chau, K.T. (1996). A software tool for learning the dynamic behaviour of power electronics circuits. *IEEE Transactions on Education*, **39**, 50–5.

Chau, K.T. and Chan, C.C. (1998). Electric vehicle technology. *International Journal of Electrical Engineering Education*, **35**, 212–20.

Chau, K.T., Chan, C.C. and Wong, Y.S. (1998). Advanced power electronic drives for electric vehicles. *Electromotion*, **5**, 42–53.

Chau, K.T., Wong, Y.S. and Chan, C.C. (1999). An overview of energy sources for electric vehicles. *Energy Conversion and Management*, **40**, 1021–39.

MathWorks (1997). *Student Edition of MATLAB Version 5 for Windows* (1st edition). Prentice Hall, New Jersey.

Wong, Y.S. (2000). *Performance Simulation and Energy Coordination for Electric Vehicles*. M.Phil. thesis, The University of Hong Kong, Hong Kong.

9 EV infrastructure

The development of EVs has advanced rapidly over the past decade. Researchers and engineers have concentrated on the improvement of EV performance through the advances in batteries, motors, converters, controllers and relevant auxiliaries. Now, it comes to the stage of commercialization. To support the publicity of EVs, an EV infrastructure is the underlying foundation, which includes the basic facilities and services to support the operation of a large number of EVs (Olsen, 1996). In order to develop a successful EV infrastructure, we should pay attention on the following aspects:

- availability of charging stations
- convenience of payment for charging
- standardization of EV batteries and charging
- regulation of clean and safe charging
- support from training and promotion and
- impacts on power utilities.

The development of EV infrastructure is a huge project, and no single party can afford the total cost and do it alone. Moreover, an EV infrastructure is a long-term investment and we need to spend many years for the development before getting any rewards. Both public and private sectors such as government, organizations, automotive companies, power utilities, battery manufacturers should work together for the development of EV infrastructure (Muntwyler, 1996; Nowicki, 1997).

Nowadays, nearly all available EVs are powered by rechargeable batteries such as the VRLA, Ni–MH, Li–Ion types. Although fuel cells are becoming applicable to EVs, there are still much to be done before they can be economically viable or commercialized. We foresee that in the near future rechargeable batteries will still dominate the market of EV energy sources. Thus, the establishment of charging systems for recharging EV batteries is a major constitution of the EV infrastructure.

A charging system for EVs poses the same functionality of a petrol station for ICEVs, but with some major differences. Firstly, an EV charging system can be either a public installation or a domestic one. An EV user can charge his EV at a public charging station or even in his own garage by simply plugging the EV charger to a power outlet. Definitely, it is rare for a person to install an in-house oil tank for refuelling his ICEV. Secondly, an EV user can have different choices on the charging period so as to get a favourable price for recharging at off-peak periods. Certainly, it is not the case for refuelling an ICEV because the oil price is generally fixed within the whole day. Thirdly, an EV charging system may bring along adverse influences to our power system such as harmonic contamination, poor power factor and high current demand. Of course, a petrol station has no impact on our power system.

The design of EV charging systems mainly depends on the level of charging currents to charge the EV batteries. There are three major current levels:

(1) Normal charging current—the EV batteries can be charged by a rather low charging current, about 15 A, and the charging period may last for over 6 h. The operation and installation costs of the corresponding charger are relatively low since the power and current ratings involved are not of critical values. This charging current usually benefits to increase the charge efficiency and to extend the battery life.
(2) Medium charging current—the EV batteries can be charged by a medium current of 30–60 A, and the charging period may last for a few hours. The operation and installation costs of the corresponding charger are relatively higher than that for normal charging current because of the necessity to upgrade the charging equipment.
(3) Fast charging current—the EV batteries can be charged up within a short time based on a high charging current of 150–400 A. In contrast to that using normal or medium charging current, the corresponding charger offers relatively low charge efficiency. Definitely, the corresponding operation and installation costs are high.

The normal charging current are adopted in both domestic and public charging infrastructures, whereas both the medium and fast charging currents are only found in the public charging infrastructure. Moreover, the fast charging current should only be adopted in those dedicated public charging stations because the corresponding current demand may cause detrimental effect on the power system network.

9.1 Domestic charging infrastructure

Charging at home is the most preferable way for people to charge their EVs since it can be done by simply plugging the EV on-board charger to the outlet that is installed or nearby the car park. People can get their EVs fully charged up after a night's parking, hence allowing for over 100-km driving range. As the batteries are slowly recharged, the power requirement is just a few kW and the charging time is from 5 to 8 h. In general, domestic charging favours an effective utilization of electricity since EVs are usually charged at night or off-peak periods. Thus, power utilities are willing to impose incentive tariff and rebate to attract EV users to charge their EVs during off-peak periods.

The basic requirement of domestic charging is the availability of a garage or a parking lot that is fed with electricity. There are two different ways:

- For those houses with a private garage, an indoor socket outlet can be installed for recharging.
- For those apartments and multi-storey buildings with car parks attached, an outdoor socket outlet can be installed. The outdoor socket outlets should have individual protection circuits, and can be independently operated. Only residents with permission can access these socket outlets for EV charging.

The payment scheme for domestic charging is rather simple because the existing revenue meters and tariff scheme can be directly adopted. An EV can simply be considered as an ordinary electric appliance. Obviously, the corresponding initial cost is relatively low since no additional installation and any other expensive meters are needed.

9.2 Public charging infrastructure

Basically, a public charging infrastructure is a distribution of public charging stations which can be readily accessed by EV drivers to recharge their EVs when necessary. It is much more difficult to develop a public charging system than a domestic one because there are many factors such as electrical safety, ergonomics, weather conditions, variation of EVs and availability of area for charging stations that should be taken into account. To ensure a public charging system to accommodate EVs of different brands and models, we should design those EV auxiliaries with a high degree of standardization. Certainly, battery chargers are the major constitutive part of a public charging infrastructure. In additional to the capability of battery charging, the infrastructure should also pose other abilities such as identification of authorized customers, notification of the battery capacity, metering and billing (Menga and Savoldelli, 1996).

9.2.1 NORMAL CHARGING STATIONS

Normal charging stations are designed to recharge EVs based on their on-board chargers using the normal charging current. These charging stations are generally located nearby the residential areas or working areas in which the EVs will be parked for 5–8 h. These charging stations are usually designed in a large scale so that many EVs can be accommodated simultaneously for normal recharging. The corresponding charging poles are usually modular designed, consisting of control terminals and delivery terminals. In practice, EV drivers simply park their EVs on the allocated areas of the charging station, then connect the cable and initiate the recharging process.

The socket outlet for normal charging stations is specific so that the plug can be locked in with protection during charging. Typically, the connection is four-wired, including live, neutral, ground and pilot wire, and it takes power from a single-phase system.

9.2.2 OCCASIONAL CHARGING STATIONS

Conceptually, EVs are expected to be fully charged at night or after 5–8 h parking in normal charging stations for normal daily operation. Practically, there are many factors affecting the usable capacity of EV batteries. For examples, some EVs may be used at both days and nights so that there are not enough time to fully charge up the batteries per single charging process, or some EVs may be desired to travel a cumulatively long distance per day. Occasional charging is a concept that EVs can be recharged whenever there are such occasions. These charging stations are normally located in those places where there are chances for EVs parking for half an hour to several hours such as parking lots, airports, train stations, hotels, hospitals, schools, shopping malls, supermarkets, convention centres and tourist attractions. For examples, parking EVs nearby lavatories may last for about half an hour, cinemas for about 1 h, restaurants for about 2 h, and malls for several hours.

Although this occasional charging concept is not limited to charge EVs using medium charging current, EV drivers should not always expect to fully charge up their EVs within the parking period. Instead, they should consider that this is a chance to increase the battery usable capacity for the upcoming travel or simply as an additional reserve. For example, the driving range of EVs may be extended by about 40 km for an hour of occasional charging using medium charging current.

9.2.3 FAST CHARGING STATIONS

Fast charging is also known as rapid charging or emergency charging which aims to recharge EV batteries within a short period that should be approaching to the time for petrol refuelling of ICEVs. Obviously, fast charging stations are the places to provide fast charging facilities for EVs. Certainly, the use of fast charging is governed by the recharging characteristics of those EV batteries. Ordinary batteries should not be subjected to fast charging as they may result in overheating when accepting a great amount of charge within a short period. The necessary time for recharging are really short, about 20 min for charging up 80% capacity. Such capability is equivalent to allow EVs travelling about 8 km further for every minute of fast charging. Thus, the total travelling distance of EVs can be greatly extended, provided that there are sufficient fast charging stations on the way.

The key of fast charging stations is obviously the off-board fast charging module, which can output 35 kW or even higher. The corresponding voltage and current ratings are 45–450 V and 20–200 A, respectively. As both power and current ratings are so high, it is preferable to install such recharging facilities in supervised stations or service centres. To avoid overcharging the EV batteries and causing excessive heat generation, there should be a communication between the fast charging module of the station and the battery monitoring circuit of the EV for information exchange so that the EV battery condition can be on-line monitored and the charging current can be on-line modified. Unlike normal charging stations, the fast charging stations adopt the cables consisting of more wires. Typically, these cables are composed of seven wires, including positive, negative, ground, pilot wire and a three-wired communication channel.

Although fast charging stations can provide the possibility of EVs offering a driving range similar to that of ICEVs, there are a number of drawbacks that need to be regulated before their widespread installation. The main drawback is their adverse impacts on our power system, namely harmonic contamination and high current demand superimposing on the peak-hour consumption, violating the principle of demand-side management. It should be noted that power utilities have played a very active role to promote the commercialization of EVs because they expect that the majority of EVs will be charged at night or off-peak periods, thus enhancing the business of power utilities and utilization of power equipment. The solution of these adverse impacts is being actively investigated. For examples, the use of new topologies of battery chargers, new arrangement of battery charging and installation of power active filters can suppress the harmonic contamination to certain extent, while higher tariffs can be adopted to offset the use of fast charging.

9.2.4 BATTERY SWAPPING STATIONS

Instead of charging the batteries immediately, we have another way to refuel the energy source of EVs—mechanically swapping the discharged batteries with those fully charged batteries. Of course, all these batteries should be owned by the service station or battery company while the EV driver is only a battery borrower. The operating principle of this battery swapping is illustrated in Fig. 9.1. An EV driver stops at a particular area and a swapping machine will replace the

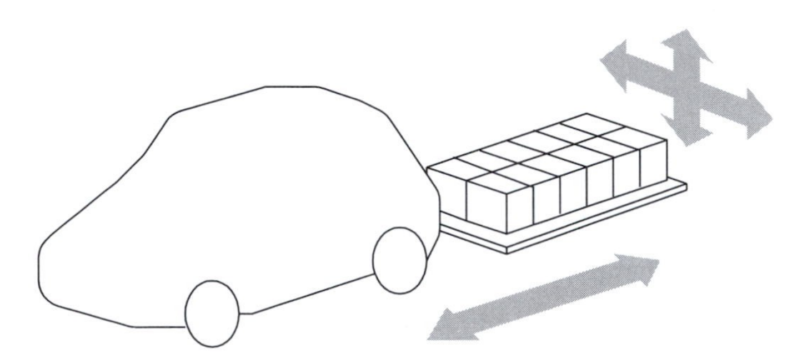

Fig. 9.1. Principle of battery swapping.

discharged batteries with fully charged ones. This machine may be a forklift that is cantilevered from its stand and has enough forward and backward adjustment to accommodate the location of the EV. It puts the fork into slots on the bottom of the battery pack, and then moves the pack to the right position. The discharged batteries will either be charged at the service station or centrally collected and charged. Since the battery swapping process involves mechanical replacement and battery recharging, it is sometimes named as mechanical refuelling or mechanical recharging. These battery swapping stations combine the merits of both normal charging stations and fast charging stations, namely recharging the EV batteries at off-peak periods while refuelling the EVs within a very short time. With the use of robotic machinery, the whole battery swapping process can be less than 10 min, directly comparable to the existing refuelling mechanism for ICEVs.

However, there are many obstacles to be solved. Firstly, the initial cost to set up this battery swapping system is very high, involving expensive robotic machinery for battery swapping and a large number of costly batteries for necessary operation. Secondly, due to the need to store both discharged and fully charged batteries, the necessary space to build a battery swapping station is much larger than that for a normal charging station or a fast charging station. Thirdly, the EV batteries need to be standardized in physical dimensions and electrical parameters before the possible implementation of automatic battery swapping. Nevertheless, this battery swapping system is being implemented in a small scale to automatically exchange Zn/Air batteries for several fleets of EVs.

9.2.5 MOVE-AND-CHARGE ZONES

The most ideal situation for charging EV batteries is to charge the vehicle while it is cruising on the road, so-called the move-and-charge (MAC). Thus, there is no need for an EV driver to find a charging station, park the vehicle and then spend time to recharge the batteries. This MAC system is embedded on the surface of a section of roadway, the charging zone, and does not need any additional space.

Both conductive and inductive types of MAC can be implemented. For the conductive MAC system, an on-board contact arch is mounted on the bottom of the EV body. By physically contacting the charging elements which have been embedded on the road surface, the arch picks up instantaneous high current. Since the EV is cruising through the MAC zone, the charging process is a kind of pulse charging. For the inductive MAC system, the on-board contact arch is replaced by inductive coils, and the embedded charging elements are replaced by high current coils which produce strong magnetic field. Obviously, the conductive MAC zone is relatively not attractive because it suffers from mechanical wears and positioning problems of the contact arch. In fact, recent investigations have only focused on the inductive type because of no mechanical contacts and large positioning tolerance. Since the cost of these MAC zones is very expensive, they are still in the experimental stage. Nevertheless, this charging concept is promising in future as people will appreciate a more convenient way to recharge their EVs.

9.2.6 PAYMENT SYSTEMS

For public charging systems, there are many possible payment schemes. Some of them may incorporate complicated tariff schemes. The following four schemes are considered to be practically viable:

(1) Charge for both parking and recharging separately—it seems to be most fair to all EV drivers because the parking time and recharging energy are separately counted or charged. None of the two components need to subsidize one another. However, both parking timer and revenue meter are necessary and the relevant management costs are relatively high.
(2) Charge for recharging only, with free parking—it can encourage EV drivers to charge the EV batteries whenever necessary. Because of free parking, those drivers park their EVs solely for recharging batteries need to subsidize the others. This scheme is particularly suitable to the public charging stations located in those rural areas where electricity is expensive. It takes the advantage that only the revenue meter is necessary.
(3) Charge for parking only, with free recharging—it can encourage EV drivers to fully charge up their EV batteries whenever possible. Because of free recharging, those EVs with high remaining capacity need to subsidize the others. This scheme is particularly suitable to the public charging stations located in those populated cities where space is very expensive. It takes the advantage that only the parking timer is necessary.
(4) One-line charge for both parking and recharging—EV drivers can pay for daily, weekly or monthly labels for parking and recharging. Thus, they can gain access to reserved parking areas and make use of those charging facilities.

In addition to the use of revenue meters for counting the power consumption, some electronic devices such as computers are necessary to record the power consumption and hence calculate the necessary payment. A mature method is to use slot machines which can provide the information of power consumption and payment so that EV drivers can make the payment by cash or credit cards. Another method is to employ smart cards which can be used to record the power consumption and also make payment automatically. The latest smart card is a proximity type which carries semiconductor chips to store the user identity and those related information.

9.3 Standardization and regulations

The EVs and EV infrastructures should be safe, reliable, easy to use and low cost (Toepfer, 1997). All these requirements involve a very important criterion—standardization. The outcome of standardization is standards (Bossche, 1997; Kosowski, 1997). According to the International Electrotechnical Commission (IEC), a standard is a document, established by consensus and approved by recognized body, that provides, for common and repeated use, rules, guidelines or characteristics for activities or their results, aimed at the achievements of the optimum degree of order in a given context. Actually, the establishment of standards, governing the dimensions, specifications and performances of both EVs and EV infrastructures, is one of the most important issues for both EV suppliers and users. After issuing those standards, both manufacturers and suppliers can follow to produce EVs and to establish the infrastructures, while drivers can use their EVs in a more safe and convenient manner.

Standards and regulations are different though they are closely related. Strictly speaking, regulations come out from standards. Standards are voluntary while regulations are compulsory. A standard can be considered as a compromise among various interests involved, which is adopted to gain mutual benefits for different parties including suppliers, manufacturers and users. The governments impose regulations and everybody should follow. Certainly, before the full implementation and commercialization of EVs, a number of standards and regulations have to be made. Most of the standards are related to the charging accessories and infrastructure of EVs, while the regulations are mainly related to safety and operation of EVs.

We can foresee that most of the standards and regulations for EVs are based on the existing ones for ICEVs. However, new standards and regulations are needed due to the facts that EVs involve electrical equipment with high-power consumption. Indeed, EVs are a product of mixed technology—a road vehicle as well as an electric appliance. As a road vehicle, the standards should follow the International Organization for Standardization (ISO); on the other hand, as an electric appliance, it should be governed by the International Electrotechnical Committee (IEC). There are different cultures on standardization for different parties. In the automobile industry, standardization is not widespread for ICEVs. Different vehicle manufacturers are likely to produce their unique products. Interchangeability of parts and components is not an important issue. The corresponding standardization is mainly covered by regulations on safety and environmental protection. However, in the electric appliance industry, standardization is highly identified. There is a great tendency to make everything being standardized, from power quality requirements to colour code for a piece of wire. To avoid duplication of work and contradictory of different parties, we should pay attention to the division of work on the standardization of EVs. As a finished product, the ISO can consider EVs as a whole for standardization. For examples, we need to have standard forms to describe the EV specifications, such as hill-climbing capability, acceleration rate and driving range. On the other hand, the IEC can impose standards on the electrical parts and components of EVs, such as power ratings and mean time between failure.

9.3.1 STANDARDS

There are two important international organizations, namely the ISO and IEC, that involve in the standardization of EVs. These organizations consist of a number of technical committees, sub-committees and working groups. The technical committees usually define their own scope of activities. In case the scope is too wide or covers different areas, sub-committees will be formed under a technical committee. Their major task is to prepare technical documents on specific subjects within their respective scopes. A technical committee or sub-committee generally appoints experts to form working groups to draft documents upon proposals of new or modified standards. Usually, these working groups will be disbanded after the completion of tasks.

Besides the above two worldwide organizations, some regional and national organizations have also played active roles for EV standardization:

- European Commission for Standardization (CEN)
- Society of Automotive Engineers (SAE) and
- Japan Electric Vehicle Association (JEVA).

In the ISO, the Technical Committee 22 (TC22) is responsible for the standardization of road vehicles. Inside the TC22, there is a sub-committee named Sub-Committee 21 (SC21) that is formed for the standardization of EVs. Within the SC21, there are two active working groups, namely WG1 and WG2. The WG1 works for the safety, operation conditions and energy storage installation of EVs while the WG2 for the EV terminology including definitions and methods of measurement. Some issued standards are the ISO/DIS 8714–1 Electric road vehicles—Reference energy consumption and range—Part 1: Test procedures for passenger cars and light commercial vehicles, and the ISO/DIS 8715 Electric road vehicles—Road operating characteristics.

In the IEC, the Technical Committee 69 (TC69) is responsible for the standardization of EVs. Inside the TC69, there are two active working groups—WG2 and WG4. The WG2 is dedicated to electric motors and their control systems, including definition of terms, integration of motors and controllers, wiring and connectors, instrumentation, diagnostics, electromagnetic compatibility (EMC), personal safety, and electric field inside the vehicle. The WG4 is responsible for power supplies and chargers, including connection of the vehicle to the ac supply, connection of the vehicle to the off-board charger, roadside energy supply, EMC, functional safety, plugs and sockets, additional supply to the vehicle for heating and cooling, communication between the battery and the charger, and inductive coupling for battery charging.

In the CEN, the Technical Committee 301 (TC301) is dedicated to the issues of EVs. There are three active working groups in the TC301, WG1, WG4 and WG5, which are responsible for the measurement of EV performances, the liaison and dialogue between EVs and charging stations, as well as the safety and miscellaneous, respectively. Moreover, the Technical Committee 69X (TC69X) is dedicated to the standardization of EVs. Its three active working groups, WG1, WG2 and WG3, are responsible for the charging design and operation, the charging environment aspects and the safety, respectively.

The SAE is a non-profit scientific and educational organization which works closely together with academia, government, industry and worldwide members. It consists of more than 600 technical committees and each committee deals with a specific subject vital to a particular mobility-related industry or engineering discipline. The SAE has issued a large number of standards concerning EVs:

- Recommended Practice for Electric and Hybrid Electric Vehicle Battery Systems Crash Integrity Testing;
- Electric Vehicle Inductive Charge Coupling Recommended Practice;
- Electric Vehicle Conductive Charge Coupler;
- Electric Vehicle Terminology;
- Electric Vehicle Acceleration, Gradeability, and Deceleration Test Procedure;
- Electric Vehicle Energy Consumption and Range Test Procedure;
- Measurement of Hydrogen Gas Emission from Battery-Powered Passenger Cars and Light Trucks During Battery Charging;
- Energy Transfer System for Electric Vehicles—Part 1: Functional Requirements and System Architectures;
- Performance Levels and Methods of Measurement of Magnetic and Electric Field Strength from Electric Vehicles, Broadband, 9 kHz–30 MHz;
- Recommended Practice for Performance Rating of Electric Vehicle Battery Modules;
- Recommended Practice for Packaging of Electric Vehicle Battery Modules;
- Performance Levels and Methods of Measurement of Electromagnetic Compatibility of Vehicles and Devices (60 Hz–18 GHz); and
- Test Procedure for Battery Flame Retardant Venting Systems.

The JEVA was established in 1976, aiming to promote the use of low-emission vehicles. It is supported by a large number of commercial organizations which include automobile manufactures, power companies and battery suppliers. In order to promote the testing, research, and practical application of EVs, it takes active participation in expositions, shows and many other events. It is also active to issue standards covering various aspects of EVs, so-called the Japan Electric Vehicle Standards (JEVS). In case the data acquisition and verification have not been completed and discussions are not mature to make the standards, the JEVA will issue technical guidelines temporarily. The issued JEVS mainly cover the dimension, construction and test procedures for lead–acid batteries, the measurements of EV performances, the symbols for controls, indicators and tell-tales, the charging systems, and the standard form of specifications (Shimizu, 1997).

Up to now, we do not have any unique standards on the input voltage and current ratings of EVs. With the advances in battery technologies, the battery current demand changes readily so that it is hard to come into any mature agreement on the voltage and current ratings for charging. Nevertheless, we can alleviate this problem by employing the latest power electronics technology. Since power converters can readily offer wide ranges of voltage and current, the charging ratings can be determined by communication between the charger and the EV.

Standard symbols are widely used in industrial equipment as they are very important for the user interface. They can overcome the barrier of languages to enable users recognizing the functionality and operation of equipment. The JEVS issued by the JEVA has defined some standard symbols for control, tell-tales and indicators of EVs as shown in Fig. 9.2.

9.3.2 REGULATIONS

The development of EVs is entering the era of commercialization. So far, we do not have dedicated regulations to govern the usage of EVs. Nevertheless, there are some regulations on emissions and harmonics that have a great influence on the development of EVs. The former is one of the major motivations for the development of EVs, whereas the latter is one of the adverse by-products due to the advent of EVs.

Symbol	Function	Symbol	Function
	State of charge of main batteries		Charge coupling
	Liquid level of main batteries	 RET	Electric brake
	State-of-charge of auxiliary batteries	READY	Operation ready
	Motor overheat		System malfunction
 VAC	Status of brake vacuum system		Reduced power

Fig. 9.2. JEVS symbols for control, indicators and tell-tales of EVs.

The Clean Air Act was first passed in 1970 which aimed to improve air quality in the United States. Under this law, the Environmental Protections Agency (EPA) sets limits on the amount of a pollutant that can be in the air. The sources of pollutant can be either stationary such as factories and power plants or mobile such as vehicles and trucks. In 1990, the California Air Resources Board established rules that 2% of all vehicles sold in the state between 1998 and 2002 should be emission-free, and 10% of vehicles put on the market must have zero emissions by 2003. In 1996, since it was discovered that there was no way for the automobile industry to meet that timetable, the Board decided to scrap the 2% mandate, but still retained the 10% mandate by 2003 (this will be reviewed in 2001). It should be noted that EVs have been identified to be the most possible technology to meet the criterion of zero emissions.

The harmonics constraints are applicable to all electric devices, including those battery chargers for EVs. One of the most important guidelines can be found in the IEEE Standard 519—1992, Recommended Practices and Requirements for Harmonic Control in Electrical Systems. Actually, there are similar codes of practices adopted in different regions and nations. These practices mainly specified the limitation on harmonic current distortions generated by an electric circuit (Arseneau *et al.*, 1997).

9.4 Training and promotion

Generally speaking, research and development on EV technology are utmost important to improve the performance of EVs, whereas training and promotion are equally important on the popularization and commercialization of EVs. Training and promotion function to educate the users (drivers) of EVs, the workers (technicians and technologists) for EVs, and the general publics relating to EVs.

9.4.1 TRAINING

As the driving and control techniques for EVs may be somewhat different from that of ICEVs, we should provide a basic training course for EV users even though they are skilful drivers of ICEVs. They should know the basic operation of EVs and their charging facilities, safety issues for EV driving as well as fundamental maintenance and performance of EVs. To be an advanced EV driver, one should also know the technology and operation of the major EV components. Moreover, it would be better if EV drivers can realize the factors that affect the travelling distance of EVs since good driving habits can maximize the corresponding driving range. Definitely, skilled workers should be available for maintenance and repairing of EVs, as well as providing technical supports to EV drivers. Multi-level training programme is necessary for a crew of qualified EV technicians and technologists. The EV technicians of the crew should be well prepared for identifying and fixing the common problems that can be found in EVs. The technologists of the crew should have advanced technical knowledge on EV technology, such as thorough understandings on the structure and operation of EVs.

Based on the aforementioned requirements, the training courses can be provided with different levels—elementary, intermediate and advanced levels. The elementary level is an introduction to EV technology and basic driving skill, which is intended for new EV drivers. The course content includes the basic issues of EVs, such as operation techniques, safety precautions and charging procedures. The intermediate and advanced levels are mainly for those technicians and technologists of EV supporting services, respectively. The intermediate level offers fundamental techniques for EV troubleshooting and repairing. The corresponding content covers the operation and performance of various EV components, procedures on troubleshooting, faulty component replacement, functional testing and hazardous material handling. The objective of the advanced level is to provide updated skill and knowledge for EV technologists. The course content covers updated EV technology, new models of EVs and new EV components (Kosak, 1997).

Besides the above in-service training courses, universities should play an active role on engineering education in EVs. The Department of Electrical and Electronic Engineering in the University of Hong Kong has recently introduced the 'Electric Vehicle Technology' course to electrical engineering students. This course aims to provide those students up-to-date and comprehensive knowledge on EV technology. Hence, the students can be ready to serve the newly born but accelerating EV industry, and can have a good foundation for further research on this technology. As shown in Table 9.1, the course syllabus is designed to cover a broad band of knowledge rather than to focus on specialized topics on EV technology.

Table 9.1 EV course syllabus

Topics	Details
Concept and impacts	Past, present and future; energy and environmental benefits; economic and social aspects.
System considerations	System configurations, parameters and specifications; all-electric and hybrid systems; conversion and ground-up design.
Electric propulsion systems	Types and characteristics of EV motor drives; selection criteria; single- and multiple-motor drives; single- and multiple-speed transmissions; motorized wheels.
Energy source systems	Types and characteristics of EV batteries, fuel cells, ultracapacitors and ultrahigh-speed flywheels; selection criteria; sole and hybrid energy sources.
Auxiliary systems	Power and electronic accessories; battery chargers; air-conditioners; power steering units; auxiliary power supplies; energy management units.

9.4.2 PROMOTION

Governments have played an important role to promote the research, development and commercialization of EVs. They have set a number of regulations and policies:

- the rule that some vehicles sold in the coming years should be emission-free;
- the policy that the first registration tax of EVs is waived;
- the incentive that the electricity tariff for recharging EVs is specifically low;
- the R&D activities on EVs are highly recommended and financially supported; and
- the infrastructure of EV charging stations is fully supported.

Apart from the governments, there are some non-profit organizations that have actively engaged in the promotion of EVs:

- World Electric Vehicle Association (WEVA)
- Electric Vehicle Association of the Americas (EVAA)
- European Electric Road Vehicle Association (AVERE) and
- Electric Vehicle Association of Asia Pacific (EVAAP).

WEVA was founded on December 3, 1990 during the 10th International Electric Vehicle Symposium in Hong Kong where C.C. Chan served as General Chairman. It is a non-profit international organization which aims to promote the development and application of EVs and HEVs worldwide, to encourage and facilitate exchange of information among the members, as well as to coordinate the scheduling of the International EV Symposium (EVS) to be hosted by its members rotationally for the furtherance of the Association's purposes. Any international organizations dedicated to the aims and purposes of WEVA and recognized as the sole representative of a region of the world may associate as a member. At present, there are three members: EVAA representing American Continents (North and South), AVERE representing Europe and Africa, and EVAAP representing Asia and Pacific.

EVAA is an industry association working to advance the commercialization of EVs in the United States, Canada and Latin America. EVAA's members include major automobile companies, power utilities, transportation companies, manufacturers of batteries and other EV components, universities, the US Department of Energy, and national EV industry associations in other countries in the American hemisphere. The Association serves as the industry's central information source about EVs. Updated information and EV activities in the Americas are available in its web site, http://www.evaa.org/.

AVERE was founded in 1978 as a European network of industrial manufacturers and suppliers for EVs. It is a non profit-making association created under the aegis of the European Community. With more than 500 members, AVERE is composed of 14 national representations. Its goal is to rationalize the efforts of its member companies in the scientific and technological development of EVs, as well as in market development and thus to promote EV use. It offers its members vast experience and network in order to disseminate the results of their activities and any other information regarding EVs. Similar to EVAA, AVERE has launched a useful web site, http://www.avere.org/, promoting and providing information on EVs in Europe. Moreover, every two months, AVERE publishes its 'Mobile' magazine promoting and providing information on EVs for both the general publics and companies.

EVAAP was established in 1990 during the 10th International EV Symposium held in Hong Kong. Its objectives are to act as a non-profit professional association to promote EVs or HEVs in the Asia-Pacific region. Members of EVAAP consist of organizations involved in activities related to the development, production and application of EVs or related technology or product from countries and places in the Asia Pacific, including Australia, China, Hong Kong, Korea, India, Japan and Singapore. To achieve its objectives, EVAAP encourages the exchange of relevant information among its members and with international organizations with similar purpose, collaborates with other international bodies in activities of similar purpose, as well as endeavours to act as a source of relevant information for public dissemination, educational and government bodies.

9.5 Impacts on power system

EVs bring both good and bad influences on power system. Positively, the batteries of EVs can be charged at off-peak periods or at night so that the overall power demand can be levelled and the utilization of power system facilities can be improved. Negatively, the EV battery chargers are nonlinear devices which generate harmonic contamination to our power system, while the battery recharging of EVs at normal or peak periods creates additional current demand burdens on our power system.

9.5.1 HARMONIC IMPACT

During the recharging process of EV batteries, the input charging current is highly distorted, leading to create significant harmonic currents. Since different harmonic currents create different voltage drops across their respective impedances, the corresponding voltage waveform is also distorted, resulting in a series of harmonic voltages. The adverse effects due to both harmonic currents and voltages are summarized below (Arseneau *et al.*, 1997; Arrillaga *et al.*, 1998; Brener *et al.*, 1997):

- Since conventional measuring devices, such as ammeters, voltmeters, wattmeters and energy meters, are calibrated at a fixed frequency (50 or 60 Hz), and generally have poor frequency response, the harmonic currents and voltages can deteriorate their accuracy to record the actual current, voltage, power and energy consumption, respectively.

- Since large capacitor banks are usually installed in our power system to improve the overall power factor, they may be damaged by excessive harmonic currents.
- As the triplen harmonic current flows through the neutral of a three-phase four-wire power system, the neutral conductor may be overloaded, causing excessive heat problems. Also, common mode noise (neutral to ground voltage) may occur at the neutral point and the live-to-neutral voltage may be affected.
- As the harmonic currents flow through the power transformer with delta connection, the triplen harmonic component can circulate within the delta loop, causing excessive heat problems.
- In the presence of harmonic currents and voltages, the over-current, over-voltage and under-voltage protection devices may suffer from false alarm or tripping.

9.5.2 HARMONIC COMPENSATION

In order to compensate the harmonic contamination on our power system, there are many possible measures proposed by researchers and engineers. Basically, these measures can be categorized into two groups—device and system levels. In the device level, many new topologies of battery chargers are being proposed in such a way that the input harmonic current distortion is aimed to be minimal. These approaches rely on the invention of new battery chargers with minimum harmonic contamination and economically viable. In the system level, it can further be divided into two subgroups—passive and active filters. The passive filters can be simply phase-shifting transformers to suppress certain low-frequency harmonics or different combinations of inductor-capacitor sets to reduce those undesirable harmonics. On the other hand, the active filters are advanced power electronic systems which can on-line measure and diagnose the system harmonics so that they can instantaneously generate the same magnitude but anti-phase harmonics to neutralize the system harmonic content. As expected, these active filters need additional power source and sophisticated real-time control technology.

Recently, a new way for compensating harmonics generated by EV chargers has been proposed (Staats *et al.*, 1997; Chan *et al.*, 1998). It is neither based on the invention of new EV chargers (device level) nor the adoption of new filters (system level). Rather than adding something to our power system, the basic idea is simply to coordinate the number of EV chargers per charging station. Since the phase angles of those harmonic currents generated by one EV charger are normally different from those by another EV charger, there is a natural effect that harmonic compensation or even cancellation may occur. The more the number of chargers are being used per station, the higher the possibility to compensate the overall harmonic currents flowing to that charging station can be resulted. However, there is a practical limitation on the number of EV chargers per station because of the availability of space.

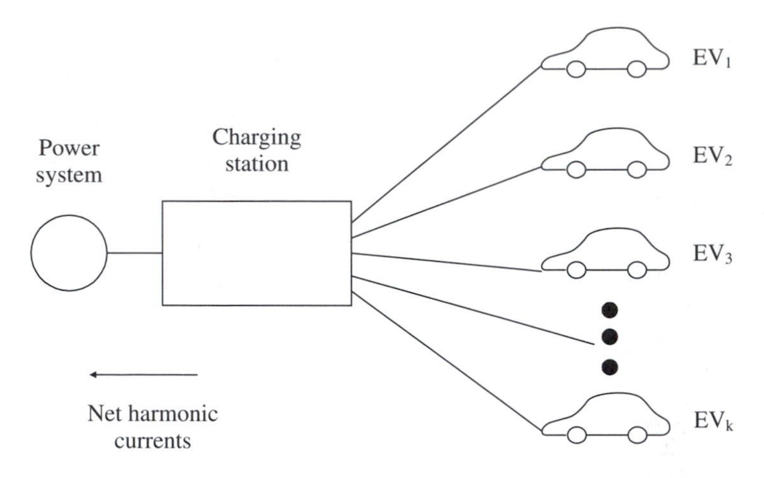

Fig. 9.3. Multiple EVs at a charging station.

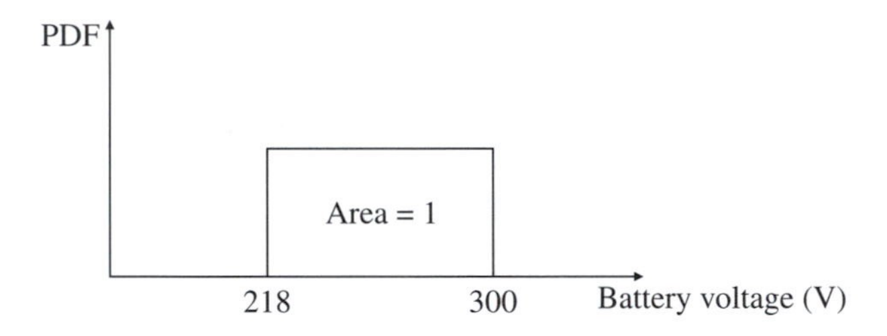

Fig. 9.4. PDF of EV battery voltages.

As illustrated in Fig. 9.3, the net harmonic current injected to the power system from a charging station with k EVs is investigated. We assign k random numbers to represent the state-of-charge (SOC) of these EVs, and represent the corresponding battery voltages by $\xi_1, \xi_2, \ldots, \xi_k$. At any instant, an EV in the station may be fully discharged, partially discharged or even fully charged (after a period of recharging). Thus, its SOC can be described by a random number with a uniform probability density function (PDF) ranging from fully discharged to fully charged. Assuming that the battery voltage is linearly related to the SOC, all EV battery voltages are represented by a uniform PDF (for example, between 218 and 300 V) as shown in Fig. 9.4. The charging current I required by each EV can be described by a Fourier series:

$$I = \sum \left\{ \frac{1}{n\pi} [\sin(n\beta_2) - \sin(n\beta_1) - \sin(n\beta_4) + \sin(n\beta_3)] \cos(n\omega t) \right\}$$
$$+ \sum \left\{ \frac{1}{n\pi} [-\cos(n\beta_2) + \cos(n\beta_1) + \cos(n\beta_4) - \cos(n\beta_3)] \sin(n\omega t) \right\},$$

where β_1, β_2, β_3 and β_4 are the conduction angles of the EV charger, and n is the order of harmonics. It can be rewritten as:

$$I = \sum c_n \cos(n\omega t + \varphi_n),$$

where $c_n = \sqrt{a_n^2 + b_n^2}$ and $\varphi_n = \tan^{-1}(b_n/a_n)$ are the magnitude and phase angle of the nth harmonic components, respectively. For various battery voltage levels, β_1, β_2, β_3 and β_4 are different. Hence, both c_n and φ_n are different for those k EVs in the same station. By summing up all generated harmonics in the complex domain, harmonic compensation at certain orders can be resulted.

Based on the Monte Carlo method (Hammersley and Handscomb, 1979; Sobol, 1994), the expected value of the total harmonic distortion (THD) of the input current flowing to the whole charging station can then be expressed as:

$$E[\text{THD}] = \frac{1}{N}\sum_{j=1}^{N}\left\{\frac{1}{k}\sum_{i=1}^{k} f(\xi_{i,j})g(\xi_{i,j})\right\},$$

where N is the total number of calculation for the Monte Carlo process, $\xi_{i,j}$ is the battery voltage of the ith EV at the jth calculation, $f(\xi_{i,j})$ and $g(\xi_{i,j})$ are the corresponding PDF and THD. Hence, the expected THD with respect to the number of EVs being charged in the same station is calculated as shown in Fig. 9.5. It is obvious that the higher the value of N (equivalent to the more the sample of calculation), the less the ripples (equivalent to the higher the degree of accuracy) can be resulted. Since the computational time significantly increases with the value of N, $N = 100$ is sufficient enough to provide the accuracy. It can be easily found that the expected THD decreases as the number of EVs increases. Although the

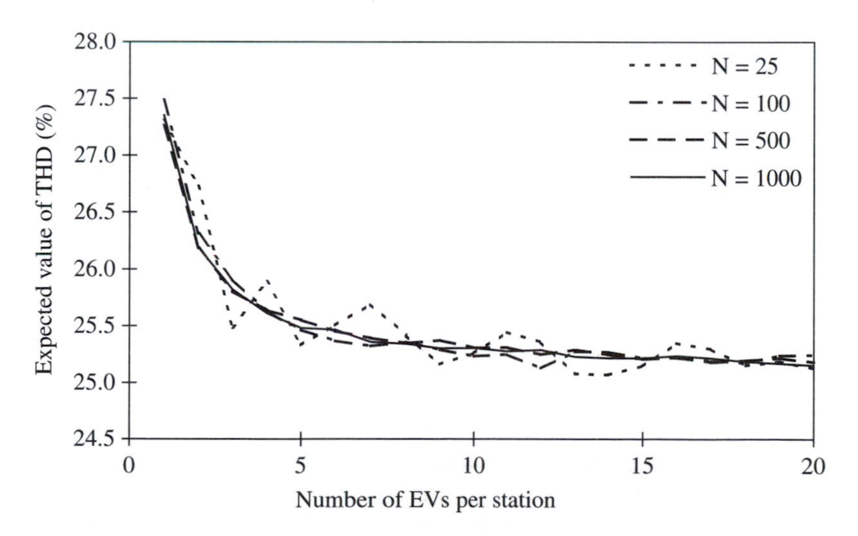

Fig. 9.5. Expected THD vs. number of EVs at a charging station.

expected THD can be minimized by encouraging infinite number of EVs to be charged at the same charging station, a finite number of about 10 EVs should be an optimal choice because the corresponding THD is near the minimum while the handling capability of about 10 EVs per charging station is realistic. It should be noted that the above numerical example is based on a typical EV charger (a diode-bridge dc-chopper charger with a constant-current charging algorithm for VRLA batteries). When EVs adopting different chargers and batteries, the resulting optimal number of EVs per station may be changed.

9.5.3 CURRENT DEMAND IMPACT

During the recharging of EVs at non-off-peak periods, the additional current demand inevitably creates burdens on the existing power system. These burdens are summarized below:

(1) In case the existing power system has been fully utilized, the additional current demand due to EV recharging at non-off-peak periods inevitably overloads the system. Otherwise, additional power plants and transmission networks are necessary to account for such additional current demand.
(2) Based on the reserve margin approach (Berrie, 1983), the product of the capacity of a reliable power plant and the number of power plants must be greater than or equal to the peak demand plus the reserve margin, typically 20%. The higher the peak current demand, the higher the generation and transmission capacities are necessary. Thus, the peak current demand governs the sizing of power plants and transmission networks, hence affecting the capital cost of the power system.
(3) In case of excessive current demand, the generators with short run-in time should be started to share the extra load or the spinning reserve should be increased to meet the maximum requirement. However, since the generators take time to run into the steady state, they should be started in advance to the occurrence of peak demand but remain running even though the peak load has passed away, hence wasting the generated energy. Similarly, the high spinning reserve also causes the power plants operating in a low-efficiency mode. Thus, the peak current demand significantly increases the running cost of the power system.

9.5.4 CURRENT DEMAND MINIMIZATION

In order to reduce the peak current demand due to the recharging of EVs, the concept of charging coordination has recently been proposed (Chan *et al.*, 1999; Chau *et al.*, 2000). The key of such concept is based on the coordination between charging current and charging time to charge a group of EVs at the same charging station, hence reducing the total maximum current demand, charging up the batteries as soon as possible and achieving a flat load profile as far as possible. Basically, there are two types of coordination approaches:

- distributive coordination and
- centralized coordination.

In case of the distributive coordination, each EV needs to install a distributive coordination controller which functions to maximize its individual charging current provided that the total current demand of the whole charging station is within the specified limit. Whenever there is any remaining current due to the charge-up or the leave of a particular EV, this unused current will be picked up by another EV based on the first-come-first-serve (FCFS) arbitration. Since each EV simply knows the total current demand and aims to grasp the unused current to shorten its individual charging time, this approach takes the advantages of simplicity and low-cost implementation. However, the FCFS arbitration cannot re-distribute the remaining current to other EVs in a balance way, thus the charging times of those EVs spread around. Also, since each EV charger knows only the total current demand of the charging station and nothing about the conditions of other EV chargers, some complicated control algorithms are not applicable to such approach.

In case of the centralized coordination, the charging station needs to install a central computer which gathers the necessary information, such as the battery capacity, SOC, current and voltage ratings as well as expected charging times, from all EV chargers. Hence, intelligent arbitration made by the central computer is adopted so that the total current demand can be minimized while the EV charging times can be equalized as far as possible. This central coordination approach takes the advantages over the distributive counterpart that the current demand fluctuation can be reduced, the spread of charging times can be optimized, and those sophisticated control algorithms can be implemented. The drawback is the increase in implementation complexity and cost, which can be well outweighed by the associated advantages.

The configuration of an EV charging station adopting centralized coordination is shown in Fig. 9.6. Based on the gathered information, a fuzzy controller is used to determine all charging currents for those EVs recharging at the station. As illustrated in Fig. 9.7, this controller is software implemented. The sequence arbitration functions to define the priority of EVs for current allocation so that an EV with a lower charging current can get a higher priority for competition. The iteration mechanism functions to avoid a large fluctuation of current demand. There are two fuzzy rules for the determination of the change of charging current ΔI_{C}, namely the Rule Set 1 is to determine the current change ΔI_{C1} for satisfying the peak current demand of the EV charging station and the Rule Set 2 is to determine the current change ΔI_{C2} for balancing the charging current of each EV. These rule sets can be described by the fuzzy relations R_1 and R_2 as defined by:

$$R_1 : \text{if } I_{d1} \text{ and } T_{\mathrm{L}} \text{ then } \Delta I_{C1},$$

$$R_2 : \text{if } I_{d2} \text{ and SOC then } \Delta I_{C2},$$

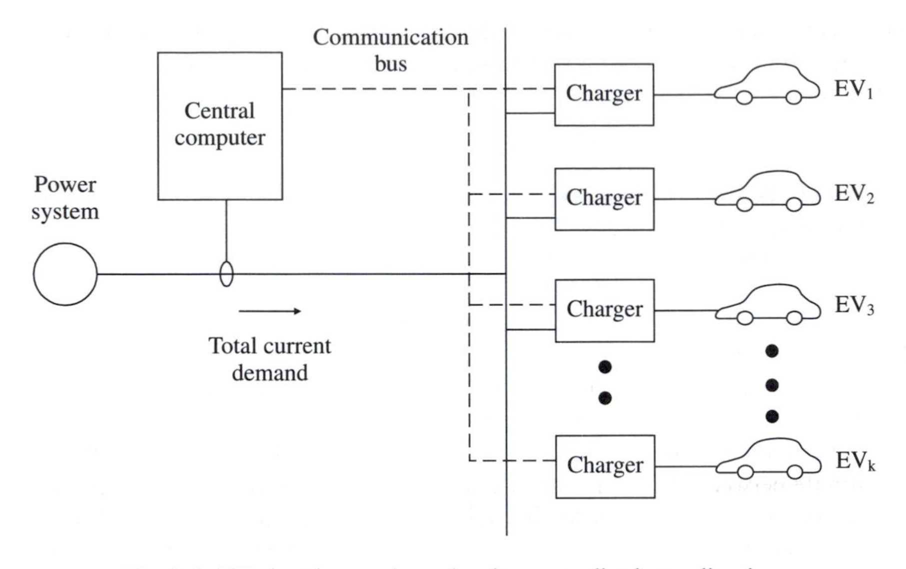

Fig. 9.6. EV charging station adopting centralized coordination.

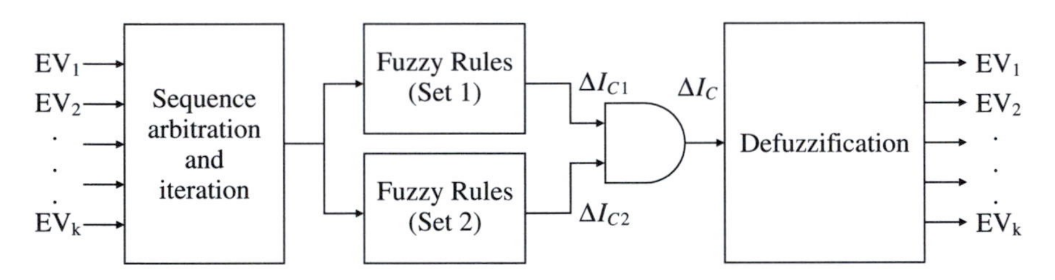

Fig. 9.7. Fuzzy controller for centralized coordination.

where I_{d1} is the deviation of the total charging current from the expected peak current demand, I_{d2} is the deviation of the individual charging current from the average charging current, T_L is the time left for charging and SOC is the battery state-of-charge. The predicates of those linguistic variables I_{d1}, I_{d2}, ΔI_{C1} and ΔI_{C2} include the negative big (NB), negative medium (NM), negative small (NS), zero (Z), positive small (PS), positive medium (PM) and positive big (PB). Since the maximum charging time is always kept to be lower than the predefined value, the negative linguistic values, NB, NM and NS, of T_L are excluded. Likewise, the set of SOC consists of no negative linguistic values. Then, the resultant current change ΔI_C is obtained by:

$$\Delta I_C = \Delta I_{C1} \wedge \Delta I_{C2},$$

which is then defuzzified to produce the crisp value by using the weighted average method. Finally, this crisp value is added to the previous charging current to derive the new charging current as given by:

$$I_C(t) = I_C(t-1) + \Delta I_C(t).$$

For exemplification, a charging station adopting centralized coordination is simulated to serve 12 EVs. For the sake of simplicity, the SOC of each EV is assumed to be zero and the EVs arrive at the charging station one-by-one with a fixed time interval ranging from 0.1 to 0.5 h for different simulations. The profiles of the total charging current with respect to different arrival time intervals are plotted in Fig. 9.8 and the corresponding charging times and peak current demands are listed in Table 9.2. It can be found that both the peak current demand and the average charging time are inversely proportional to the arrival time interval. The reason is due to the fact that the larger the arrival time interval, the higher the flexibility to coordinate between charging currents and charging times of the EVs is resulted. In case the centralized coordination is voided, the peak current demand will rise up to 720 A, indicating that the coordination can significantly minimize the peak current demand of the EV charging station. Also, it can be seen that the charging times of the EVs are nearby the average values, particularly when there is a large arrival time interval. This equalization merit is actually due to the feature of centralized coordination, which cannot be offered by distributive coordination.

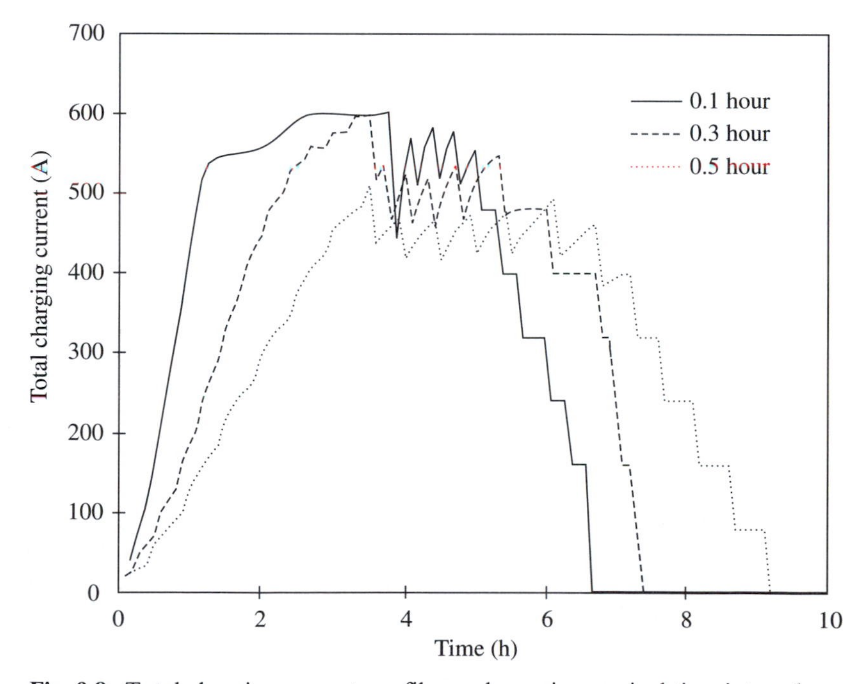

Fig. 9.8. Total charging current profiles under various arrival time intervals.

Table 9.2 Charging times and peak current demands under various arrival time intervals

	Arrival time intervals				
	0.1 h	0.2 h	0.3 h	0.4 h	0.5 h
Charging time of EV_1 (h)	3.8	3.6	3.6	3.6	3.6
Charging time of EV_2 (h)	3.8	3.6	3.6	3.6	3.6
Charging time of EV_3 (h)	4.0	3.6	3.6	3.6	3.6
Charging time of EV_4 (h)	4.2	3.7	3.6	3.6	3.6
Charging time of EV_5 (h)	4.4	4.0	3.7	3.7	3.6
Charging time of EV_6 (h)	4.6	4.3	4.0	3.9	3.8
Charging time of EV_7 (h)	4.8	4.6	4.4	4.2	3.9
Charging time of EV_8 (h)	5.0	5.0	4.8	4.3	3.9
Charging time of EV_9 (h)	5.3	5.1	4.7	4.2	3.8
Charging time of EV_{10} (h)	5.5	5.0	4.5	4.1	3.8
Charging time of EV_{11} (h)	5.7	4.9	4.4	4.1	3.8
Charging time of EV_{12} (h)	5.6	4.8	4.2	4.0	3.8
Average charging time (h)	4.725	4.350	4.092	3.908	3.733
Peak current demand (A)	602	599	597	548	510

References

Arrillaga, J., Bradley, D.A. and Bodger, P.S. (1998). Power System Harmonics. John Wiley & Sons, New York.

Arseneau, R., Heydt, G.T. and Kempker, M.J. (1997). Application of IEEE Standard 519–1992 harmonic limits for revenue billing meters. *IEEE Transactions on Power Delivery*, **12**, 346–53.

Berrie, T.W. (1983). Power System Economics. Stevenage: Peregrinus.

Bossche, P.V. (1997). The need for standards in an emerging market—the case of the electric vehicle. *Proceedings of the 14th International Electric Vehicle Symposium*, CD-ROM.

Brener, G.M. and Frearson, K.R. (1997). The influence of 220kV capacitor banks on power system harmonics. *Proceedings of Cigre Regional Meeting South East Asia and Western Pacific*, 259–68.

Chan, C.C., Chau, K.T. and Chan, M.S.W. (1998). Impact of electric vehicles to power system. *Proceedings of the 15th International Electric Vehicle Symposium*, CD-ROM.

Chan, M.S.W., Chau, K.T. and Chan, C.C. (1999). Coordination of charging stations for electric vehicles. *Proceedings of the 16th International Electric Vehicle Symposium*, CD-ROM.

Chau, K.T., Chan, M.S.W. and Chan, C.C. (2000). Current demand minimization by centralized charging coordination. *Proceedings of the 17th International Electric Vehicle Symposium*, CD-ROM.

Hammersley, J.M. and Handscomb, D.C. (1964). *Monte Carlo Methods*. Chapman and Hall, London.

IEEE Standard 519–1992 Recommended Practices and Requirements for Harmonic Control in Electrical Systems. IEEE, New York. 1993.

Kosak, R. (1997). Electric vehicle training. *Proceedings of the 14th International Electric Vehicle Symposium*, CD-ROM.

Kosowski, M.G. (1997). Current status of electric vehicle codes and standards. *Proceedings of the 14th International Electric Vehicle Symposium*, CD-ROM.

Menga, P. and Savoldelli, A. (1996). Methodology for economical and managerial assessment of electric vehicles recharging infrastructures. *Proceedings of the 13th International Electric Vehicle Symposium*, **I**, 748–55.

Muntwyler, U. (1996). Cost of an infrastructure for electric vehicles. *Proceedings of the 13th International Electric Vehicle Symposium*, **I**, 383–89.

Nowicki, J.J. (1997). Building an electric vehicle infrastructure. *Proceedings of the 14th International Electric Vehicle Symposium*, CD-ROM.

Olsen, J.M. and Edison, D. (1996). National electric vehicle infrastructure working council health & safety committee update. *Proceedings of the 13th International Electric Vehicle Symposium*, **II**, 164–70.

Shimizu, K. (1997). Status of Japanese EV standardization activities in '97. *Proceedings of the 14th International Electric Vehicle Symposium*, CD-ROM.

Sobol, I.M. (1994). *A Primer for the Monte Carlo Method.* CRC Press, Boca Raton.

Staats, P.T., Grady, W.M., Aprapostathis, A. and Thallam, R.S. (1997). A statistical method for predicting the net harmonic current generated by a concentration of electric vehicle battery chargers. *IEEE Transactions on Power Delivery*, **12**, 1258–66.

Toepfer, C.B. (1997). The EV as a vehicle for advancing electrical safety. *Proceedings of the 14th International Electric Vehicle Symposium*, CD-ROM.

10 Energy, environment and economy

Energy is the source of our lives, environment is the reliance of our lives, and economy is the operation of our lives. EVs can contribute to energy, environment and economy, hence beneficial to our lives (Chan, 2000). Their contributions are due to the improvement in energy security and energy efficiency for transportation, the alleviation of air pollution and noise pollution caused by ICEVs, and the creation of new businesses for supplementing or even replacing ICEVs. In fact, the first impetus for the redevelopment of EVs was due to the energy crisis in the 1970s, whereas the second impetus for the promotion of EVs was due to the goal of environmental protection in the 1990s. It is anticipated that the third impetus for the commercialization of EVs will be due to the economic opportunity in the 2000s.

10.1 Energy

Energy plays a very important role in the evolution of our society. Among various energy resources, oil is the major one. The exclusively wide acceptance of ICEVs for road transportation has created a vast market for oil business. Any disruption of oil supply will definitely cause a large turbulence in our society. In 1973, the oil price increased dramatically when there was an embargo on oil. This oil crisis let us realize that we have been too dependent on oil, and it will inevitably be exhausted in the future. Figure 10.1 shows a scenario of global energy

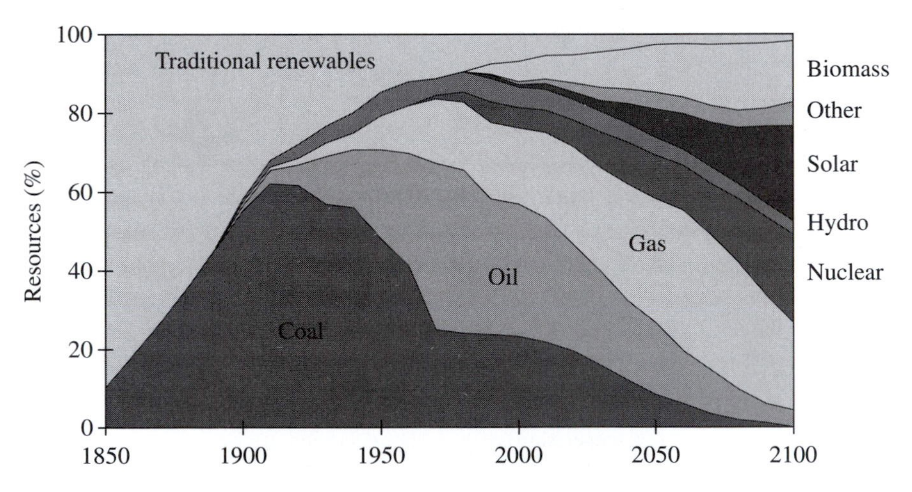

Fig. 10.1. Global energy consumption.

consumption based on the available energy resources (Nakićenović *et al.*, 1998). Therefore, continual development of alternative energy resources and efficient utilization of oil have recently become key issues in most government strategic planning around the world.

10.1.1 ENERGY DIVERSIFICATION

Deriving from oil, petrol and diesel are essentially the exclusive fuels for ICEVs. So, our present transportation means are heavily dependent on oil. EVs are an excellent solution to regulate this unhealthy dependence because electricity can be generated by almost all energy resources in the world. Figure 10.2 illustrates the merit of energy diversification for EVs in which electrical energy can be obtained from thermal power generation, nuclear power generation, hydropower generation, tidal power generation, wave power generation, wind power generation, geothermal power generation, solar power generation, chemical power generation, and biomass power generation.

Thermal power generation is to combust fossil fuels, hence producing heat energy, heating up water to form steam, driving the steam turbine, and finally generating electricity. Three common fossil fuels for this power generation are coal, oil and natural gas. Coal is the most abundant fossil fuel. As shown in Fig. 10.3, coal contributes about 91% of global fossil reserves, whereas oil and natural gas are only of 4% and 5%, respectively. Currently, mineable coal reserves exceed one trillion tonnes which are expected to be large enough to last over 200 years while the current reserves of oil may last 45 years and natural gas 70 years at current rates of consumption. Moreover, different from oil and natural gas reserves that are concentrated in specific regions, coal is widely distributed around the world. World coal production totals about 4.6 billion tonnes per year, and

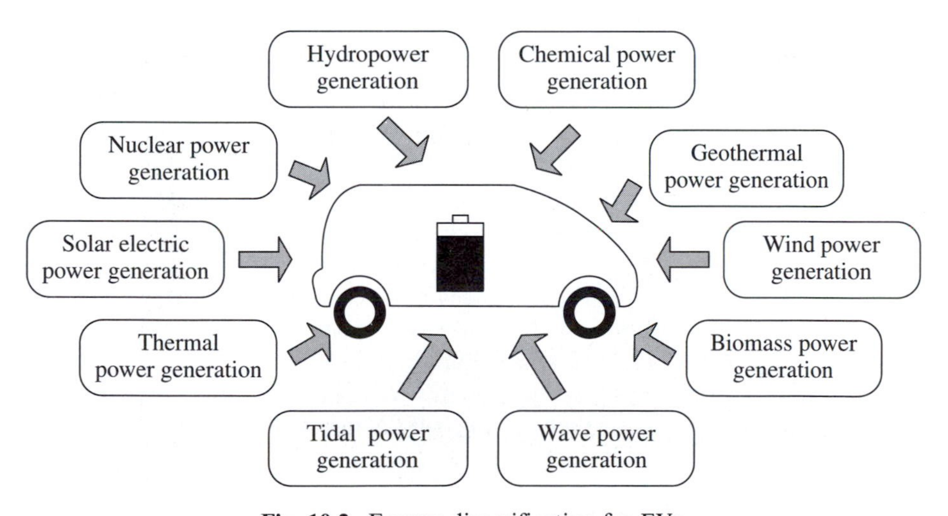

Fig. 10.2. Energy diversification for EVs.

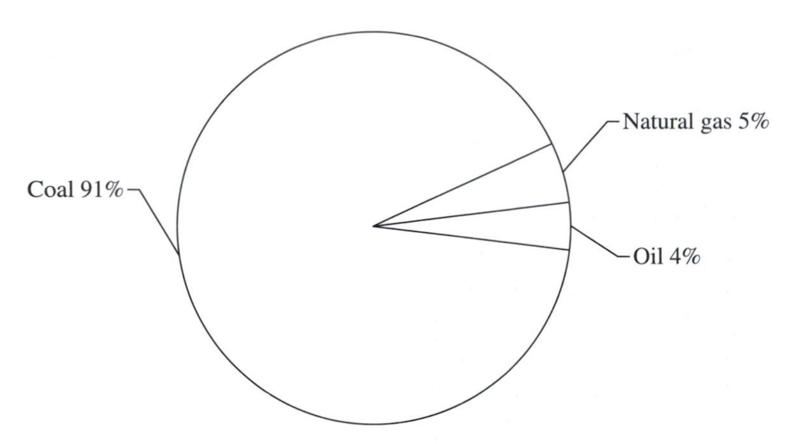

Fig. 10.3. Fossil fuel thermal power generation.

about 70% of the production is being used to generate about 40% of the world's electricity. Thus, the use of coal to generate electrical energy for EVs is not only economically advantageous, but also greatly alleviates the dependence on oil for transportation.

Nuclear power generation is similar to thermal power generation except that the heat energy is resulted from nuclear reactions rather than fossil-fuel combustion. At present, all commercial nuclear power is obtained by fission of very heavy atoms, which are generally extracted from uranium or thorium ores. Nuclear power is a main source for electricity generation. In 1996, nuclear power plants supplied about 23% of the electricity production for countries with nuclear units, and about 17% of the total electricity generated worldwide. Global and regional operable nuclear capacities are shown in Fig. 10.4 in which the power levels are only for indicative purposes. The nuclear generating capacity continues to grow in some countries especially for those of the Far East. It is estimated that more fraction of world's new nuclear capacity will be anticipated in Asia. However, nuclear power generation faces a complex set of issues including economic competitiveness, social acceptance, and the handling of nuclear waste.

Hydropower generation is basically the conversion from potential energy to kinetic energy of falling water, hence driving the water turbine and finally generating electrical energy. The operation of hydropower plant is clean, produces no emissions nor greenhouse gases, and leaves behind no wastes. Because of simple operation and no combustion, hydropower can generate electricity with an efficiency as high as 90% which is much higher than that of combustion thermal power generation. In general, the production cost of hydropower is about one-third the cost of using fossil fuels or nuclear. Moreover, hydropower does not experience rising or unstable fuel costs, and its operating cost generally grows at less than the rate of inflation. The main impacts arising from hydropower generation are intrusion to local environment and changes to local ecology.

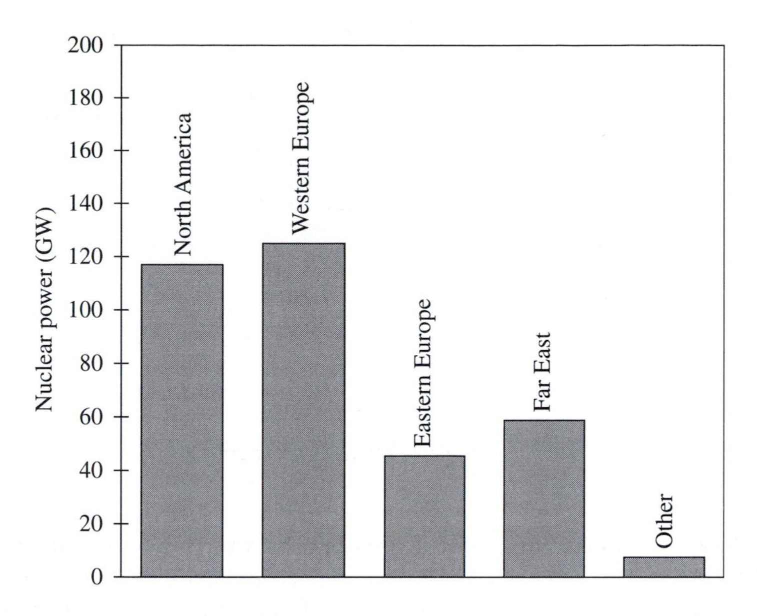

Fig. 10.4. Nuclear power generation.

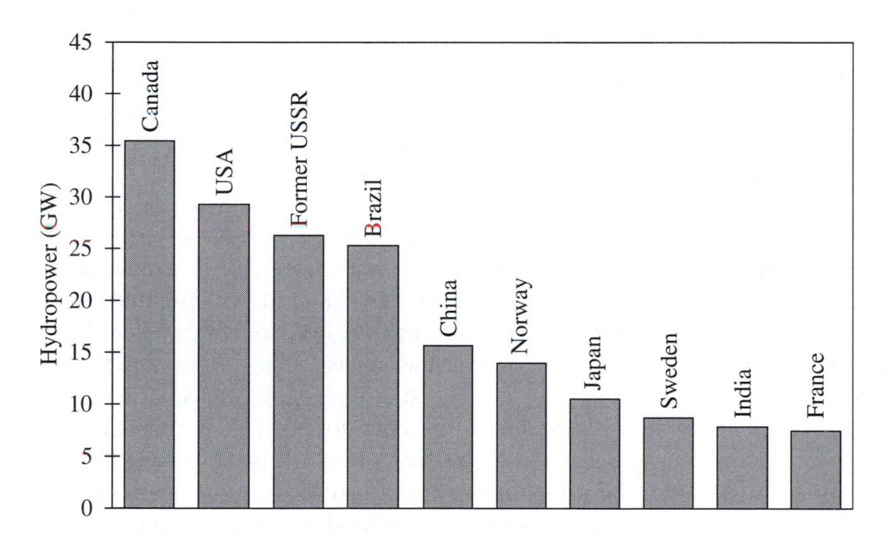

Fig. 10.5. Hydropower generation.

Naturally, hydropower generation suffers from geographical constraint. About 20% of worldwide electricity is generated by hydropower. Those top hydropower generation countries are shown in Fig. 10.5 in which the power levels are only for indicative levels. It is interesting to note that Norway produces more than 99% of its electricity with hydropower.

Tidal power generation is the most well developed technology to extract ocean power. It operates on similar principles to hydropower generation. Tidal power is governed by weather conditions, slope of the continental shelf and the coast. The total potential tidal power of the world has been estimated to be about 64 GW. This source is non-depletable and pollution free. Some tidal power generation systems have been constructed in the world. The Rance tidal power plant in France is the largest one in the world, generating about 240 MW power. The main drawback of tidal power generation is the ecological problem which may be caused by the disruption of the environment and change of habitats.

Wave power, similar to tidal power, is a kind of ocean power as it takes the advantage of ocean waves caused primarily by interaction of winds with the ocean surface. Since wave power is an irregular and oscillating low-frequency energy resource, many types of wave energy converters have been developed to extract energy from surface waves, from pressure fluctuations below the water surface, or from the full wave. It has been estimated that the wave power along California coastline is of 4–10 MW/km. The technology is still at the experimental stage.

Wind power generation basically employs wind turbine technology to generate electricity. It is economical and pollution free, but its security is constrained by intermittence and unreliability. It has been estimated that the total physical potential of wind power may accommodate 20% of the global electricity demand. At present, Germany has already installed the wind power capacity over 1300 MW and Denmark is more than 1000 MW. The largest wind power plant, having 112 wind turbines and capable of generating electricity of about 35 MW, has been installed in the United States.

Geothermal power is heat energy contained within the Earth, such as hot springs, fumaroles, geysers and volcanoes that can be recovered and put to useful work. Geothermal power generation is to convert those high-temperature (above 150 °C) geothermal resources to electricity using steam or organic vapour turbines. Similar to other power generation using renewable resources, it is economical and pollution free. The United States offers the largest geothermal power generation in the world (over 2700 MW), followed by Philippines with about 890 MW, Mexico with about 700 MW, Italy with about 545 MW, and New Zealand with about 460 MW. Other countries such as Iceland, Japan and Russia have also installed geothermal power plants.

Solar power generation is generally classified into two categories, namely solar-thermal and solar-electric (photovoltaic). Solar-thermal power generation makes use of the solar radiation incident on the collector to heat a fluid, commonly water, to its boiling point so that the resultant steam drives a turbine to generate electricity. At present, more than 350 MW of electricity is generated by commercial solar-thermal power plants in the United States. On the other hand, photovoltaic power generation converts the solar photons to electricity directly. The four countries with the greatest installed photovoltaic power capacity are Germany, Italy, Japan and the United States. The world largest photovoltaic power plant is located in Italy with electricity capacity of about 3.3 MW.

Chemical power generation is generally based on fuel cell technologies. A fuel cell is the chemical system in which a fuel (commonly hydrogen) combines with oxygen to produce electricity with pure water as by-product. Among various types of fuel cells, the PAFC, MCFC and SOFC are relatively more attractive for power utilities. The corresponding maximum power levels are of about 5 MW, 100 MW and 100 MW, respectively. It should be noted that the preferred fuel cell is the SPFC for EV on-board applications. Its maximum power level is of about 1000 kW.

Biomass power refers to the energy created by firewood, agricultural residues, animal wastes, charcoal and other derived fuels. Biomass can be used as a substitute for conventional fossil fuels for thermal power generation. The corresponding heat is generated by direct combustion of biomass or biomass-derived products. Those gaseous or liquid biofuels are derived by using thermal conversion, biological conversion and biochemical conversion. Biomass power generation usually operates on a small to medium scale (under 100 MW), and those biomass power plants are likely installed in rural locations that are connected to the distribution network close to the final consumers.

Although various energy resources are widespread over different regions and different countries, electrical energy can be obtained by using the aforementioned technologies of power generation. Therefore, the use of electrical energy as the source for road transportation essentially solves the problems of regional and national dependence on particular energy resources, especially oil. Moreover, with the continual development of energy diversification for power generation, the energy security for road transportation can be achieved.

10.1.2 ENERGY EFFICIENCY

Besides the definite merit of energy diversification resulting from the use of EVs, another advantage is the high energy efficiency offered by EVs. In order to compare the overall energy efficiency of EVs with ICEVs, their energy conversion flows from crude oil to road load are illustrated in Fig. 10.6, where the numerical data are only for indicative purposes and may have deviations for different refineries, power plants, power transfers, batteries, EV types, ICEV types or driving cycles. By taking crude oil as 100%, the total energy efficiencies for EVs and ICEVs are 18% and 13%, respectively. Therefore, even when all electricity are generated by oil-fuelled power plants, EVs are more energy efficient than ICEVs by about 40%. It should be noted that since ICEVs presently spend over 60% of oil demand in advanced countries, the use of EVs can significant save the consumption of oil, hence saving in both energy and money.

Moreover, EVs possess a definite advantage over ICEVs in energy utilization, namely regenerative braking. As shown in Fig. 10.7, EVs can recover the kinetic energy during braking and utilize it for battery recharging, whereas ICEVs wastefully dissipate this kinetic energy as heat in the brake discs and drums. With this technology, the energy efficiency of EVs is virtually boosted up by further 20%.

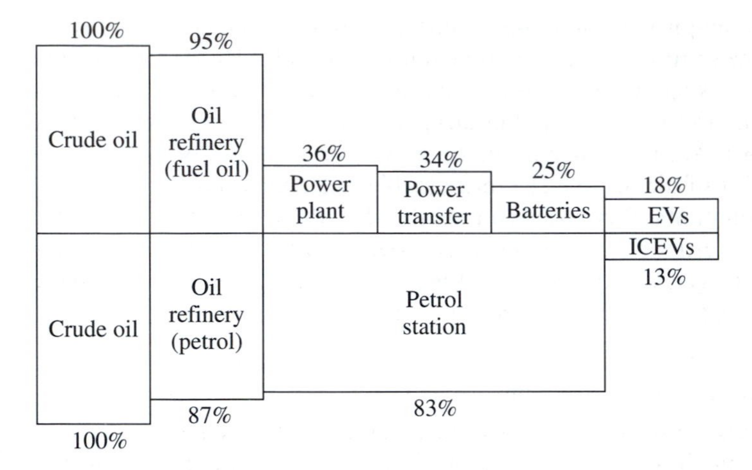

Fig. 10.6. Energy efficiencies of EVs and ICEVs.

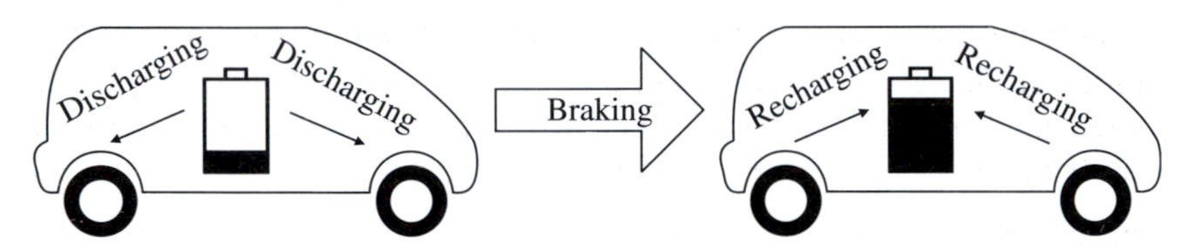

Fig. 10.7. Energy saving by regenerative braking.

10.2 Environment

Nowadays, ICEVs have been exclusively used for road transportation. Since the energy source of ICEVs is solely based on oil-derived fuels such as petrol and diesel, the combustion of these fuels for propulsion inevitably liberates harmful pollutants to air and generates severe noise to the surroundings. As more and more people prefer to own a car instead of using public transportation, the pollution due to road transportation continues to grow and reaches an unhealthy condition in many advanced cities (Houghton, 1996). It is the time for us to take action on the use of EVs for road transportation.

10.2.1 TRANSPORTATION POLLUTION

Among various environmental impacts due to road transportation, air pollution is the major issue. ICEVs not only emit toxic gases and dust that pollute our local environment, but also release greenhouse gases which cause unnecessary global warming. Those toxic gases and particulate are directly harmful to our health, causing respiratory and lung diseases as well as stimulating disastrous cancers.

The next major pollution due to road transportation is noise. It not only distracts our concentration and causes annoyance, but is also harmful to our hearing sense.

The first step on solving the air pollution problem due to road transportation is to identify those harmful pollutants generated by ICEVs. In the United States, the Clean Air Act in 1970 authorized Environmental Protection Agency (EPA) to issue a list of air toxics, and then established regulation to control the spread of those pollutants. Based on the potential health and environmental hazards, a list of 189 hazardous air pollutants has been identified, including carbon monoxide (CO), nitrogen oxides (NO_x), sulphur oxides (SO_x), volatile organic compounds (VOCs) and lead molecules. Some hazardous pollutants are commented below:

- CO is odourless, colourless and highly toxic gas which can reduce the oxygen-carrying capability of haemoglobin in our blood, thus causing unconscious death.
- NO_2 and SO_2 gases are harmful to our respiratory passages and lung. When dissolving in water or forming acid rain, they cause the impairment of forest as well as the contamination of lakes and rivers.
- VOCs such as benzene, ethylene, formaldehyde, methyl chloroform and methylene chloride are cancerogenic substances, thus very harmful to human and plants.
- Lead molecules in our environment are very harmful to human, especially the children, because they cause permanent damages to our brain, nervous and digestive systems. In some countries or cities, leaded petrol is becoming prohibited.

Smog is a mixture of fog, smoke and other pollutants, which can be easily realized whenever there is a reduction in visibility. The influence of smog is not restricted to urban areas, but also spreads to rural areas. Figure 10.8 illustrates the

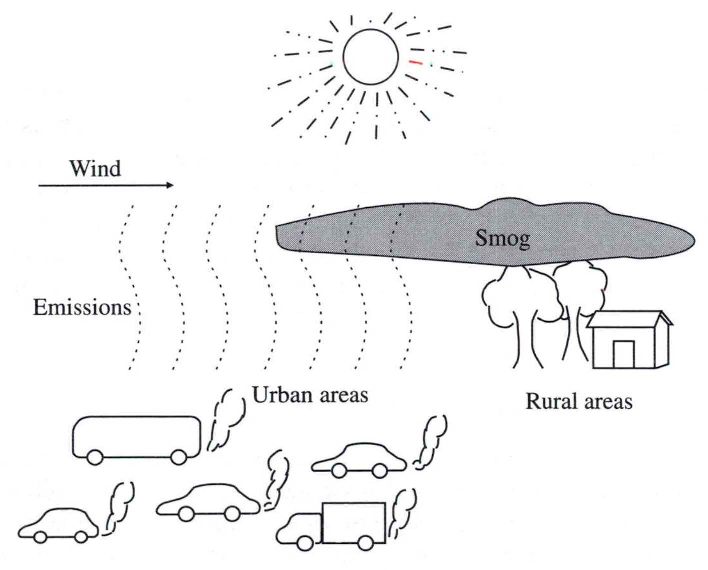

Fig. 10.8. Principle of smog formation.

formation of smog. ICEVs are the major source of smog-forming pollutants. As wind may blow the pollutants away from their sources, they perform photochemical reactions under direct sunlight. These reactions combine the mixture of hydrocarbons (HC) and NO_x to form ground-level ozone, which is the primary constitution of smog. This ground-level ozone is so highly oxidizing and chemically active that it is corrosive and irritating to our respiratory system. Besides this ground-level ozone, smog consists of other pollutants (such as VOCs and NO_2) that are detrimental to our health, namely the reduction of infection resistance as well as the irritation of asthma and eyes.

Apart from the pollutants mentioned before, there are other emissions that are classified as greenhouse gases. The major ingredients of greenhouse gases include carbon dioxide (CO_2), methane (CH_4), nitrous oxide (N_2O) and chlorofluorocarbon (CFC). The main influence of greenhouse gases is the problem of global warming. Initially, the global temperature of the Earth can be kept constant if the solar energy absorbed by the Earth is always equal to the energy re-radiated back to the space. However, the greenhouse gases above the Earth surface prevent the re-radiated energy from escaping the Earth and hence trap the heat in atmosphere. As more and more greenhouse gases are emitted, the global temperature continues to increase, causing permanent changes in the sea level and hence damages to our global ecology. Among all greenhouse gases, CO_2 plays the most important role as it is exclusively dominant over the others. Furthermore, among various sources of CO_2 emissions, road transportation is the major source.

Another adverse influence of road transportation to our environment is noise pollution. Inside the engine of ICEVs, the mixture of air and fuel is burnt in burst so that the burning gas can push the pistons drastically to provide the desired mechanical driving force. Such kind of combustion produces a severe noise problem to our surroundings. Additionally, practical imperfection of gears, belts and axles along the mechanical transmission of ICEVs is another major source of annoying noise. Although the noise pollution seems to be less detrimental than the air pollution, this noise problem is definitely harmful to our health in terms of concentration and hearing sense.

10.2.2 ENVIRONMENT-SOUND EVS

Nowadays, in many metropolises, ICEVs are responsible for over 50% of hazardous air pollutants and smog-forming compounds. Although the engine of ICEVs is continually improved to reduce the emitted pollutants, the increase in the number of ICEVs is much faster than the reduction of emissions per vehicle. Hence, the total emitted pollutants due to ICEVs continues to grow in a worrying way.

In order to alleviate or at least control the growth of air pollution due to road transportation, we need to introduce a new concept of vehicles—zero-emission vehicles. In October 1990, the California Air Resources Board established rules that 2% of all vehicles sold in the state between 1998 and 2002 should be

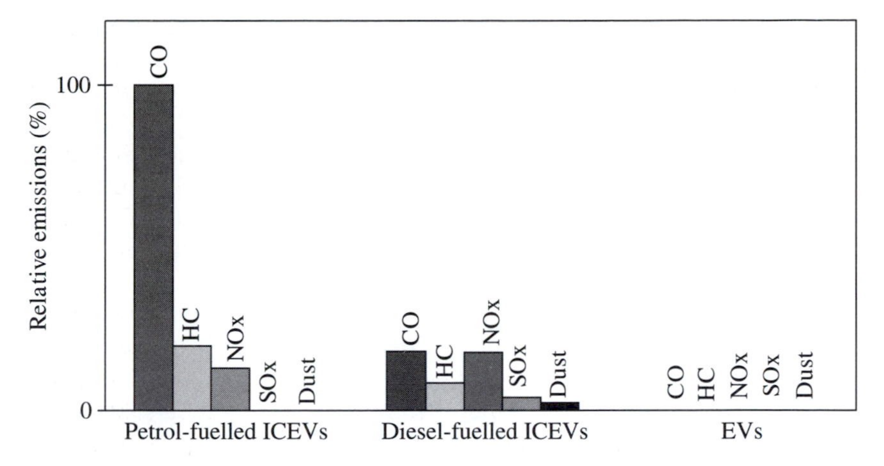

Fig. 10.9. Comparison of local harmful emissions.

emission-free, and 10% of vehicles put on the market must have zero emissions by 2003. Although the Board in 1996 decided to scrap the 2% mandate because of the difficulty for the automobile industry meeting that timetable, it retained the 10% mandate by 2003. In the foreseeable future, EVs are the most possible choice to meet the zero-emission criterion. It should be noted that this zero-emission criterion is applied to local emissions rather than global emissions. Figure 10.9 shows a comparison of those local harmful emissions generated by petrol-fuelled ICEVs, diesel-fuelled ICEVs and EVs. As expected, EVs offer no local emissions at all. On the other hand, petrol-fuelled ICEVs emit remarkable toxic gases, especially CO gas, whereas diesel-fuelled ICEVs produces serious air pollution, particularly dirty dust. Taking into account the emissions generated by refineries to produce petrol and diesel for ICEVs as well as the emissions by power plants to generate electricity for EVs, a comparison of those global harmful emissions is shown in Fig. 10.10. It can be found that the global harmful emissions of EVs are much lower than those of ICEVs. Notice that the indicative data in those comparisons may vary with the fuel used, the efficiency of power plants and the driving cycle of vehicles.

In additional to low global emissions, the use of EVs offers the possibility of further reducing the air pollutants by centralized treatment. Since the corresponding global emissions are produced during electricity generation, the pollutants can be easily collected for centralized treatment. The commonest treatment is to add filters to adsorb the dirty dust. It should be noted that this centralized treatment is ill-suited to ICEVs because it is practically impossible to apply expensive treatment to each ICEV.

Besides the reduction of global harmful emissions, the use of EVs can benefit the reduction of CO_2 gas which is the major component of greenhouse gases. Figure 10.11 shows an indicative comparison of those CO_2 gas globally emitted by petrol-fuelled ICEVs, diesel-fuelled ICEVs and EVs. Increasingly, the effective

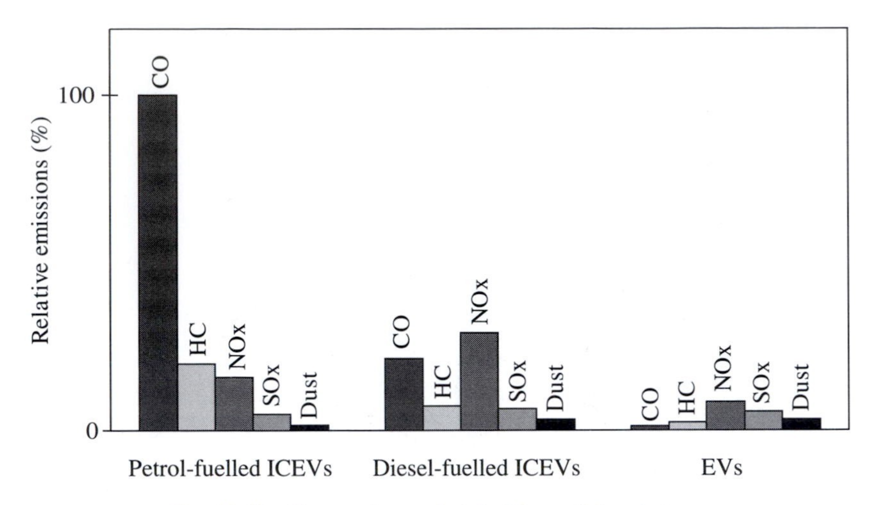

Fig. 10.10. Comparison of global harmful emissions.

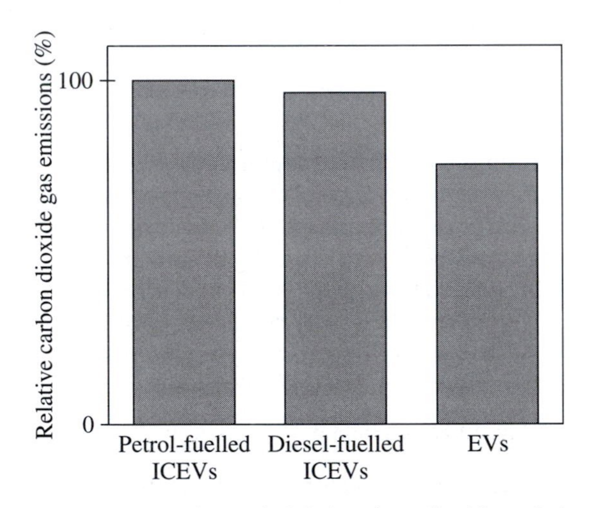

Fig. 10.11. Comparison of global carbon dioxide emissions.

emissions of CO_2 gas due to the use of EVs can be further suppressed by collecting the gas from the power plants for recycling. As shown in Fig. 10.12, a CO_2 recycling system has been adopted by Kansai Electric Power in Japan. Firstly, CO_2 gas is recovered from the power plant flue gas by chemical absorption method. It can be extracted with a recovery rate of about 90%. The collected CO_2 gas is then liquefied for storage and transportation. Secondly, it is experimentally used to synthesize methanol (CH_3OH) by the catalytic hydrogenation method as described by:

$$CO_2 + 3H_2 \rightarrow CH_3OH + H_2O.$$

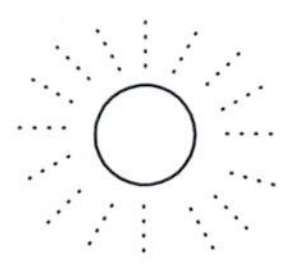

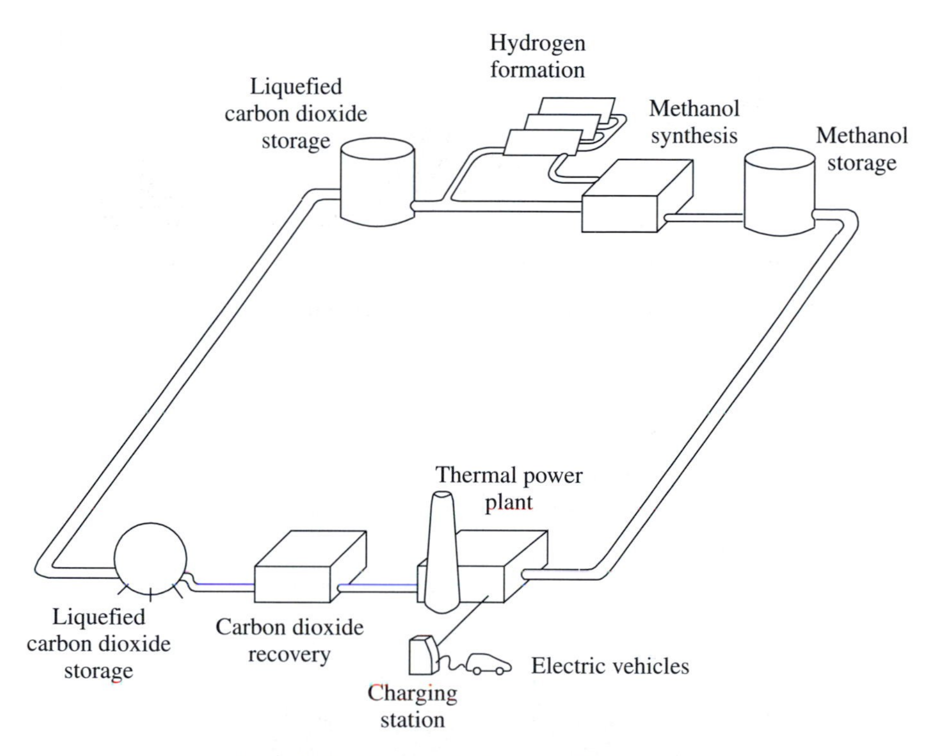

Fig. 10.12. Carbon dioxide recycling system.

The required hydrogen (H_2) gas can be produced by utilizing micro-organisms existing in the natural environment. As illustrated in Fig. 10.13, green algae produces starch from CO_2 by photosynthesis in cultivation vessels, then the resulting starch is decomposed to form organic acid in dark anaerobic fermentation vessels, hence H_2 gas is generated when the organic acid is decomposed by photosynthetic bacteria in the presence of sunlight. Finally, CH_3OH resulting from the whole CO_2 recycling process is used as an alternative fuel for thermal power generation.

Certainly, the above-mentioned CO_2 recycling scheme is not the only way to reduce greenhouse gases. In fact, the popularity of EVs can stimulate energy diversification for electricity production and many renewable energy sources such as hydropower, tidal power, wave power, wind power and solar power produce zero harmful emissions nor greenhouse gases.

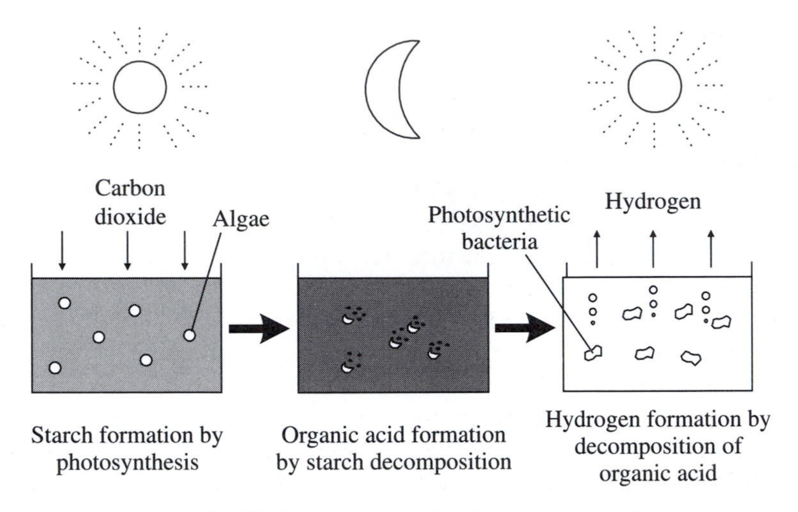

Fig. 10.13. Hydrogen production by micro-organisms.

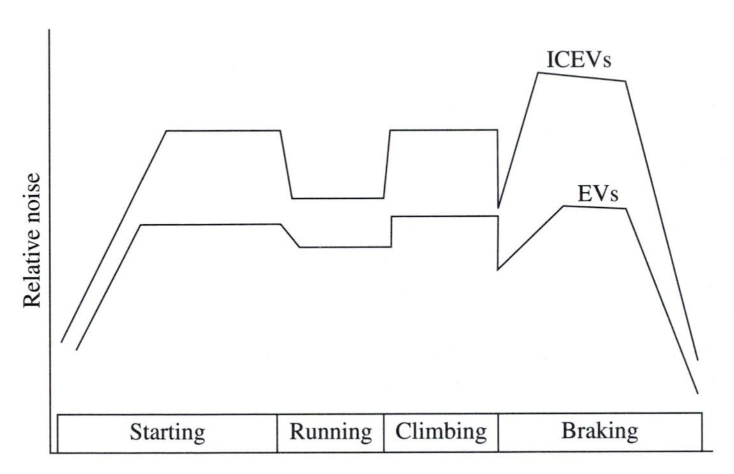

Fig. 10.14. Comparison of noise by ICEVs and EVs.

EVs have another definite advantage over ICEVs, namely the suppression of noise pollution. Different from ICEVs that their combustion engine and complicated mechanical transmission produce severe noise problems to our surroundings, EVs are powered by electric motors operating with very low acoustic noise. Moreover, EVs offer either gearless or single-speed mechanical transmission so that the corresponding annoying noise is minimum. Actually, the operation of EVs is too silent that it may not be easy to detect the approaching EVs. Figure 10.14 gives a relative comparison of noise created by ICEVs and EVs during starting, running and braking.

10.3 Economy

As introduced before, it is anticipated that the coming impetus for the commercialization of EVs will be due to the economic opportunity in the 2000s. In response to the latest California rule that a mandate of 10% of vehicles put on the market must have zero emissions by 2003, it implies that Californian consumers will buy approximately 400 thousand EVs in 2003. Optimistically, if the other states adopt the same rule, the total EV sales in the USA will be almost 1800 thousands in 2003 (which is based on the estimation that the total vehicle sales in the USA will be of 18 millions in 2003). In case other states may not enforce the same rule or may adopt a slower schedule, a conservative estimation of the total EV sales in the USA should be about 800 thousands in 2003. Taking into account some unforeseeable adverse factors, this amount may be reduced to 100 thousands. Apart from the USA, Europe and Asia also actively promote the EV markets. It has been forecasted that Europe and Asia will create a market size of 50 thousand EVs. It should be noted that the California mandate will be reviewed periodically, therefore the above estimation may be reviewed in due course. Also, the ultralow-emission vehicles, HEVs and FCEVs will have certain market share in the coming years.

On 16 February 2001, the Electric Vehicle Association of the Americas (EVAA) reported as follows: '22 states throughout the USA have introduced legislation this year that would provide financial and/or non-financial incentives for electric drive technologies. Examples of innovative financial incentives include legislation introduced in Connecticut that would provide economic incentives to manufacturers who build new, or retool existing, plants to manufacture electric or solar-powered cars; and Georgia has legislation pending that would require, beginning in Model Year 2003, that 5% of the state-owned motor vehicles be alternatively-fuelled. This percentage would ramp up to 75% by Model Year 2012 and thereafter. In order to encourage the use of neighbourhood EVs (NEVs), Hawaii has introduced a bill that would create a task force within the department of business, economic development, and tourism to develop state policy to convert the state motor pool to EVs, including NEVs. Finally, providing access to high occupancy vehicle (HOV) lanes to single-occupant drivers of battery EVs, hybrids and/or other alternative fuel vehicles is one of the non-financial incentives being introduced in state houses across the country. Arizona, Florida, Georgia, Texas, Utah and Washington all have introduced bills that would provide single-occupant drivers of some or all of these types of vehicles access to HOV lanes. Moreover, there are a variety of incentives that states are considering to promote the purchase and use of battery, hybrid electric and fuel cell cars, trucks and buses throughout the USA'.

In face of such potential markets, new economic opportunities must be enormous. Hence, the global economy can be benefited from the sustainable growth of zero-emission and ultralow-emission vehicles. Since different countries have different vehicle mileages, petrol/diesel prices, electricity prices and energy scenarios, the corresponding economic benefits may be different. We expected that the most beneficial businesses are automobile companies, power utilities, battery manufacturers, electric motor manufacturers, and other EV auxiliary manufacturers. It should be noted that the worldwide development and markets of existing vehicles have been monopolized by several countries for many decades, and it is hardly possible or at least very difficult for other countries to chase the technological development or to share a reasonable amount of markets. The new EV markets open up this possibility because the development of EVs has not yet been monopolized. As experienced by the developers of ICEVs, a successful developer of EVs can bring the country enjoying a drastic growth of economy. Therefore, it is anticipated that the coming impetus for the commercialization of EVs will be due to economy.

The key issues of successfully commercializing and promoting EVs lie in how to produce low-cost, good performance EVs; how to leverage the initial investment; and how to provide an efficient infrastructure. The overall strategy should take into account how to fully utilize the competitive edge, to share the market and resources, and to produce EVs that can meet the market demand. Although specific strategies may vary with different manufacturers and countries, and are very complex, the key element is the willingness and commitment of manufacturers and their industrial partners, governments and public authorities, electric utilities and users. It is negative to say, 'Let the market decide'. Market forces are just abstract concept, derived from individuals who have different interests. Therefore, a wise initial strategy and agreement on market penetration are essential.

The key of success is two integrations. First is the integration of society strength, which includes government's policy support, financing and venture capital's interest, incentives for industry, and technical support from academic institutions. Second is the integration of technical strength, that is the effective integration of the state-of-the-art technologies of automobile, electrical, electronic, chemical and material engineering.

At the beginning, EVs cannot compete with ICEVs in every application. Therefore, it is important to identify the niche markets that are feasible, consequently to identify the required technical specifications, and to adopt the system integration and optimization. In order to achieve cost effectiveness, a unique design approach and a unique manufacturing process should be developed. Excellent after-sales service and effective infrastructure are also essential. Figure 10.15 shows the basic considerations of government, industry and market. Although the viewpoint and emphases of these three parties are not identical, if these three parties are willing to cooperate and commit, a common interest and awareness can be achieved. This implies that the corresponding common area will be bigger, and hence the chance of success will be greater.

Since EVs are not traditional vehicles, innovative marketing strategies and programmes should be developed so that the financier, manufacturer and customer can cooperate together at the win–win situation. As mentioned before, the major obstacles of marketing EVs are the short range and high initial cost.

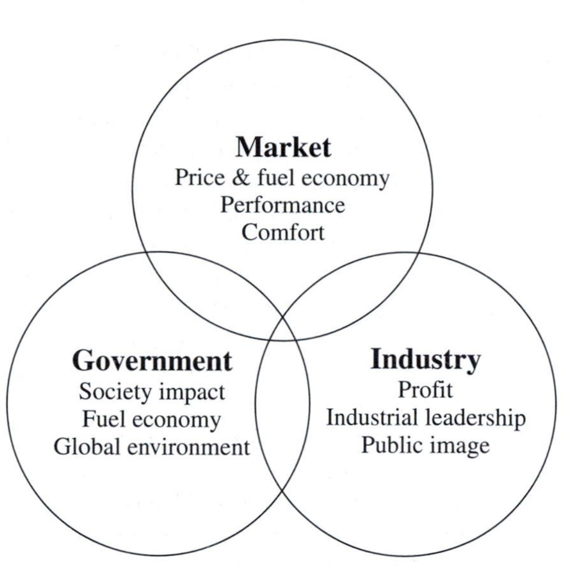

Fig. 10.15. Government, industry and market.

However, we should look into the overall economic analysis throughout the life cycle of EVs. The total cost of EVs including two parts: initial cost and operating cost. The initial cost of EVs is much higher than that of ICEVs due to the use of batteries as the energy storage device, while the energy storage device in ICEVs, namely the fuel tank, represents only a minor fraction of the total vehicle cost. In order to relieve the burden of the EV customer, it is proposed to lease the battery, so that the initial cost of the EVs mainly only consists of the cost of the chassis, body and propulsion system. Thus, the initial cost of EVs can be comparable or even cheaper than that of ICEVs if mass produced. Then, the operating cost of EVs will include three parts: maintenance cost, fuel cost and battery rental cost. According to the statistic of EV operation in various countries, the maintenance cost of EVs accounts only 25–50% of that of ICEVs. The ratio of fuel cost of EVs to ICEVs varies in different countries, since the oil price depends on the energy resource, energy policy and tax system of different countries. In general, the oil price in the USA is much cheaper than in some Asian countries, such as China. Again, the electricity price depends on the primary energy used and price, efficiency of the power system and the energy policy of different countries. Generally, the ratio of fuel cost of EVs to ICEVs is much cheaper in China as compared to that in the USA. The author has made an overall cost comparison of electric public minivans and diesel-engine public minivans in Hong Kong based on annual operation of 50 000 km and using the battery leasing programme. Taking into account a reasonable profit of the battery leasing company, the total annual operating cost of electric minivans is cheaper than that of diesel minivans. The saving of the operating cost is sufficient to compensate the higher initial cost of electric minivans within a duration of several years. This is because even without the battery, the initial cost of electric minivans may still be more expensive than that of diesel minivans, due to the absence of mass production.

The battery leasing company may be a joint venture among the battery manufacturer, battery dealer, electric power utility and oil company. This company owns the battery, leases the battery, provides charging and other services to the customer, and is responsible for the recycling of the battery. Since such a large quantity of batteries is managed by a leading company, it must be more cost-effective. The abovesaid programme will not only release the initial financial burden of the EV customer, but also relieve the psychology burden of the EV customer. Whenever the customer has a problem, he or she just needs to go to the service station. Depending on the customer's situation, the company may adopt fast charging, occasional charging, normal charging or battery swapping (the battery will be recharged during the off-peak demand of the power system), thus it can leverage the energy demand and hence increase the overall energy efficiency.

Another example of innovative marketing programme is by the introduction of innovation-bond proposed by Abt (Abt, 2000). The innovative manufacturer must have a well proven product, know the investment required for mass production, know the price for large scale of sales, and know the production period. The bank issues the innovation-bond with its price, the same as the EV price, its interest rate as low as government bonds, and its duration valid until the delivery of EV. The buyer will receive interest from the bond, and will receive good product or money back after expiration of the bond. The buyer can sell the bond to someone else. The procedure of marketing is as follows: the customer buys the innovation-bond from the bank, while the bank grants credit to the innovative manufacturer, and the manufacturer supplies good product at low price to the customer. Thus, the innovative manufacturer can benefit from minimized investment risk, can get lower interest rates of loan from the bank and can predict the market. The customer has no risk, since he or she is paid with interest rates comparable to government bonds by purchasing the bond and waiting the product. The bank is the biggest winner of all, because the bank will charge the manufacturer for issuing the bond, and will charge the customer a handling fee. The major profit of the bank may come from the interest gain, since the bank will charge a higher interest rate than government bonds for the credit granted to the manufacturer, and the risk of granting such loan is minimized. Although there are inevitable difficulties in implementing the idea of innovation-bond, creative EV marketing schemes should be developed to make commitment to the financier, manufacturer and customer with the support of government's policy. Other innovative EV marketing programmes with the aid of information technology should also be developed to suit different situations in different countries and regions.

The present EV market situation can be described as the following adverse circle of chain reaction: high initial price → low consumer satisfaction → low demand → lack interest of investment → no mass-production lines → no large-scale sales → high initial price . . .

This situation must be changed by the following favourable circle of chain reaction: keen interest of investment → mass production lines → large-scale sales → low initial price → high consumer satisfaction → high demand → keen interest of investment . . .

We should strive for a Clean, Efficient, Intelligent and Sustainable transportation means for the 21st Century.

References

Abt, D. (2000). The innovation-bond: an instrument for the broad market introduction of electric vehicles. *Proceedings of the 17th International Electric Vehicle Symposium*, CD-ROM.

Assessment of Electric Vehicle Impacts on Energy, Environment and Transportation Systems. International Energy Agency, 1999.

Chan, C.C. (2000). *The 21st Century Green Transportation Means—Electric Vehicles*. National Key Book Series in Chinese, Tsing Hua University Press, Beijing.

Houghton, J.T. (1996). *Climate Change 1995: The Science of Climate Change*. Cambridge University Press, Cambridge.

Electric Vehicle. Energia, 1996.

Electric Vehicle. Kansai Electric Power, 1996.

Electric Vehicle. Kyushu Electric Power, 1996.

Electric Vehicle—Current Situation and Perspectives. VARTA Batteries, 1996.

EV Town 21. Japan Electric Vehicle Association, 1996.

Http://eia.doe.gov/.

Http://www.epa.gov/.

Http://www.hydro.org/.

Http://www.iea.org/.

Nakićenović, N., Grübler, A. and McDonald, A. (1998). *Global Energy Perspectives*. Cambridge University Press, Cambridge.

Preserving The Global Environment—Carbon Dioxide Recycling Research Facility. Kansai Electric Power, 1996.

State Legislative Alert. EVAA Leaflet, 2001.

Index